Matrices

AND

Graphs

Stability Problems in Mathematical Ecology

Dmitrii O. Logofet
Laboratory of Mathematical Ecology
Russian Academy of Sciences
Moscow, Russia

CRC Press
Boca Raton   Ann Arbor   London   Tokyo

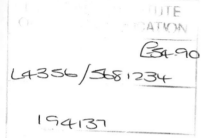

Library of Congress Cataloging-in-Publication Data

Logofet, Dmitriĭ Olegovich
    Matrices and graphs: stability problems in mathematical ecology / author,
Dmitriĭ O. Logofet
    p.   cm.
    Includes bibliographical references and index.
    ISBN 0-8493-4246-5 ✓
    1. Ecology—Mathematical models.   2. Matrices.   3. Stability.   4. Graph
theory.   I. Title.
QH541.15.M3L64   1992
574.5'01'1–dc20   92-20183
                                                                      CIP

This book was formatted with LaTeX by Archetype Publishing Inc., 15 Turtle Pointe
Road, Monticello, IL 61856.

Direct all inquiries to CRC Press, Inc., 2000 Corporate Blvd., N.W., Boca Raton, Florida,
33431.

© 1993 by CRC Press, Inc.

International Standard Book Number 0-8493-4246-5

Library of Congress Card Number 92-20183

Printed in the United States of America   1 2 3 4 5 6 7 8 9 0

Printed on acid-free paper

Preface

Here is one more book about stability, stability in ecological systems... Yet the vast variety of ideas, concepts, and methods which are nowadays included under the fuzzy term "mathematical ecology" have at least one unambiguous feature in common: all of them are related, to a greater or lesser extent, with mathematical models. Stability, understood generally as an ability of a living system to persist in spite of perturbations, when observed in reality, must be in models too. Once we realize the specific kind of mathematical model that is available to us, we can then define stability in formal terms, choosing among those definitions of stability which are relevant to the mathematical apparatus in hand a practical definition that accommodates the stable functioning of the ecological system concerned. This analytical framework will then both enable us to observe how a system responds to perturbations in the computer model (made easier by the current boom in personal computers and accessible software) and to speculate on the role that any particular feature of the system plays with regard to the system's stability or lack of stability.

In this projection to the "plane" of stability problem, the eternal question of whether the exogenous or endogenous causes are responsible for particular effects in the quantitative behavior of a system becomes reduced to the question of what are the effects of system structure and functional mechanisms, the fundamentals of the system. In general, these are matrices and graphs which have long served as a mathematical means of representing those fundamentals, and in particular, these are the community matrices and the graphs of trophic and other relations which are considered in this book as a means of studying stability in model ecosystems.

One more question is relevant, which is both of more recent origin and yet still "perennial," whether a specific pattern revealed by computer simulation should be ascribed to nature or considered to be nothing else than an artifact of the mathematical description or the mathematical software routines which are used beyond their correct scopes of application? I consider this question, which arises in particular cases of interest, to have a greater chance of being answered within mathematical ecology, by means of what is now called the "thinking of ecological problems in mathematical terms."

Thinking of stability problems in mathematical terms such as matrices and graphs requires that we know constructive criteria for a matrix or graph to possess a particular mathematical property. Fortunately or unfortunately, mathematics has its own unsolved problems too, and it is often a hard problem to find a constructive criterion, or a characterization, of the property in terms of matrix elements. That is why a knowledge of the general hierarchy of stability notions, as well as stability matrix subsets, may greatly facilitate our treating particular matrices of known structure. Even if an exact characterization of the property in question is not yet known, the hierarchy may give some insight into the kind of restructuring or "perestroyka," that can be expected in the ecosystem structure in response to perturbation.

There are three types of thinkers about "ecological problems in mathematical terms": the field or laboratory ecologist, the theoretical ecologist, or "modeler" (terms which have now become almost synonymous), and the applied mathematician. Reading this book may give the first of the three a somewhat more "indicative" than "informative" knowledge of the kinds of problems that can be formulated, the second might accept or argue about the inferences from the formulations, and the third could see (and would not object to, I hope) the ways by which such formulations have resulted in those inferences. For all three, the inferences might, in particular, increase or decrease their trust in the outcomes of particular simulation models.

I hope, therefore, that the book will be of interest and use to the whole community of mathematical ecologists. For the sake of readability, I have tried to provide as many illustrations as possible, believing that even a plain idea is better perceived in a visualized form; mathematical proofs are gathered as a rule in appendices to each chapter, although some proofs of possible general interest or originality are commented on in the main text. Introductory courses in linear algebra and differential equations will have provided sufficient background to understand the mathematics, while further references are proposed for deeper insight into the subjects treated.

The book can hardly serve as the sole textbook for an introductory course in mathematical ecology, unless the course is especially designed to cover only multispecies or multicomponent issues. For such issues the book may well be referred to, especially in courses given for mathematics students. I optimistically think that the time will come when "a mathematically trained student of biology" will no longer be a rarity, and this is the kind of reader who merits the author's highest appreciation.

Dmitriĭ O. Logofet
Moscow, Russia
The Commonwealth of Independent States

Acknowledgments

I am grateful to all my colleagues in the Laboratory of Mathematical Ecology who patiently survived the local disaster called a Scientist Who Writes the Monograph. Professor John Maybee made several critical remarks that have improved the text. My special thanks go to Mrs. T. M. Guseva and Dr. L. K. Makarova for their invaluable help in preparing the manuscript. Many people of CRC Press, both known and unknown to me, as well as Ms. Lori Pickert of Archetype Inc., deserve my sincere thanks for their careful attitude to the author's manuscript and for the high quality of the finished book.

"Don't worry about your mathematical difficulties;
I can assure you that mine are even greater..."

Albert Einstein

"When you don't know what to do, apply what you do know."

Richard Bellman

Contents

CHAPTER 1

Stability Concepts in Ecology and Mathematics

Understood generally as an ability of a living system to persist in spite of perturbations, stability can be defined in many ways, both in verbally descriptive and formal terms, either in ecology or mathematics. While neither one of these "ecological" meanings of stability can now be recognized as being more fundamental than the other, mathematics has the greater advantage, since it has given rise to the notion of Lyapunov stability, which appears to be inherent in or important to any further conceptions of stability—at least, within the theory of dynamical systems. To see whether and how this mathematical substantiality has produced any consequences in ecology, we concentrate, in this introductory chapter, on the way this notion is applied in mathematical theory of population, community, or ecosystem dynamics, or more shortly, in the proper chapters of mathematical ecology.

I. BASIC APPROACHES TO THE NOTION OF STABILITY

In spite of appearing to be intuitively clear, "an ability to persist in spite of perturbations" can scarcely be defined in a unique and unambiguous way. The reason for this difficulty is that both the "persistence" and the "perturbations" parts of the idea need further clarification, to say nothing of the scale factors of the system being considered. What is understood by "an ability to persist" and what kind of "perturbations" is relevant? Different answers to these basic questions and a variety of stability concepts have been proposed and discussed in the ecological literature,[1-2] (yet only a few of them have been given proper mathematical attention[3-4]). Perhaps this is why stability in general lacks—and should apparently not have—any "stable" definition.

A. Species Diversity as a Measure of Stability

It has long been a paradigm of theoretical ecology that the more complex the structure of a biological community is and the more abundant the number of constituent species in it, the more stable the community will be in response to perturbations.[5-6] The argument in favor of this thesis—first proposed apparently by MacArthur[7] and

1

Elton[8]—is briefly as follows. Various species have different ranges of adaptation to changes in their environment. A wider species composition may therefore respond to various environmental changes more successfully than a community with a lower number of species. The former is hence considered to be more stable than the latter. This is probably the main reason why a variety of species diversity indices (in particular, the entropy of information theory or its analogs) are used to characterize the community's stability.[6]

The most popular index is the Shannon information entropy measure:

$$D = - \sum_{i=1}^{n} p_i \ln p_i, \qquad p_i = N_i/N, \qquad N = \sum_{t=1}^{n} N_i, \qquad (1.1)$$

where n is the number of species in a community, N_i is the population size of the ith species. A somewhat different measure, but of the same genus, is the index

$$D_1 = 1 - \sum_{i=1}^{n} \frac{N_i(N_i - 1)}{N(N - 1)} . \qquad (1.2)$$

It can be shown that D_1 represents the probability that two individuals, when chosen randomly from a set of N ones, do not belong to the same species. For large values of N and uniform enough distribution of N among species, the proportion of species i can be assumed to approximate $p_i \approx 1 + \ln p_i$, whereby it can be shown that $D_1 \approx D$.

Other definitions of this kind (e.g., those using R. A. Fisher's measure of information) have also been proposed as stability measures, yet all of them rely upon notions originating from theoretical physics or information theory (see a survey by Goodman[6]). There is no doubt that diversity measures do carry some objective information about the properties of a system. E. Odum[9] has shown, for instance, that there are certain relationships among the diversity level, the structure, and the functioning of energy flows in an ecosystem. However, using such diversity indices as stability measures leads to some drawbacks, or paradoxes, in the theory,[10,4] where a typical "route" to a drawback can be traced in the following argument.

Under the "maximum diversity–maximum stability" assumption, it would be logical to suppose further that maximum stability is attained at equilibrium, if this state is ever reached. A simple exercise in finding a conditional maximum of a function like $D(\mathbf{p})$ over the space of frequencies p_i will then readily generate a fixed pattern of equilibrium distribution, which can hardly be interpreted in ecological terms. For example, the maximum of D (1.1) is attained at $\mathbf{p}^* = [1/n, 1/n, \ldots, 1/n]$, i.e., at uniform species abundance. This excludes any species domination or quantitative hierarchy in the community, whereas the empirical evidence is just the opposite: communities which exist long enough to be considered stable do feature dominating species, which carry out the major part of the work to provide for the matter and energy turnover through the ecosystem.[11] In other words, a quantitative hierarchy, rather than uniformity, is normal for a real system.

Other diversity indices may, of course, generate hierarchical distributions too, but the causes of the hierarchy will always be fixed to a particular mathematical form of the diversity function to be maximized, rather than to real properties of the system under study. That is why the use of diversity indices as measures of stability can scarcely be accepted as a faultless approach.

On the other hand, an increase in diversity may really be observed in many natural and laboratory communities, particularly at the early stages of their evolution to an equilibrium state. To all appearances, the diversity measures are still able to characterize a community in some dynamic respects, although falling short of general applicability.

The reason for such paradoxes probably arise because the models of theoretical physics and information theory are formally applied to systems to which they are not applicable. Both the Boltzmann entropy in statistical physics and the information entropy in information theory make sense only for ensembles of weak interactions among particles or other objects. Introducing an entropy measure is grounded quite well for such ensembles. But once we turn to systems whose elements interact strongly, the entropy measure can no longer suffice. Biological communities and ecosystems represent systems with strong interactions, because it is mainly the interactions themselves among constituent species or ecosystem components that form the structure of the system.

As far as a steady equilibrium can be regarded as the final outcome of the functioning of structure, quite explainable may be progress in the application of entropy measures at early stages of community evolution. The point is that, these stages being far from equilibrium, the competition or other interactions among species are still weak, interspecies pressures are still low, and the community may well be considered as a system of weak interactions.

So, is the diversity "only the spice of life or is it a necessity for the long life of the total ecosystem comprising man and nature?"[12] Although the diversity measures have the obvious advantage of being determined through observable and measurable characteristics of a real ecosystem, the question of a cause–effect relationship between diversity and stability has no simple answer.

B. Model Approach to Definitions of Stability

In contrast with intuitive understanding of stability typical the "stability vs. diversity" speculations, the model approach can provide for quite formal, mathematically rigorous definitions. Mathematical theory of stability has a long history of development and application in science and engineering, where any issue relies essentially upon a *mathematical model* of the system under study, i.e., a mathematical description which is adequate to the purposes posed in the study. So, if we have a "good" enough (seen from the viewpoint of adequacy and descriptive completeness) mathematical model of a biological community or ecosystem, then the stability properties of a real system can be deduced by investigating the model of the system, using mathematical techniques of stability theory.

But mathematics also has many notions—and corresponding formal definitions—of stability. The task is to decide what kind of model behavior should correspond to a stable functioning of the real system and to choose those mathematical definitions of stability which are adequate both to a meaningful, "ecological," perception of stability and to the mathematics of the model.

Speaking of ecosystem stability, E. Odum,[11] for example, distinguishes between two major types, namely, a *resistant* and a *resilient* stability. *Resistant* stability is the ability of an ecosystem to resist perturbations (or disturbances), while keeping its structure and functioning unchanged. *Resilient* stability is the ability of an ecosystem to restore itself after its structure and functioning have been disturbed. Both types are observed in nature, and the evidence is growing that an ecosystem can scarcely develop both types of stability, probably due to the fact that a particular type of the environment (whether it is permanently favorable or temporarily unfavorable) is likely to promote either the resistant or the resilient type of stability.[11]

An ideal mathematical analog to the above distinction is the relation between the *trajectory* and *structural* types of stability in dynamical systems, the former referred to most often as just *stability*. Formalized as a system of differential (or difference, or integral-differential, etc.) equations, a model is generally considered *stable* if a particular solution of interest (e.g., an equilibrium, or steady state) is stable in a certain mathematical sense, which generally means that small deviations of the initial state will not result in a great difference between the new and the reference solutions. For example, the well-known Lotka–Volterra pair of prey-predator equations (Section 4.III), that has become classic in mathematical ecology, should be regarded as stable, even having a set of concentric ovals as its phase portrait (Figure 12): a small initial perturbation shifts the system to yet another oval, but still close to the initial one.

Suppose now that the structure of this ideal model is disturbed by the introduction of self-limiting in the prey population. In a laboratory prototype it might be any change in the set-up, interpretable as competition for a limiting factor, while in the model equations it corresponds to a new negative term appearing in the prey equation, or a negative perturbation of the corresponding zero element of the community matrix. As will be seen later on, this disturbance—no matter how small it is—changes the whole pattern of dynamic behavior in the model: concentric ovals now turning into spirals converging to the center (Figure 13).

This is an illustration of a *structurally unstable* system, whereas *structural stability* is defined as the ability to keep the phase pattern unchanged at least for small perturbations of coefficients in the equations. For example, the prey-predator pair with self-limitation is already a structurally stable system, since perturbation of any coefficient will neither change the spiral into another pattern, nor affect the converging nature of the spiral unless the perturbation is great enough.

Also, this is an illustration to the fact that a mathematical system, unlike ecological systems after E. Odum,[11] may possess the trajectory and structural stability/instability properties in various combinations: the stable Lotka–Volterra ovals are structurally unstable, the structurally stable Lotka–Volterra spiral is trajectory stable too, whereas the zero solution to the equation of exponential population growth,

$$dN/dt = \mu N,$$
(1.3)

is unstable for any positive value of the *Malthusian parameter* μ. Yet it will always be difficult to decide whether the behavior of a real system after a perturbation should be interpreted in terms of resilient stability to the structural perturbation, or in terms of trajectory-type stability to a perturbation which is so great as to affect the structure. Chaparral ecosystems near the California seaside are known to recover soon after burning in brushfires,[11] but whether we have structural stability with respect to eradicating the links the shrubbage has with other components of the ecosystem, or a stability with respect to even as great a perturbation as the state becoming zero, will strongly depend on what particular system we consider as the model.

Investigations of stability thus rely on a particular mathematical model or a class of particular models, assuming them to be adequate enough to the system under study. As far as the assumption is true, the model approach has an obvious advantage in its prognostic ability, as well as in its capability of relating stability to other system properties such as the structure and particular mechanisms of functioning. Expressed in formal terms, stability conditions of the model promote a formulation of hypotheses concerning the functioning of the real system; their tests in theory and practice may generate new knowledge of the system and give more insight into modeling issues. Besides, the adequacy assumption itself may always be questioned and sometimes answered—at least, in qualitative terms—also from the outcome of stability analysis.[13−16]

Further sections of the chapter will discuss how these views and ideas are realized in the Lyapunov and other concepts of stability for multicomponent systems of ecological modeling, such as age-structured populations and multispecies communities.

II. LYAPUNOV AND "ECOLOGICAL" STABILITY IN ECOSYSTEM DYNAMICS

When using the model approach to studying stability, it should be always borne in mind that stability makes sense only with respect to a specified class of perturbations, while "an ability to persist" is identified with the model variables being within a specified region of the state space. If, for example, the perturbation is reduced to a single change in the initial state of a dynamical system, the well-known *Lyapunov stability* concept arises—either *local* or *global*, *asymptotic* or *neutral*, depending on whether the perturbations are considered sufficiently small or definitely finite and whether the perturbed trajectory is assumed to converge or just to be close enough to the unperturbed one. If in addition, sufficiently small although permanent perturbations are expected to affect the right-hand sides of the model equations, then this leads to the notion of stability under *permanently acting perturbations*. The intimate connection that exists between these two notions in the general theory of stability of motion also finds a proper expression in the population equations.[17−18]

When the matter concerns a system of equations to describe the population or community dynamics, we usually consider the stability of steady-state solutions, such that the population size of age groups or constituent species neither vanishes, nor tends to infinity. These solutions are said to be *nontrivial* or *positive equilibrium* (or sometimes just *equilibrium*). Stability of such solutions implies, in particular, that the number of species will be preserved in the community despite the fact that the initial equilibrium distribution of population sizes has been somehow shifted. Thus, elimination of one or several species from the community can naturally be interpreted as a loss of its stability. The set of remaining species may well comprise a stable composition, that will however be a different community.

On the other hand, the list of species can be preserved—at least in the model— even if the equilibrium is Lyapunov unstable but neither of the state variables vanishes. This weaker model property seems to be more adequate to an ecological meaning of stability than the Lyapunov concept, and the distinction is considered below in more detail.

A. Formal Definitions and Interpretations

Let a model population, community, or ecosystem be represented by a system of differential equations written in vector form as

$$dN/dt = f(N), \qquad (1.4)$$

where the vector of state variables $N = [N_1, N_2, \ldots, N_n]$ consists of population sizes (or densities), $N_i(t)$, of n groups composing a population, or of n species composing a community, or of n components composing an ecosystem. Let also the right-hand side of the equation, $f(N)$, which reflects our hypothetical knowledge of the object under study, be mathematically correct, that is, meet all technical conditions for a solution, $N(t)$, to exist and to be unique, corresponding to a given initial condition $N(t_0) = N_0$. While, in general, a definition of stability has to be relevant to any particular, say, *reference*, solution $N^r(t)$, *a* proper change of variables can always reduce the problem to the stability of a so-called *equilibrium* solution, that is, a constant vector N^{eq} nullifying the right-hand side of equations (1.4).

Definition 1.1 *Solution* N^{eq} *is said to be* locally Lyapunov stable *if for any small* $\varepsilon > 0$ *there exist* $\delta > 0$ *such that the inequality* $\|N_0 - N^{eq}\| < \delta$ *results in* $\|N(t; N_0) - N^{eq}\| < \varepsilon$ *for any* $t \geq t_0$, *where* $N(t; N_0)$ *designates the solution corresponding to the initial state* N_0. *If, in addition,* $N(t) \to N^{eq}$ *as* $t \to \infty$, *then* N^{eq} *is asymptotically* Lyapunov stable; *if, moreover, this is true for any initial state of a certain domain* G *in the state space, then* N^{eq} *is* asymptotically stable globally in G, *or just* globally asymptotically stable.[19]

In theoretical mechanics, a locally Lyapunov stable motion is traditionally interpreted as a small "tube" of trajectories propagating along the time direction: any perturbed trajectory remains within the ε-tube unless the initial perturbation has shifted it out of the ε-tube. The Lyapunov asymptotic stability then means that the "thickness" of the tube tends eventually to zero, while the global stability turns the tube rather into a sharpened "funnel." Phase portraits of population systems, shown

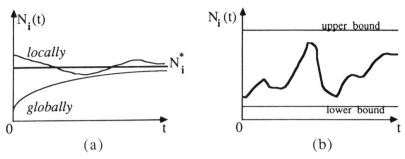

Figure 1.   Examples of a Lyapunov stable (a) and ecologically stable (b) trajectories in ecosystem models.

in Figures 12 and 13, can also justify these images if the time axis is imagined in the direction perpendicular to the phase plane.

In terms of a biological community composed by n interacting species, Lyapunov stability means quite a special mode of "an ability to persist in spite of perturbations": if an impact of any kind has shifted somehow the equilibrium distribution, N^{eq}, of population sizes to a new state N_0 (but has not affected the mechanisms governing the dynamics), then the state must return, in the course of time, close to the equilibrium again; the versions of the Lyapunov concept modify the meaning only in what concerns the magnitude of the initial shift and the closeness of the return. (The disputable question of what should be considered as an equilibrium in the community remains entirely within the scope of the model adequacy problems.)

As a partial consequence, it follows that the species composition, i.e., the list of n constituent species, remains unchanged in a Lyapunov stable community. At the same time, this represents the ecological meaning of stability: if neither species is eliminated, the community should be considered stable. But, in model terms, nonelimination can be achieved in quite a number of patterns of dynamic behavior, of which the Lyapunov stable equilibrium is just one particular instance. So, a formal definition of ecosystem stability might be weakened to the requirement that model trajectories neither grow infinitely, nor turn to zero, but vary within a certain bounded region of the state space. In what follows, this kind of stability will be referred to as *ecological stability*. Its contrast to the Lyapunov concept is illustrated in Figure 1 and a formal definition is given below.

Definition 1.2 *Let Ω_0^n and Ω^n be closed bounded domains in the interior of the positive orthant, $\mathbb{R}_+^n$, of the state space and $\Omega_0^n \subseteq \Omega^n$. A model ecosystem (1.4) is called* ecologically stable *(or* ecostable*) if any trajectory originated from Ω_0^n goes never out of Ω^n, or in symbolic terms,*

$$\forall N_0 \in \Omega_0^n, \quad \forall t \geq t_0 : N(t; N_0) \in \Omega^n \subset \text{Int } \mathbb{R}_+^n.$$

Domain Ω_0^n is then called the domain of ecostability.

Figures 12 and 13 may illustrate the notion, although the ecostability in these models is nothing but a consequence of their Lyapunov stability. Any oval in Figure 12 comprises a domain of ecostability $\Omega_0^2 = \Omega^2$, while in Figure 13 the domain is given by the whole positive quadrant $\mathbb{R}_+^2$.

Clearly, the domain Ω_0^n has to be *contensive* enough, i.e., it must cover a region of the state space within which the model still provides a representative picture of the ecosystem behavior, if it does so in the whole state space. It is also clear that the question whether or not a particular domain of ecostability is contensive leads to the issue of model adequacy and cannot be answered by a mathematical study alone.

Note that, in an ecostable model, an equilibrium, if perturbed, may not necessarily be restored, or there may be no equilibrium at all. Thus, a model may be ecologically stable, even having no Lyapunov stable equilibrium, so that the Lyapunov stability is a sufficient but not a necessary condition of ecological stability. Further on (Chapter 4) we will see, however, that there is a whole class of community models where the both stability notions become equivalent. These are the so-called *Volterra conservative* and *dissipative* systems; conservative are, in particular, the classical Lotka–Volterra pairs of prey-predator species, while dissipative are the "horizontal-structured" communities of competing species organized by the niche overlap principle (Chapter 7).

Further arguments in favor of Lyapunov stability analysis in population and community models may cite a method proposed below to investigate ecological stability in simple models.

B. A General Method for Simple Models

Among several mathematically formalized properties that contribute to the correctness of a model (1.4), the simplest one is probably that the positive orthant $\mathbb{R}_+^n$ is invariant with respect to the system. Since all $N_i > 0$ in the Int $\mathbb{R}_+^n$, one can always define the following change of variables:

$$\xi_i = \ln\,(N_i/N_j^0), \qquad i = 1, \ldots, n. \tag{1.5}$$

Substitution of ξ_i into (1.4) yields

$$d\xi_i/dt = \varphi_i(\xi_i, \ldots, \xi_n; N_1^0, \ldots, N_n^0) \tag{1.6}$$

with the initial conditions $\xi_i(0) = 0$, $i = 1, \ldots, n$. Obviously, if $\mathbf{N}(t; \mathbf{N}_0) \to \infty$, then $\boldsymbol{\xi}(t) \to +\infty$, but if $\mathbf{N}(t; \mathbf{N}_0) \to 0$, then $\boldsymbol{\xi}(t) \to -\infty$ (for finite $\mathbf{N}_0$). A solution to system (1.6), as a function of parameters $[N_1, \ldots, N_n]$ is thus defined in the entire state space $\mathbb{R}_\xi^n$ (rather than in its positive orthant).

As $\varphi_i(0, \ldots, 0; N_1^0, \ldots, N_n^0)$ is generally not zero, i.e., $\boldsymbol{\xi}^0(t) \equiv \mathbf{0}$ is generally not a solution to system (1.6), it follows that (1.6) can be transformed into

$$d\xi_i/dt = \Phi_i(\xi_1, \ldots, \xi_n; N_1^0, \ldots, N_n^0) + B_i, \qquad i = 1, \ldots, n. \qquad (1.7)$$

where

$$\begin{aligned} \Phi_i &= \varphi_i(\xi_1, \ldots, \xi_n; N_1^0, \ldots, N_n^0) - \varphi_i(0, \ldots, 0; N_1^0, \ldots, N_n^0), \\ B_i &= \varphi_i(0, \ldots, 0; N_1^0, \ldots, N_n^0), \qquad \mathbf{B} = [B_1, \ldots, B_n]. \end{aligned}$$

It is clear that if B_i are considered as *permanently acting perturbations* (PAP) of system (1.6), then $\boldsymbol{\xi}^0 = \mathbf{0}$ is a solution to the *unperturbed* system $d\xi_i/dt = \Phi_i$. We can now make use of the stability problem for the motion under PAP.[20]

Definition 1.3 *Solution* $\boldsymbol{\xi}^0 \equiv \mathbf{0}$ *to the system of equations (1.7) with* $\mathbf{B} = \mathbf{0}$ *is called* stable to PAP, *if for any small positive* ε *there exist positive numbers* $\eta_1(\varepsilon)$ *and* $\eta_2(\varepsilon)$ *such that for any initial value* $\xi(t_0)$ *not exceeding* $\eta_1(\varepsilon)$ *in its norm and for any PAP* $\mathbf{B}$ *not exceeding* $\eta_2(\varepsilon)$ *in their norm, the norm of solution* $\xi(t)$ *to the perturbed system (1.7) does not exceed* ε *for any* $t \geq t_0$. *In symbolic terms,*

$$\forall \, \varepsilon > 0 \; \exists \; \eta_1(\varepsilon), \, \eta_2(\varepsilon) > 0 \text{ such that } \forall \, \boldsymbol{\xi}(t_0), \|\boldsymbol{\xi}(t_0)\| < \eta_1(\varepsilon),$$

$$\forall \, \mathbf{B}, \|\mathbf{B}\| < \eta_2(\varepsilon), \forall \, t \geq t_0 : \|\boldsymbol{\xi}(t)\| < \varepsilon.$$

In other words, the perturbed solution must be close enough to the unperturbed one if both the initial value and the PAP are sufficiently small.

If we now solve the PAP-stability problem for a particular value of parameter $\mathbf{N}^0 > \mathbf{0}$ and determine a region of such values in the original model state space, then certain sufficient conditions of ecological stability in system (1.4) will arise. The proof of this statement, relying on Definition 1.3 and the properties of mapping $\mathbb{R}_+^n \to \mathbb{R}_\xi^n$, is given in the Appendix.

By the Chetaev–Malkin theorem,[20] the trivial solution $\boldsymbol{\xi}^0 \equiv \mathbf{0}$ is stable to PAP if it is Lyapunov asymptotically stable in the unperturbed system and the perturbations B_i are sufficiently small. Unfortunately, there are no effective and sufficiently general means to evaluate the smallness of B_i. Nevertheless, the ecostability problem for model ecosystem (1.4) is thus reduced—at least theoretically—to the Lyapunov stability problem for a transformed system.

This method has appeared capable of giving, if not the exact ranges, then at least meaningful enough estimates of the ecostability domain in $\mathbb{R}_+^n$ for simple particular cases of population models, where the dynamics can also be described by direct analysis.[18] This is quite logical, since the Lyapunov asymptotic stability of the trivial solution represents only a sufficient condition, rather than a criterion, of its stability to PAP. This indicates also the close tie between these two types of stability, of which the Lyapunov concept is more fundamental and mathematically much more developed.

III. MATRIX PROPERTIES AS A KEY TO STABILITY ANALYSIS

Progress in the theory and applications of the Lyapunov concept of stability was predetermined by two fundamental theorems of A. M. Lyapunov: on stability by the Lyapunov function of certain properties and on stability by linear approximation.[21] If a Lyapunov function is found for a particular system (or a particular class of systems), then one can judge the domain of stability in the state space of the model, while linear approximation can establish the local stability only. Unfortunately, there is no general method to construct a Lyapunov function, whereas the second Lyapunov method is more universal, reducing the stability problem to verification of certain properties of a certain matrix. Both methods are extensively used in ecological modeling, though the second one provides the major gate to the matrix- and graph-theoretic approaches.

A. Stability of Linear and Linearized Systems

Let a system of linear ordinary differential equations be written in the form

$$dx_i/dt = \sum_{j=1}^{n} a_{ij}x_j, \qquad i = 1, \ldots, n, \tag{1.8}$$

or in vector and matrix notation,

$$dx/dt = A\,x, \tag{1.8'}$$

where x denotes a column vector $x = [x_1, \ldots, x_n]^T$ and $A = [a_{ij}]$ is a matrix of constant entries and size $n \times n$. Its counterpart in difference equations assumes the form

$$x(t + 1) = A\,x(t). \tag{1.9}$$

The solution theory is completely developed for the case of linear systems, reducing the task of finding the general solution to that of finding all eigenvalues of matrix A. The solution is then to be constructed as a linear combination of time-dependent exponential curves with the exponents to be given by the eigenvalues (in case of difference equations the eigenvalues provide the bases to the exponential functions).

Consequently, the stability problem for the *zero solution* $x(t) \equiv 0$ to system (1.8) is reduced to the following well-known criterion.[19]

Theorem 1.1 *The zero solution of a linear system (1.8) is Lyapunov asymptotically stable if and only if all the eigenvalues of matrix A have negative real parts:*

$$\mathrm{Re}\ \lambda_i(A) < 0, \qquad i = 1, 2, \ldots, n.$$

If, on the contrary, there is an eigenvalue whose real part is positive, then the solution is unstable.

The mathematician would be jobless if all the eigenvalues of a matrix, i.e., its *spectrum*, $\Lambda(A)$, could generally be expressed explicitly in terms of the matrix entries. Actually, the stability problem for a matrix, in a more general context, is a kind of *eigenvalue localization* problem, and to study the problem requires naturally all elementary properties of the eigenvalues to be known. The basic property is that $\lambda(A)$ is neither an additive, nor a multiplicative function in the space of $n \times n$-matrices: $\lambda(A + B) \neq \lambda(A) + \lambda(B)$ and $\lambda(AB) \neq \lambda(A)\lambda(B)$ in the general case, although for the *identity matrix* $I = \text{diag}\{1, \ldots, 1\}$, the identity

$$\lambda(aA + bI) \equiv a\lambda(A) + b$$

holds true for any numbers a and b, any $n \times n$-matrix A, and each of its eigenvalues λ.

Also, if a matrix A is nonsingular, so too is its inverse, A^{-1}, whose eigenvalues are inverse to those of A. Hence, both must be stable or unstable simultaneously. The same is true for a matrix and its transpose, although the spectrum is now the same for both.

The strong analogy that exists between solutions of differential and corresponding difference equations, extends also to the stability criterion, resulting in a similar theorem with the only difference being that negativity of the eigenvalue real parts is replaced by the condition that all the eigenvalues be less than one in modulus.[22]

It is important to note that if a solution to the linear system of differential equations is locally asymptotically stable, then it is globally stable too, which is generally not true in nonlinear systems. Also, in the case of linear difference equations, the statement is already not so simple, requiring, as we shall see further in Chapter 2, additional constraints on matrix A.

Consider now a nonlinear system (1.4) which, however, admits linearization at an equilibrium point $\mathbf{N}^{\text{eq}}$. In other words, consider a system that, after the change of variables

$$\mathbf{x}(t) = \mathbf{N}(t) - \mathbf{N}^{\text{eq}} \tag{1.10}$$

and calculation of the *Jacobian matrix*

$$A = \left[\left| \frac{\partial f}{\partial N_j} \right|_{\mathbf{N}^{\text{eq}}} \right], \tag{1.11}$$

takes on the form

$$d\mathbf{x}/dt = A\mathbf{x} + \mathcal{O}(\mathbf{x}), \tag{1.12}$$

with vector function $\mathcal{O}(\mathbf{x})$ obeying some nonrestrictive technical conditions. Then the properties of solutions $\mathbf{N}(t)$ are obviously equivalent to those of $\mathbf{x}(t)$ and the Lyapunov asymptotic stability of the equilibrium $\mathbf{N}^{\text{eq}}$ can be verified by the fundamental theorem on stability *in the first (linear) approximation.*[19]

Theorem 1.2 *If all the eigenvalues of the Jacobian matrix A have negative real parts, then the zero solution of a nonlinear system (1.12) is asymptotically stable.*

If, on the contrary, there is an eigenvalue whose real part is positive, then the solution is unstable.

Note that the "only if" part inherent in Theorem 1.1 has now disappeared from the formulation, leaving the asymptotic behavior in the critical cases with a variety of possible stable or unstable patterns. But Theorem 1.2 still gives a powerful method to establish sufficient conditions for an equilibrium to be stable in a non-linear model. Similar to the fact in real analysis that the linear term bears the major part of the function increment, the linear stability analysis reveals the major tendency, if any exists, in the dynamical behavior of the nonlinear system.

For example, in multispecies Lotka–Volterra systems of population (nonlinear!) equations, where the Jacobian matrix appears to be in a certain fixed relation with the matrix of interaction coefficients (see Chapter 4), and in more general cases too, it opens the way to speculate on the effects that the structure of interactions causes in the set of matrix eigenvalues, the matrix *spectrum*, hence in the patterns of stability behavior. At the same time, this is a way in which the power of matrix and graph theory finds application to stability problems of theoretical ecology.

B. Indecomposability of a Matrix and Strong Connectedness of Its Associated Graph

Among the basic properties of a square matrix, *indecomposability* is one that plays an important role in the analysis of its *spectral* characteristics. For example, it is well known in the theory of matrix models for age-structured populations (Chapters 2 and 3), that the well-known Perron–Frobenius theorem on nonnegative matrices guarantees that the projection matrix have the maximal-in-modulus positive eigenvalue λ_1, the *dominant eigenvalue* (which appears to be related to the Malthusian parameter $r = \ln \lambda_1$), and the positive eigenvector corresponding to λ_1 (the *dominant vector*), which is interpreted as a steady-state age structure. What is less known though is that indecomposability is a necessary premise for the theorem to be actually applied to the projection matrix. It also proves to be in strong connection with a basic structural property such as *strong connectedness* of the associated graph.

The formal definition of indecomposability requires the notion of matrix permutation. A permutation of matrix rows is equivalent to its being multiplied from the left by the so-called *row permutation matrix* $P = [p_{ij}]$, whose entries p_{ij} are equal to 1 if the permutation places the jth row on the ith one, and to 0 if otherwise. On the contrary, a *column permutation matrix* has a unit entry p_{ij} if the permutation places the ith column on the jth one, and zero entries if otherwise, while a permutation of matrix columns is equivalent to its being multiplied from the right by the column permutation matrix. Clearly, a permutation matrix has only one nonzero entry in each row and column.

Two matrix equalities below illustrate these statements for permutation

$$\begin{pmatrix} 1 & 2 & 3 \\ 2 & 3 & 1 \end{pmatrix}$$

of a matrix 3×3:

$$
\begin{bmatrix} 0 & 0 & 1 \\ 1 & 0 & 0 \\ 0 & 1 & 0 \end{bmatrix}
\begin{bmatrix} a & b & c \\ d & e & f \\ g & h & i \end{bmatrix}
=
\begin{bmatrix} g & h & i \\ a & b & c \\ d & e & f \end{bmatrix},
$$

$$
\begin{bmatrix} a & b & c \\ d & e & f \\ g & h & i \end{bmatrix}
\begin{bmatrix} 0 & 1 & 0 \\ 0 & 0 & 1 \\ 1 & 0 & 0 \end{bmatrix}
=
\begin{bmatrix} c & a & b \\ f & d & e \\ i & g & h \end{bmatrix}.
$$

The row and column permutation matrices for any one permutation are obviously transposed with respect to each other, so that the result of a simultaneous permutation of matrix A's rows and columns can be calculated as $P^{T}AP$. For example,

$$
\begin{bmatrix} 0 & 0 & 1 \\ 1 & 0 & 0 \\ 0 & 1 & 0 \end{bmatrix}
\begin{bmatrix} a & b & c \\ d & e & f \\ g & h & i \end{bmatrix}
\begin{bmatrix} 0 & 1 & 0 \\ 0 & 0 & 1 \\ 1 & 0 & 0 \end{bmatrix}
=
\begin{bmatrix} i & g & h \\ c & a & b \\ f & d & e \end{bmatrix}.
\tag{1.13}
$$

A remarkable property of permutation matrices is their being *orthogonal*, i.e., nonsingular matrices satisfying the condition $P^{T} = P^{-1}$. Thus, after a simultaneous permutation, we have

$$
A \sim P^{-1}AP
\tag{1.14}
$$

which is a formal expression that two matrices, A and $B = P^{-1}AP$, are *similar*. Similar matrices can be shown to have identical spectra of their eigenvalues, so that any simultaneous row-and-column permutation of a matrix keeps its eigenvalues unchanged.

Note that a simultaneous permutation of matrix rows and columns means renumbering the model variables (of species if we deal with a community matrix, or population subclasses if we deal with a projection matrix of an age-structured population). Renumbering should certainly not affect the properties of the model, in general, and its asymptotic behavior, in particular. Since the latter is essentially a function of the eigenvalue spectrum, the above statement gives just a formal justification to this clear conjecture.

Definition 1.4 *A matrix $A = [a_{ij}]$ of dimensions $n \times n$ is called* decomposable *if there exists a simultaneous permutation of matrix rows and columns that brings it to the form*

$$
\begin{bmatrix} B & O \\ C & D \end{bmatrix} \sim A
\tag{1.15}
$$

where B and C are square blocks of the sizes $p \times p$ and $q \times q$ respectively ($p+q = n$), while O is a $(p \times q)$-block of zeros. Otherwise, a matrix is called indecomposable,[23] *or* irreducible *in some other texts.*[24-25]

The matrix of example (1.13) becomes decomposable if, for instance, $b = h = 0$.

If a decomposable matrix is presented in its definitive form (1.15), then, by a proper Laplace decomposition of its characteristic determinant, the spectrum of the matrix can be shown to consist of the spectra of its diagonal blocks B

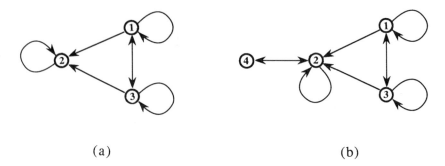

(a)   (b)

Figure 2.   Examples of the associated graph for a 3 × 3 (a) and 4 × 4 (b) matrices.

and D. Since permutation does not affect eigenvalues, the eigenvalue problem for a decomposable matrix reduces to the same problem individually for each of its definitive diagonal blocks of lower dimensions, whereas indecomposable matrices can no longer admit such a reduction.

To check whether there exists a proper permutation in the problem of indecomposability for a given matrix A is a task of combinatorial awkwardness. Fortunately there always exists another formulation of the problem, which often helps us to see the result quite simply. The formulation is in terms of the so-called *associated graph*, $D(A)$, the directed graph (or *digraph*) which is associated with matrix A by the following rule: the graph has n vertices numbered in the very same way as rows and columns of the matrix are (usually by integers 1 through n) and each nonzero matrix entry of subscripts i, j represents an *arc* (or *link*) directed from vertex j to vertex i.

For example, a 3 × 3-matrix of (1.13) with $b = h = 0$ and the rest of the entries being nonzero has its associated graph, as shown in Figure 2a; the digraph in Figure 2b is the associated graph for the 4 × 4-matrix of (1.16) below.

Any square matrix has a unique associated graph, whereas there exists a whole class of $n \times n$-matrices for which a given digraph is the associated graph: these are matrices having the same pattern of nonzero entries, while differing in any way in numerical values of those entries. Within that class there is a unique matrix, $A(D)$, that has all its nonzero entries equal to one. Its transpose, $[A(D)]^T$, is known in graph theory as the *adjacency matrix* of a digraph D: the element a_{ij}^T indicates whether or not vertex j is *adjacent* to vertex i by the link $j \rightarrow i$ (or by the directed path of length 1).

Since transposing does not affect decomposability,[24,26] we may stick to the same rule in constructing the adjacency matrix as in above construction of the associated graph, i.e., the link $j \rightarrow i$ corresponds to $a_{ij} = 1$, which appears more convenient in our further considerations. So, for the digraphs in Figure 2, we have

$$A(D_a) = \begin{bmatrix} 1 & 0 & 1 \\ 1 & 1 & 1 \\ 1 & 0 & 1 \end{bmatrix}, \quad A(D_b) = \begin{bmatrix} 1 & 0 & 1 & 0 \\ 1 & 1 & 1 & 1 \\ 1 & 0 & 1 & 0 \\ 0 & 1 & 0 & 0 \end{bmatrix}. \quad (1.16)$$

Definition 1.5 *A directed graph is called* strongly connected *(or* strong*) if for each pair of vertices k ≠ l there exists a directed path from k to l.*[26]

The property of being strongly connected is obviously independent of how the vertices are numbered. Moreover, it is well known (and does not require much mathematical effort to show) that if a matrix is indecomposable then its associated graph is strongly connected and, conversely, if a digraph is strong then any matrix for which the digraph is the associated graph, is indecomposable. The digraph in Figure 2a, for example, is not strong, as there is no path from vertex 2 to any other one, and the matrix is thus decomposable, as was noted above. The digraphs in Figure 3 of Chapter 2 are obviously strong and we shall discuss later on what follows from their matrices being indecomposable.

If a digraph D is not strong then it must consist of (a finite number of) *strong components*, the maximal subsets of the vertex set which are strongly connected in D. Strong components of a digraph associated with a decomposable matrix A correspond to its indecomposable diagonal blocks. For example, the digraph in Figure 2a has two strong components: the first one is formed on the subset $\{1, 3\}$ by links $1 \rightarrow 1$, $1 \rightarrow 3$, $3 \rightarrow 1$, and $3 \rightarrow 3$, while the second one on the subset $\{2\}$ by link $2 \rightarrow 2$. Figure 2b has apparently two strong components, too.

There is also an algebraic method to identify strong components in a digraph or to verify its strong connectedness when the number of strong components equals one. The method is based upon analysis of the so-called *reachability matrix R*, whose entry of subscripts i, j equals one if there is a finite directed path in D that reaches vertex i starting from j, and equals zero if, otherwise, no path from j reaches vertex i.

The reachability matrix for a digraph D can be calculated by the formula

$$R(D) = [I + A(D) + A(D)^2 + \cdots]_B, \tag{1.17}$$

where I is the identity matrix, $A(D)$ is the adjacency matrix for D, and subscript B indicates the Boolean nature of the algebraic operations.[26] Each term $A(D)^l$ in (1.17) shows which vertices are connected in D by a directed path of length l ($l = 1, 2, \ldots$). The maximal length of a path (with no self-crossing) in D is obviously not greater than n (equals n for a loop spanning all n vertices), so that the sum in (1.17) is actually finite. In practice, the summing should proceed until a subsequent summand can no longer change the sum. Equal rows of R will then indicate those vertices in D which belong to the same strong component, while the number of different rows will give the number of strong components.

Thus, for the digraphs in Figure 2, we have

$$R(D_a) = \begin{bmatrix} 1 & 0 & 1 \\ 1 & 1 & 1 \\ 1 & 0 & 1 \end{bmatrix}, \qquad R(D_b) = \begin{bmatrix} 1 & 0 & 1 & 0 \\ 1 & 1 & 1 & 1 \\ 1 & 0 & 1 & 0 \\ 1 & 1 & 1 & 1 \end{bmatrix}. \tag{1.18}$$

By (1.17) matrices $R(D)$ appear quite computable in general, though too high dimensions may cause pure technical problems. Either verified visually in the associated graph, or established by computation of its reachability matrix, inde-composability serves as a formal analog to some structural integrity in the system under study.

Traditionally considered by matrix theory in connection with nonnegative ma-trices, indecomposability does not require, by definition, any sign constraints on nonzero entries. In the sequel, it is treated both for nonnegative matrices (e.g., pro-jection matrices in age-structured models) and for more general ones (e.g., commu-nity matrices in multispecies models). Each area brings about further development of indecomposability issues as well as further interpretation of indecomposability in more meaningful terms.

IV. HOMAGE TO LYAPUNOV OR WHY THERE ARE SO MANY STABILITY PROBLEMS

The world of mathematical ecology is essentially a nonlinear world where models, even those which are simple in structure, may reveal complex types of dynamic behavior. Unless these types result in extinction or outbreaks of any population, all are covered by the concept of ecological stability, which appears simple only in its general formulation. Recent advances in modern theory of dynamical sys-tems, which include such fascinating phenomena as stochasticity, strange attractors, and catastrophes, find highly illustrative applications in the field of mathematical ecology,[27] expanding the variety of models used to describe the phenomena under study.

The idea of multiple equilibria, for instance, has grown (see the survey by Levin[28]) from the understanding that regular oscillations around an equilibrium are quite far from providing a representative model of ecologically stable behavior. Depending on perturbations, a system may pass from one domain of attraction of an equilibrium into another or, under small stochastic PAP, it may demonstrate irreg-ular transitions among different steady states, spending the major part of the time in their respective vicinities.[29-30] This kind of dynamic behavior has been general-ized into the notion of resilience for ecological models.[31-32] It can be shown[33] that resilience is generally a weaker property than the Lyapunov equilibrium stability, though a stronger one than just ecological stability.

The latter kind of stability also covers the behavior of some models in topo-logical catastrophe theory.[34-36] It appears that the dynamics that can be classified as a catastrophe in topology,[37] may be quite far from being a catastrophe in the ecological sense,[33,38] causing just some reorganization in the ecosystem rather than the loss of its stability.

Related to the problem of ecological stability are also the issue of "chaos" and the existence of so-called "strange attractors" in models of population and com-munity dynamics. After Lorenz[39] had discovered a chaotic, or "pseudo-stochastic" regime in a deterministic system of equations, this behavior was related to the tur-bulence phenomenon by the concept of the "strange attractor";[40] similar regimes were discovered in nonlinear first-order difference equations,[41-42] which stimulated further efforts in the field of population models—both in nonlinear difference equations and systems of ordinary differential equations of dimension $n \geq 3$ (surveys can be found, e.g., in Yatsalo[43] and Shaffer[44]).

The fact that even a simple discrete analog of the logistic model gives rise to a chaotic behavior[45-46] put certain methodological difficulties in our path toward forecasting the dynamics of biological populations. It is important, however, that attempts to fit single-species equations with data on both field[47-48] and laboratory[49] populations often results in nonchaotic regimes. The cause is simply that estimates obtained for critical population parameters fall outside the corresponding regions of chaotic behavior in the model. As a result, some ecologists are prone to consider the potential to chaos inherent in difference equations as a peculiarity of this kind of mathematical description rather than a property of real dynamics, where the irregularities have to be related with stochasticity in external perturbing factors.

In multispecies models, however, the conditions giving rise to chaos and strange attractors sometimes prove to be natural and interpretable in ecological terms.[50-54] As a matter of fact, the question of whether a population behave stably in a stable environment or whether it is the law of its own dynamics which implies chaos, cannot be definitely solved in the framework of the model alone, because it may always happen that factors responsible for a particular type of dynamic behavior are either neglected or averaged in the model's formulation.[55]

All the above-mentioned alternatives to the classical Lyapunov notion of stability logically expand the class of mathematical models that are capable of giving an adequate description of population or ecosystem dynamics. Unfortunately, the alternative analysis in particular models often encounters a lack of adequate analytical methods. Nowadays, some successful attempts are known which apply these concepts to certain phenomena of ecological interest (see e.g. a survey in Svirezhev[56]), but neither of the concepts have yet achieved a stage in its development where it could consequently be applied to all basic types of ecological models at a satisfactory level of generality.

On the other hand, powerful mathematical apparatus of the Lyapunov stability—in spite of the limiting sense of the concept—provides a potential to develop a stability theory for ecological models, to pose some stability problems anew and to find new results (particularly, in multispecies formulations) of dynamical theory of ecological systems. This is the Lyapunov notion which serves as a basis to develop further notions of stability, and this seems to be a promising way to combine theoretical ecology with mathematical theories of matrices and graphs, to which this book is devoted.

In matrix models of (isolated) age-structured populations (Chapters 2 and 3), the difference-equations counterpart to Lyapunov Definition 1.1 and the mathematics of linear algebra and matrix theory give a potential tool to relate the pattern of asymptotic behavior and population age structure with the demographic parameters and the pattern of the life cycle.

Developed in multispecies models, further notions of matrix stability constitute a hierarchy of community matrix properties, each having its own ecological meaning (Chapter 4). Through the Lyapunov concept, the notion of qualitative stability in model ecosystems is formalized into sign stability of the community matrix, that has now been developed up to a completed theory of sign-stable structures and their finite description in terms of the types of intra- and inter-species relations (Chapter 5). The latter theory has found an application in another theory concerned with a more particular class of ecological systems, namely, trophic chains, that also relied upon the Lyapunov stability in a series of special kinds of equilibria (Chapter 6).

Legitimately connected with (although, in some speculations, omitted from) the issue of equilibrium stability is the existence problem for an equilibrium with all positive components. In some cases it is exactly this problem whose solution may generate a meaningful theory of stable structures such as, for instance, the theory of competition communities which are organized by the niche overlap principle (Chapter 7).

A traditional theme in stability issues of theoretical ecology is the stabilization effect of spatial distribution in models and reality. Though not being a keystone of stability studies in spatially distributed models, the Lyapunov concept still promotes insight into the effect of space dimension by studying the so-called "box" models of spatial systems, where it proves to be able to generate interpretable conditions for stabilizing or destabilzing effects of migration on the dynamics of the whole system (Chapter 8).

In addition, the Lyapunov stability analysis always represents a preliminary and necessary step in studying more complex types of ecological stability, such as resilience,[31-33] Hopf bifurcations,[57] "hierarchical" stability,[38] as well as in studying the "stability vs. complexity" problem of theoretical ecology.[3,58]

Besides these theoretical aspects, there is also an "empirical" evidence in favor of Lyapunov analysis in ecological models, that is backed by studies on random construction of community matrices (see e.g. Pimm,[59] Cohen et al.,[60] and references therein). When some properties observable in a statistical ensemble of community matrices such as the number of trophic levels, connectedness, the number of "omnivory" links, etc., are compared with those in a representative collection of real communities, it turns out that real structural properties are reproduced with much greater probability within the subset of stable matrices than over the whole ensemble.

To summarize, various stability problems are even more numerous in mathematics than stability meanings in ecology. However, it is encouraging to observe that the Lyapunov concept, which supports certain general features, is quite relevant to both areas. That is why it makes sense to investigate the Lyapunov and other stabilities in ecological models and to develop interpretations pertinent to the major types of such models; and this is what the further chapters of the book are mostly devoted to.

Appendix

Statement A1.1.   The ecostability problem for system (1.4) can be reduced to a stability problem under permanently acting perturbations (PAP).

Proof: Since the change of variables (1.4) is continuous and monotone, the domains $\Omega_0^n \subseteq \Omega^n$, as bounded and closed in $\mathbb{R}_+^n$ will be so in the space $\mathbb{R}_\xi^n$ too, $\xi(\Omega_0^n) \subseteq$

$\xi(\Omega^n)$, containing also the origin. Then, in the definition of ecostability for the transformed system (1.7), the domains $\xi(\Omega_0^n)$ and $\xi(\Omega^n)$ can be replaced merely by spheres of sufficiently small and sufficiently large radii ρ_0 and P:

$$S_0 = \{\boldsymbol{\xi} : \|\boldsymbol{\xi}\| \leq \rho_0\} \subset \xi(\Omega_0^n), \tag{A1.1}$$

$$S = \{\boldsymbol{\xi} : \|\boldsymbol{\xi}\| \leq P\} \supset \xi(\Omega^n). \tag{A1.2}$$

The formal definition thus becomes equivalent to the following condition:

$$\exists \; \rho_0 > 0 : \forall \; \boldsymbol{\xi}(t_0), \|\boldsymbol{\xi}\|(t_0)\| \leq \rho_0, \text{ and } \forall \; t \geq t_0 : \|\boldsymbol{\xi}(t; \mathbf{N}^0\| \leq P. \tag{A1.3}$$

If the parameter $\mathbf{N}^0$ varies, the unique solutions to system (1.7) starting at the origin form a family as a function of $\mathbf{N}^0$, the whole family staying within the finite sphere of radius P.

Comparing Definition 1.3 of stability under PAP and the condition (A1.3), one can see that if $\boldsymbol{\xi}^0$ is stable to PAP, then for sufficiently small perturbations $\mathbf{B}$ the radii $P = \varepsilon = \eta_2^{-1}(\|\mathbf{B}\|)$ and $\rho_0 = \eta_1(\varepsilon)$ guarantee the condition (A1.3) to hold true, thus confirming the ecological stability. The domain of ecostability in the orthant $\mathbb{R}_+^n$ will then be given by $\Omega_0^n = \xi^{-1}(S_0)$. ∎

Additional Notes

To 1.I. Further comments on diversity measures in relation with the concept of ecosystem *homeostasis* can be found in Trojan.[61]

Particular examples of the model approach to stability analysis are so numerous in the current literature that it can hardly make sense to survey all of them. For students of mathematics the books by Jeffries[62] and Hofbauer and Sigmund[63] can be useful respectively as an elementary and a deeper introduction to the "elementary" theory of dynamical systems and its application to the theory of evolution; the latter book contains also an extensive bibliography of related works of the 1970s and 1980s.

To 1.II. The trivial observation that environmental conditions may often vary in time and modify the dynamic behavior of an ecosystem, can be accounted for in models by resorting to the concept of stability to PAP. Clearly, the outcome depends essentially on which parameters of the model are assumed to undergo the respective variations and there are several possible ways to include this kind of perturbations (namely, PAP) into a model, particularly when the perturbations are stochastic.[64−66]

Ecological stability of population models was sometimes referred to as *Lagrange stability*, a particular case of orbital stability in theoretical mechanics. Under the names of *Lagrange stability*[18,67−68] and *boundedness*,[69] some kinds of ecological stability were analyzed in few publications, but there are still no methods sufficiently well developed to make general inferences.

Among a number of related concepts, the closest one—which has produced some elegant mathematical findings—is *permanence*;[70−73] this requires that any trajectory, if started in the interior of the positive orthant, eventually attains and then forever remains in a compact subset within the orthant; in particular, the boundary faces of the orthant must act as a *repellor* set. For this reason, for example, there is no permanence in the conservative Lotka–Volterra prey-predator system, which should certainly be regarded as ecologically stable. Note also, that both the necessary and the sufficient conditions of permanence in Lotka–Volterra systems[73] reduce the problem to investigating the equilibria in the interior and the boundary of the positive orthant, thus revealing again the close tie that permanence has with the Lyapunov concept.

To 1.III. The theory of solutions to linear systems of ODEs with constant co-efficients is explained in any introductory textbook on ODEs, although sampling at examinations has revealed that exactly these highly practical chapters of the theory are really not as well known as they should be, even among students of mathematics. Even more confusion appears—even in some published manuals—due to the disappearance of the "only if" part in the formulation of Theorem 1.2. In fact, a variety of nonlinear stability patterns has appeared just because of this "disappearance."[56−57]

Although omitted sometimes from standard courses of linear algebra and ma-trix theory, indecomposability is of great importance in applications. Despite *re-ducibility* sounding much easier than *decomposability*, we have reserved the term for another notion. In favor of decomposability one can only add that not only the matrix itself is *composed* of its diagonal blocks, but also its spectrum is *composed* of the spectra of those blocks.

To 1.IV. Each of the dynamic properties mentioned has certainly a variety of meanings and corresponding definitions in the mathematical theory of dynamical systems. But even the same notions do not escape the multi-name fate. For example, after R.Thom,[34] the term *catastrophe* became quite popular both in theory and applications,[35] although Russian mathematicians[36] prefer making use of the original meaning, which is *Whitney's singularity in smooth mappings*.

Homage to the linear approach may sound strange nowadays, given today's background of rapidly developing nonlinear techniques already mentioned. Men-tioned in relation with stability theorems, the linearity-versus-nonlinearity dialectics is also realized in a broader mathematical context,[74] and the dialectics can probably be recognized by any reader who is absorbed in problems arising in any particular area of ecological modeling.

REFERENCES

1. Lewontin, R. C., The meaning of stability, in *Diversity and Stability in Ecological Systems, Brookhaven Symposium in Biology* No. 22. National Bureau of Standards, U. S. Department of Commerce, Springfield, Va., 1969, 13–24.

2. Usher, M. B. and Williamson, M. H., Eds., *Ecological Stability*, Chapman & Hall, London, 1974.

3. May, R. M., *Stability and Complexity in Model Ecosystems* (Monographs in Population Biology, Vol. 6), 2nd ed., Princeton University Press, Princeton, NJ, 1974, 265 pp.

4. Svirezhev, Yu. M. and Logofet, D. O., *Stability of Biological Communities* (revised from the 1978 Russian edition), Mir Publishers, Moscow, 1983, 319 pp.

5. Margalef, R., *Perspectives in Ecological Theory*, University of Chicago Press, Chicago, 1968, 112 pp.

6. Goodman, D., The theory of diversity-stability relationships in ecology, *Quart. Rev. Biol.*, 50, 237–266, 1975.

7. MacArthur, R. H., Fluctuations of animal populations and a measure of community stability, *Ecology*, 36, 533–536, 1955.

8. Elton, C. S., *The Ecology of Invasions by Animals and Plants*, Methuen, London, 1958, 181 pp.

9. Odum, E. P., Diversity as a function of energy flow, in *Unifying Concepts in Ecology* (*Proc.* 1st *Int. Congress of Ecology*), van Dobben, W. H. and McConnel, R. H., Eds., W. Junk Publ., The Hague, 1976, 11–14.

10. Logofet, D. O. and Svirezhev, Yu. M., Stability in models of interacting populations, in *Problemy Kibernetiki* (*Problems of Cybernetics*), Vol. 32, Yablonsky, S. V., Ed., Nauka, Moscow, 1977, 187–202 (in Russian).

11. Odum, E. P., *Basic Ecology*, Saunders College Publishing, Philadelphia, 1983, Chaps. 2 and 3.

12. Odum, E. P., *Fundamentals of Ecology*, 3rd ed., Saunders, Philadelphia, 1971, p. 256.

13. Logofet, D. O., What is mathematical ecology? in *Mathematical Models in Ecology and Genetics*, Svirezhev, Yu. M. and Passekov, V. P., Eds., Nauka, Moscow, 1981, 8–17 (in Russian).

14. Logofet, D. O. and Alexandrov, G. A., Modelling of matter cycle in a mesotrophic bog ecosystem. II. Dynamic model and ecological succession, *Ecological Modelling*, 21, 259–276, 1983/1984.

15. Alexandrov, G. A. and Logofet, D. O., A dynamic model of the combined nitrogen and organic matter cycle in the biogeocenosis of a transient bog, in *Mathematical Modeling of Biogeocenotic Processes*, Svirezhev, Yu. M., Ed., Nauka, Moscow, 1985, 80–97 (in Russian).

16. Logofet, D. O. and Alexandrov, G. A., Interference between mosses and trees in the framework of a dynamic model of carbon and nitrogen cycling in a mesotrophic bog ecosystem, in: Mitsch, W., Straškraba, M., and Jorgensen, S. E., Eds., *Wetland Modelling* (Developments in Ecological Modelling, Vol. 12). Elsevier, Amsterdam, 1988, 55–66.

17. Logofet, D. O. and Svirezhev, Yu. M., Stability concepts for biological systems, in *Problems of Ecological Monitoring and Ecosystem Modelling*, Vol. 5, Izrael, Yu. A., Ed., Gidrometeoizdat, Leningrad, 1983, 159–171 (in Russian).

18. Logofet, D. O. and Svirezhev, Yu. M., Ecological stability and Lagrange stability. A novel view on the problem, in *Problems of Ecological Monitoring and Ecosystem Modelling*, Vol. 7, Izrael, Yu. A., Ed., Gidrometeoizdat, Leningrad, 1985, 253–258 (in Russian).

19. Barbashin, E. A., *Introduction to Stability Theory*, Nauka, Moscow, 1967, Chap. 1 (in Russian).

20. Malkin, I. G., *Theory of Motion Stability*, Nauka, Moscow, 1966, Chap. 6 (in Russian).

21. Malkin, I. G., *ibid*, Chaps. 1 and 2.

22. Mirolyubov, A. A. and Soldatov, M. A., *Linear Homogeneous Difference Equations*, Nauka, Moscow, 1981, 208 pp., Chaps. 2, 3 (in Russian).

23. Gantmacher, F. R., *The Theory of Matrices*, Chelsea, New York, 1960, Chap. 13.

24. Berman, A. and Plemmons, R. J., *Nonnegative Matrices in the Mathematical Sciences*, Academic Press, New York, 1979, 316 pp., Chap. 2.

25. Horn, R. A. and Johnson, C. R., *Matrix Analysis*, Cambridge University Press, Cambridge, 1990, Chap. 6.2.

26. Harary, F., Norman, R. Z., and Cartwright, D., *Structural Models: an Introduction to the Theory of Directed Graphs*, John Wiley, New York, 1965, Chap. 7.

27. Svirezhev, Yu. M. and Logofet, D. O., Complicated dynamics in simple models of ecological systems, in: *Lotka-Volterra Approach to Cooperation and Competition in Dynamical Systems. Proc. 5th Meeting UNESCO Working Group on System Theory, Wartburg, Eisenach, GDR, March 5–9, 1984*. Ebeling, W. and Peschel, M., Eds. (Mathematical Research, Vol. 23), Academie-Verlag, 1985, 13–22.

28. Levin, S. A., Multiple equilibria in ecological models, in *Proc. Int. Symp. Problems of Mathematical Modeling in the Man-Environment Interaction (Telavi, September, 1978)*, Vol. 1. Computer Center, USSR Acad. Sci., Moscow, 1979, 164–230.

29. Freydlin, M. I. and Svetlosanov, V. A., On the effect of small random perturbations on stability of states in ecological systems, *Zhurnal Obshchei Biologii* (Journal of General Biology), 37, 715–719, 1976 (in Russian).

30. Sidorin, A. P., Behavior of a population with several steady states in a random environment, in *Mathematical Models in Ecology and Genetics*, Svirezhev, Yu. M. and Passekov, V. P., Eds., Nauka, Moscow, 1981, 71–74 (in Russian).

31. Holling, C. S., Resilience and stability of ecological systems, *Annual Rev. Ecol. and Syst.*, Vol. 4, Palo Alto, CA, 1973, 1–23.

32. Grümm, H. R., Definitions of Resilience. Research Report RR-76-5, IIASA, Laxenburg, Austria, 1976, 20 pp.

33. Logofet, D. O., Towards stability problem of ecological systems, in *Proc. Int. Symp. Problems of Mathematical Modeling in the Man-Environment Interaction (Telavi, September, 1978)*, Vol. 2. Computer Center, USSR Acad. Sci., Moscow, 1981, 61–75.

34. Thom, R., Topological models in biology, *Topology*, 8, 315–335, 1969.

35. Poston, T. and Stewart, I., *Catastrophe Theory and Its Applications*, Pitman, London, 1978.

36. Arnold, V. I., Varchenko, A. H., and Guseyn-Zadeh, S. M., *Singularities of Differentiable Mappings*, Nauka, Moscow, 1982, 304 pp. (in Russian).

37. Jones, D. D., The application of catastrophe theory to ecological systems. Research Report RR-75-15, IIASA, Laxenburg, Austria, 1975, 57 pp.

38. Logofet, D. O. and Svirezhev, Yu. M., Ecological modelling, catastrophe theory and some new concepts of stability, in *EURO IFIP 79. European Conference on Applied Information Technology, London, 25–28 September, 1979*. Samet P. A., Ed., North-Holland, Amsterdam, 1979, 287–293.

39. Lorenz, E. N., Deterministic nonperiodic flow, *J. Atmos. Sci.*, 20, 130–141. 1963.

40. Ruelle, D., Takens, F., On the nature of turbulence, *Commun. Math. Phys.*, 20, 167–192, 1971.

41. Sharkovsky, A. N., Coexistence of cycles in a continuous mapping of the straight line into itself, *Ukrainian Math. Journal*, 16, 61–71, 1964 (in Russian).

42. Li, T.-Y. and Yorke, J. A., Period three implies chaos, *Amer. Math. Monthly*, 82, 982–985, 1975.

43. Yatsalo, B. I., Complicated Dynamical Regimes in Models of Closed Ecological Systems. Candidate of Sciences thesis, Moscow State University, Moscow, 1984, 112 pp.

44. Shaffer, W. M., Order and chaos in ecological systems, *Ecology*, 66, 93–106, 1985.

45. May, R. M., Biological populations obeying difference equations: stable points, stable cycles, and chaos, *J. Theor. Biol.*, 51, 511–524, 1975.

46. Svirezhev, Yu. M. and Logofet, D. O., *Stability of Biological Communities*, Mir Publishers, Moscow, 1983 (translated from the 1978 Russian edition), Chap. 2.

47. Hassel, M. P., Lawton J. H., and May R. M., Patterns of dynamical behavior in single-species populations, *J. Anim. Ecology,* 45, 471–486, 1976.

48. Bellows, T. S., The descriptive properties of some models for density dependence, *J. Anim. Ecology*, 50, 139–156, 1981.

49. Thomas, W. R., Pomcrantz, M. J., and Gilpin, M. E., Chaos, asymmetric growth, and group selection for dynamical stability, *Ecology*, 61, 1312–1320, 1980.

50. Alekseev, V. V., Biophysics of communities of living organisms, *Uspekhi Fizicheskih Nauk* (Advances in Physical Sciences), 120, 647–676, 1976 (in Russian).

51. Alekseev, V. V. and Kornilovsky, A. N., Autostochastic processes in biophysical systems, *Biofizika* (Biophysics), 27, 890–894, 1982 (in Russian).

52. Alekseev, V. V. and Kornilovsky, A. N., Ecosystem stochasticity model, *Ecol. Modelling*, 28, 217–229, 1985.

53. Dombrovsky, Yu. A. and Markman, G. S., *Spatial and Temporal Ordering in Ecological and Biochemical Systems*, Rostov University Press, Rostov-na-Donu, 1983, 118 pp. (in Russian).

54. Svirezhev, Yu. M., *Nonlinear Waves, Dissipative Structures and Catastrophes in Ecology*, Nauka, Moscow, 1987, 368 pp., Chap. 10 (in Russian).

55. Nisbett, R. M. and Gurney, W. S. C., *Modelling Fluctuating Populations*, John Wiley, New York, 1982.

56. Svirezhev, Yu. M., 1987, *ibid.*

57. Marsden, J. E. and McCracken, M. *The Hopf Bifurcation and Its Applications* (Applied Mathematics Series, Vol. 19), Springer, Berlin, 1976.

58. Svirezhev, Yu. M. and Logofet, D. O., *Stability of Biological Communities* (revised from the 1978 Russian edition), Mir Publishers, Moscow, 1983, Chap. 9.

59. Pimm, S. L., *Food Webs*, Chapman & Hall, London, 1982, 219 pp.

60. Cohen, J. E., Briand, F. and Newman, C. M., Community food webs: data and theory, *Biomathematics*, Springer-Verlag, Berlin, 20, 1990, 308 pp.

61. Trojan, P., Ecosystem Homeostasis, PWN—Polish Scientific Publ., Warsaw, 1984, 132 pp.

62. Jeffries, C., Mathematical Modeling in Ecology: a Workbook for Students, Birkhäuser, Boston, 1988, 193 pp.

63. Hofbauer, J. and Sigmund, K., *The Theory of Evolution and Dynamical Systems*, Cambridge University Press, Cambridge, 1988, 341 pp.

64. Levins, R., The effects of random variations on different types of population growth, *Proc. Nat. Acad. Sci. U.S.A.*, 64, 1061–1065, 1969.

65. Krapivin, V. F., *Persistence Theory of Complex Systems in Conflict Conditions*, Nauka, Moscow, 278 pp. (in Russian).

66. Svirezhev, Yu. M., 1987, *ibid*, Chaps. 11, 12.

67. Thornton, K. W. and Mulholland R. J., Lagrange stability and ecological systems, *J. Theor. Biol.*, 45, 473–485, 1974.

68. Adzhabyan, N. A., Ecological stability of a predator-prey system, in *Problems of Ecological Monitoring and Ecosystem Modelling*, Vol. 10, Izrael, Yu. A., Ed., Gidrometeoizdat, Leningrad, 1988, 162–171 (in Russian).

69. Brauer, F., Boundedness of solutions of predator-prey systems, *Theor. Popul. Biol.*, 15, 268–273, 1979.

70. Hutson, V. and Law, R., Permanent coexistence in general models of three interacting species, *J. Math Biol.*, 21, 285–298, 1985.

71. Jansen, W., A permanence theorem for replicator and Lotka–Volterra systems, *J. Math Biol.*, 25, 411–422, 1987.

72. Hofbauer, J., Hutson, V, and Jansen, W., Coexistence for systems governed by difference equations of Lotka–Volterra type, *J. Math Biol.*, 25, 553–570, 1987.

73. Hofbauer, J. and Sigmund, K., *The Theory of Evolution and Dynamical Systems*, Cambridge University Press, Cambridge, 1988, 341 pp., Chap. 19.

74. Casti, J., Linear methods for nonlinear problems, in *Linear Algebra and Its Role in Systems Theory. Proc. Summer Research Conference held on July 29–August 4, 1984* (*Contemporary Mathematics*, Vol. 47), Brualdi, R. A., Carlson, D. H., Datta, B. N., Johnson, C. R., and Plemmons, R. J., Eds., American Mathematical Society, Providence, RI, 1985, 65–80.

CHAPTER 2

Leslie Matrix Model: A Challenge to Linear Algebra

This chapter deals with the Leslie matrix model for the dynamics of a population with discrete age structure. The model is based upon several simplifying assumptions, whose oversimplicity has long been invoking widespread theoretical discussions, while narrowing the area of practical applications of the model in its "pure" Leslie form. On the one hand, linearity of the model has allowed all the power of linear algebra to give an exhaustive picture of model behavior. On the other hand, applications require the rigid assumptions to be somehow attenuated, resulting in various modifications of the classical doctrine. However, the fact that the mathematics of the classical Leslie model has been thoroughly investigated makes many of its modifications equally amenable to analysis, rather than to computation alone.

I. CLASSICAL LESLIE MODEL: CONSTANT SCHEDULE OF THE LIFE SPAN

Assuming that the population is a homogeneous group of individuals often turns out too coarse to deal with many issues of both theoretical and applied ecology. To study population dynamics of an individual species it is necessary, as a rule, to account for the biology of various age groups or stages of development in the life cycle of organisms. If these groups or stages are distinguishable enough and if we consider their population sizes at consecutive discrete moments of time, then formalizing the dynamics of the population brings about a model which is discrete both in time and in the number of state variables.

Let the population of a species under study be divided into a certain number, say, n, of age groups. The way how we define these groups is usually determined by the biology of the species and specificity of the problem with which we are concerned. The simplest postulates on how population sizes of age groups depend upon each other and upon time, result in the model which is commonly referred to as the *Leslie matrix* model[1-2], although similar matrices to describe the age structure dynamics were proposed in even earlier papers.[3-4]

Let $x_i(t)$ denote the size (in absolute numbers or relative population densities) of the i-th age group ($i = 1, 2, \ldots, n$) if we do not take sex partition into account, or the size of the female part of the i-th group if sex partition is essential to the problem under concern. The time $t = 0, 1, 2, \ldots$ counts down the discrete moments of transition from any one age group to the next one. In other words, all age groups are supposed to be the same in duration, which is equal to the interval between consecutive moments of time. This is probably the main constraint that hinders the application of the Leslie model to a wide class of real populations, while still giving enough space to both practical examples and theoretical research.

Let the birth-rate function $b_i(x_1, \ldots, x_n)$ indicate, in general, the average number of offspring (or of new-born females) produced per one individual of age i during one interval of time and assume it, in particular, to be a constant *birth rate* b_i, i.e.,

$$b_i(x_1, \ldots, x_n) = b_i, \qquad i = 1, 2, \ldots, n, \tag{2.1}$$

Then the size of the first age group, made up of the offspring from all age groups together, is given by

$$x_1(t+1) = \sum_{i=1}^{n} b_i x_i(t). \tag{2.2}$$

Assume also the functions $s_i(x_1, \ldots, x_n)$, which show what portion of individuals aged i survive to the age $i + 1$, to be constant too, i.e.,

$$s_i(x_i, \ldots, x_n) = s_i, \qquad i = 1, 2, \ldots, n - 1, \tag{2.3}$$

where the *survival rates* s_i obviously meet the condition $0 < s_i \leq 1$. Then we have

$$x_{i+1}(t+1) = s_i x_i(t), \qquad i = 1, 2, \ldots, n - 1, \tag{2.4}$$

for any i-th age group, excluding the n-th one. Assumptions (2.1) and (2.3) mean that we do not take into account the parameter variations induced by environmental conditions and neglect any population density effects on birth and death processes.

If we denote by $\mathbf{x}(t)$ the column vector built up of the sizes of all age groups, it follows from (2.2) and (2.4) that

$$\mathbf{x}(t+1) = L\mathbf{x}(t), \tag{2.5}$$

where matrix L of size $n \times n$ takes on the form

$$L = \begin{bmatrix} b_1 & b_2 & \cdots & b_{n-1} & b_n \\ s_1 & 0 & & & \\ & s_2 & 0 & & \mathbf{0} \\ & & & \ddots & \ddots \\ \mathbf{0} & & & s_{n-1} & 0 \end{bmatrix}, \tag{2.6}$$

with $b_i \geq 0$ ($i = 1, 2, \ldots, n$), $0 < s_j \leq 1$ ($j = 1, 2, \ldots, n - 1$), and is called a *Leslie matrix*. In more general cases of equation (2.5), matrix L is also termed the *transition* or *projection matrix*, while its entries are often referred to as *vital rates* or *demographic parameters*.

Equation (2.5) represents a system of n first-order linear difference equations with constant coefficients. The solution that corresponds to an initial distribution $\mathbf{x}(0)$ of age group sizes, can be written as

$$\mathbf{x}(t) = L^t \mathbf{x}(0), \quad t = 1, 2, \ldots, \tag{2.7}$$

where L^t is the t-th power of matrix L. Vector $\mathbf{x}(t)$ shows a distribution of the total population size among the age groups in absolute, rather than relative, values. So the age distribution $\mathbf{x}(t)$ is not a distribution in the sense of probability theory or statistics, with the normalizing condition $x_1(t) + \cdots + x_n(t) = 1$. The latter would make no sense for the Leslie model in general, as its matrix may not be respectively stochastic. (In botanical literature, the relative age distribution is sometimes called the *age spectrum*,[5-6] but we reserve the term *spectrum* for the set of all eigenvalues of a matrix.)

Matrix L defines a linear operator in the n-dimensional Euclidean space, and we call it also the *Leslie operator*. Since the variables $x_i(t)$ stand for group sizes, they are nonnegative, and we are interested in observing how the Leslie operator acts on the positive orthant $\mathbb{R}^n_+$ of n-dimensional space. Since all the elements of matrix L are nonnegative (the matrix itself is then called *nonnegative*), the Leslie operator can clearly bring no vector of the positive orthant beyond the orthant boundaries. In other words, any trajectory $\mathbf{x}(t)$ $(t = 1, 2, \ldots)$ initiating in $\mathbb{R}^n_+$, remains within $\mathbb{R}^n_+$. All the subsequent properties of the Leslie model ensue from matrix L being nonnegative and having a special pattern of its nonzero entries allocation.

The asymptotic behavior of solutions to equation (2.5) is principally bound up with spectral properties of matrix L and the most essential properties are given by the well-known Perron–Frobenius theorem. To use this theorem one has to be sure that the nonnegative matrix L is *indecomposable* (or *irreducible* in some texts), or, equivalently, that its associated digraph is strongly connected. In Figure 3 the digraphs are shown which are associated with the Leslie matrix examples (2.49) and (2.50).

From the definitions given in Chapter 1 it follows immediately that for a Leslie matrix to be indecomposable it is necessary and sufficient that

$$b_n \neq 0. \tag{2.8}$$

This condition corresponds to n being the oldest reproductive, rather than the maximal possible, age of the organisms. It is clear that the size of postreproductive groups could influence the age structure of younger groups if only a limiting effect of population density were involved. Since the classic Leslie model does not account for effects of this kind, constraint (2.8) is acceptable as long as we take into account only the reproductive groups.

A human population is a nearest example where it is not true and the sizes of postreproductive groups are still of interest. One can easily calculate them from

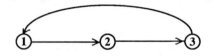

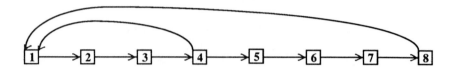

Figure 3.   Associated graphs for Leslie matrices. The link $n \to 1$ affords strong connectedness.

the dynamics of the last reproductive group $x_n(t)$ by the following recursion:

$$
\begin{aligned}
x_{n+1}(t+1) &= s_n x_n(t) \\
x_{n+2}(t+1) &= s_{n+1} x_{n+1}(t) \\
&\cdots \\
x_{n+k}(t+1) &= s_{n+k-1} x_{n+k-1}(t),
\end{aligned}
$$

which results in

$$
x_{n+j}(t+1) = s_n s_{n-1} \ldots s_{n+j-1}(t-j), \qquad j = 1, 2, \ldots, k
$$

(k is the number of postreproductive age groups). The Leslie matrix should then have the following form:

$$
L = \begin{bmatrix} L_r & \mathbf{0} \\ X & S \end{bmatrix} \tag{2.9}
$$

with an $n \times n$ indecomposable Leslie matrix L_r for n reproductive groups, a $k \times k$ survival matrix

$$
S = \begin{bmatrix}
0 & & & & \\
s_{n+1} & 0 & & \mathbf{0} & \\
& s_{n+2} & 0 & & \\
\mathbf{0} & & \ddots & \ddots & \\
& & & s_{n+k-1} & 0
\end{bmatrix} \tag{2.10}
$$

for k postreproductive groups, and $k \times n$ matrix X having the only nonzero entry s_n in its upper right corner.

Matrix S is *nilpotent*, i.e., there exists a natural m such that $S^m = 0$ (apparently $S^k = 0$). With any indecomposable upper left block, nilpotent lower right, and nonzero lower left, matrix (2.9) represents a standard form of the so-called *almost indecomposable* matrices, which exhibit the same long-term behavior in the model as its indecomposable part. Thus we concentrate, in what follows, on indecomposable Leslie matrices, whose order is equal to the number of the oldest reproductive group.

Expanding the characteristic determinant of matrix (2.6) with respect to elements of the first row, or in another way, one can get the characteristic equation

$$
\begin{aligned}
p(\lambda) &= \det(\lambda I - L) \\
&= \lambda^n - b_1\lambda^{n-1} - b_2 s_1 \lambda^{n-2} - \cdots - b_n s_1 \ldots s_{n-1} = 0.
\end{aligned}
\tag{2.11}
$$

Since $b_n \neq 0$, the free term in (2.11) is nonzero too. The equation has thereby no zero roots, hence matrix L is *nonsingular*, or *invertible*, in the sense that $\det A \neq 0$. Thus, invertibility of a Leslie matrix is equivalent to its being indecomposable, which is certainly not true for a general matrix.

II. SPECTRAL AND ASYMPTOTIC PROPERTIES OF THE LESLIE OPERATOR

These are the spectral properties of matrix L, or of the corresponding operator L, which determine the asymptotic behavior of trajectories in the matrix model.

A. Leslie Operator of Simple Structure

A great amount of matrix algebra appears more simple if we deal with an operator of *simple structure*, that is, an operator which has n linearly independent eigenvectors. Since eigenvectors corresponding to different eigenvalues are linearly independent, a simple structure occurs when the eigenvalues of a given operator are all different. But this is not a necessary condition in general: there exist linear operators of simple structure with multiple roots of their characteristic polynomials. However, the Leslie operator has a simple structure if and only if its eigenvalues are all different.[7] The proof (given in Appendix, Theorem A2.1) relies basically on the fact that the *minimal polynomial* $\psi(\lambda)$ of the Leslie matrix (i.e., a nullifying polynomial of the minimal power) coincides with its characteristic polynomial $p(\lambda)$.

Note that a Leslie matrix of a dimension greater than 2 may indeed have multiple eigenvalues, i.e., may not be of simple structure, in spite (or probably because) of a quite special allocation of nonzero matrix elements.

The above-mentioned coincidence is of great methodological importance as many results of matrix theory used below rely entirely on properties of the minimal polynomial. In particular, several more statements about the operator L are deduced below to establish the pattern of asymptotic behavior in the Leslie model.

B. Invariant Polynomials and the Jordan Form

The fact that the minimal polynomial of L has the same power as the dimensionality of the space itself, means that the space is *cyclic* with respect to the operator, i.e., there exists a vector $\mathbf{e}$, such that n vectors $\mathbf{e}, L\mathbf{e}, L^2\mathbf{e}, \ldots, L^{n-1}\mathbf{e}$ are linearly independent, while the vector $L^n\mathbf{e}$ is their linear combination (as is any vector of an n-dimensional space). For instance, $\mathbf{e} = [1, 0, \ldots, 0]^T$ can be such a vector.

Invariant polynomials for a matrix A are, by definition, the polynomials

$$
\begin{aligned}
i_1(\lambda) &= \psi(\lambda) = \frac{D_n(\lambda)}{D_{n-1}(\lambda)}, \\
i_2(\lambda) &= \frac{D_{n-1}(\lambda)}{D_{n-2}(\lambda)}, \ldots, \\
i_n(\lambda) &= \frac{D_1(\lambda)}{D_0(\lambda)}
\end{aligned}
\tag{2.12}
$$

where $D_k(\lambda)$ denotes the greatest common divisor (g.c.d.) for all k-order minors of the characteristic matrix $\lambda I - A$ ($D_0(\lambda) \equiv 1$ reduced[7] to the following:

$$
i_1(\lambda) = \psi(\lambda) = D_n(\lambda), \qquad i_2(\lambda) = 1, \ldots, i_n(\lambda) = 1.
\tag{2.13}
$$

The normal Jordan form of any matrix is known to represent decomposition of the vector space into cyclic subspaces corresponding to elementary divisors of all the invariant polynomials. In combination with equalities (2.13), this enables one to state that if the spectrum of a Leslie matrix consists of m different values λ_j ($j = 1, 2, \ldots, m \leq n$) with multiplicities k_j ($k_1 + k_2 + \cdots + k_m = n$), then the normal Jordan form of the matrix has exactly m cells of sizes $k_j \times k_j$ and entries λ_j in their principal diagonals. Therefore, all eigenvalues that are equal come into a single Jordan cell. This is obviously not so for a general matrix, whose multiple eigenvalues may belong to different cells.

So, a Leslie matrix can generally be represented as

$$
L = P \, \mathrm{diag}\{J_1, J_2, \ldots, J_m\} P^{-1},
\tag{2.14}
$$

where J_j is a $k_j \times k_j$ Jordan cell

$$
J_j =
\begin{bmatrix}
\lambda_j & 1 & & \mathbf{0} \\
& \lambda_j & \ddots & \\
& & \ddots & 1 \\
\mathbf{0} & & & \lambda_j
\end{bmatrix}.
\tag{2.15}
$$

In case all the eigenvalues λ_j are different, the columns of the (nonsingular) transformation matrix P are the eigenvectors corresponding to different eigenvalues. In a general case, these are the vectors which constitute the so-called Jordan basis.

Since all equal eigenvalues enter the same Jordan cell, vectors $\mathbf{e}_1, \mathbf{e}_2, \ldots, \mathbf{e}_k$ of that basis, corresponding to an eigenvalue λ of multiplicity k, can be successively obtained from the relations

$$(L - \lambda I)\mathbf{e}_1 = \mathbf{0}, \quad (L - \lambda I)\mathbf{e}_2 = \mathbf{e}_1, \ldots, (L - \lambda I)\mathbf{e}_k = \mathbf{e}_{k-1}. \tag{2.16}$$

A function of a matrix, in particular, any natural power of it, is defined by the values the function takes over the spectrum, so that

$$L^t = P \, \text{diag}\{J_1^t, J_2^t, \ldots, J_m^t\}P^{-1}, \tag{2.17}$$

where power t of the Jordan $k \times k$ cell corresponding to an eigenvalue λ of multiplicity k takes on the following form:

$$J^t(\lambda) = \begin{bmatrix} \lambda^t & \frac{t}{1!}\lambda^{t-1} & \frac{t(t-1)}{2!}\lambda^{t-2} & \cdots & \frac{t(t-1)\cdots(t-k+2)}{(k-1)!}\lambda^{t-k+1} \\ 0 & \lambda^t & \frac{t}{1!}\lambda^{t-1} & \cdots & \frac{t(t-1)\cdots(t-k+3)}{(k-2)!}\lambda^{t-k+2} \\ \cdots & \cdots & \cdots & \cdots & \cdots \\ & & & & \frac{t}{1!}\lambda^{t-1} \\ 0 & 0 & 0 & \cdots & \lambda^t \end{bmatrix}. \tag{2.18}$$

When the eigenvalues λ_j are all different, formulas (2.17) and (2.18) reduce to

$$L^t = P \begin{bmatrix} \lambda_1^t & & & \mathbf{0} \\ & \lambda_2^t & & \\ & & \ddots & \\ \mathbf{0} & & & \lambda_n^t \end{bmatrix} P^{-1}. \tag{2.19}$$

C. Perron–Frobenius Theorem and Convergence to the Limit Structure

By formulas (2.17)–(2.18) we can calculate, for any initial age distribution $\mathbf{x}(0)$, the distribution $\mathbf{x}(t)$ at any time step t and study the asymptotic behavior of model trajectories as $t \to \infty$, i.e., examine the limit

$$\lim_{t \to \infty} L^t \mathbf{x}(0). \tag{2.20}$$

From (2.18) it is clear that if $|\lambda_j| < 1$ for all λ_j, then $\lim_{t \to \infty} L^t \mathbf{x}(0) = \mathbf{0}$. If there exists a λ_j such that $|\lambda_j| > 1$, then $\lim_{t \to \infty} L^t \mathbf{x}(0) = \infty$. Thus a nontrivial limit in (2.20) can exist only if the maximal absolute eigenvalue equals unity. In the above two cases however, it would be interesting to find the limit

$$\lim_{t \to \infty} \frac{L^t \mathbf{x}(0)}{|\lambda_1^t|}, \tag{2.21}$$

where λ_1 is an eigenvalue with the largest modulus (named also the *dominant eigenvalue*).

For a nonnegative indecomposable matrix A the Perron–Frobenius theorem states that:

1. A has a real positive eigenvalue r (the largest eigenvalue), which is a simple root of the characteristic equation. If λ_j is any other eigenvalue of A, then $|\lambda_j| \leq r$.

2. There exists a positive eigenvector corresponding to r.

3. If A has h eigenvalues which are equal to r in modulus, then they are all different and represent the roots of the equation $\lambda^h - r^h = 0$; h is known as the *index of imprimitivity* (or *Frobenius index*, or *index of cyclicity*) of matrix A (the matrix is called *primitive* when $h = 1$ and *imprimitive* otherwise).

4. If $\lambda_1, \lambda_2, \ldots, \lambda_n$ is the set of all eigenvalues of matrix A and $\theta = \exp(i2\pi/h)$, then $\lambda_1\theta, \lambda_2\theta, \ldots, \lambda_n\theta$ coincides, to within indexing, with $\lambda_1, \lambda_2, \ldots, \lambda_n$.

A Leslie matrix with $b_n \neq 0$ is indecomposable and so it has the largest eigenvalue $r > 0$, which is a simple root of the characteristic polynomial (2.11). By the Descartes "rule of signs"[8] it is furthermore the only positive root of that polynomial as there is only one change of sign among the polynomial coefficients. The corresponding eigenvector is also the only one positive (to within multiplying by a positive scalar).

Are there any other eigenvalues equal in modulus to r, or in other words, do there exist imprimitive Leslie matrices? The answer is positive by the Theorem A2.2 proved in the Appendix. The imprimitivity index for a Leslie matrix is equal to the greatest common divisor for the numbers of the age groups with nonzero natality. Now one can easily construct both primitive and imprimitive Leslie matrices. In particular, for a Leslie matrix to be primitive it is sufficient that either b_1 be positive, or b_js be positive in some two consecutive age groups, i.e., there exists j such that $b_j \neq 0$ and $b_{j+1} \neq 0$.

Since the Jordan form is defined within renumbering of its cells, L can be presented in the form of

$$L = P \, \text{diag} \, \{\lambda_1, \lambda_2, \ldots, \lambda_n, J_{n+1}, \ldots, J_m\} P^{-1}, \tag{2.22}$$

where $\lambda_1 = r, \lambda_2, \ldots, \lambda_h$ are the h eigenvalues that equal r in modulus. It follows from the relations $LP = P \, \text{diag} \, \{\ldots\}$ and $P^{-1}L = \text{diag} \, \{\ldots\}P^{-1}$ that the first h columns of matrix P and the first h rows of matrix P^{-1} are, to within multipliers, the column and the row eigenvectors, respectively, corresponding to the eigenvalues $\lambda_1 = r, \lambda_2, \ldots, \lambda_n$.

If $\mathbf{C}(\lambda)$ stands for the column eigenvector and $\mathbf{R}(\lambda)$ for the row eigenvector, both corresponding to λ, then, as a nice exercise in linear and polynomial algebra, one can show that

$$\mathbf{C}(\lambda) = \sqrt{\frac{r^{n-1}}{p'(r)}} \left[\frac{c_1}{1}, \frac{c_2}{\lambda}, \dots, \frac{c_n}{\lambda^{n-1}} \right]^T, \tag{2.23}$$

$$\mathbf{R}(\lambda) = \sqrt{\frac{r^{n-1}}{p'(r)}} \left[\frac{q_1(\lambda)}{c_1\lambda^n}, \frac{q_2(\lambda)}{c_2\lambda^{n-1}}, \dots, \frac{q_n(\lambda)}{c_n\lambda} \right], \tag{2.24}$$

where

$$c_i = \begin{cases} 1, & \text{if } i = 1, \\ s_1 s_2 \dots s_{i-1}, & \text{if } i = 2, \dots, n; \end{cases}$$

$q_i(\lambda)$ are "tails" of the characteristic polynomial $p(\lambda)$, truncated at its consecutive terms:

$$q_i(\lambda) = \sum_{j=i}^{n} b_j c_j \lambda^{n-j}.$$

Formulas (2.23–24) contain no incorrectness since for any complex number z we have

$$\sum_{t=1}^{n} q_i(z) = zp'(z) - np(z),$$

and hence for $z = r$

$$p'(r) = \frac{1}{r} \sum_{i=1}^{n} q_i(r) > 0.$$

The structure of vectors (2.23) and (2.24), with the real coefficient they have in common, provide the following conditions for their scalar products:

$$\langle \mathbf{C}(\lambda_i), \mathbf{R}(\lambda_i) \rangle = 1, \quad \langle \mathbf{C}(\lambda_i), \mathbf{R}(\lambda_j) \rangle = 0, \quad i \neq j; \quad i, j = 1, 2, \dots, h. \tag{2.25}$$

Therefore, the first h columns of matrix P and the first h rows of P^{-1} in (2.22) are respectively $\mathbf{C}(\lambda_1), \mathbf{C}(\lambda_2), \dots, \mathbf{C}(\lambda_h)$ and $\mathbf{R}(\lambda_1), \mathbf{R}(\lambda_2), \dots, \mathbf{R}(\lambda_h)$.

Since $|\lambda_{h+1}| < r, \dots, |\lambda_n| < r$, it follows from (2.18) that for the corresponding Jordan cells we have

$$\lim_{t \to \infty} \frac{J_j^t}{r^t} = 0, \quad j = h+1, \dots, m.$$

Moreover, the Perron–Frobenius theorem gives

$$\frac{\lambda_l}{r} = \cos \frac{2\pi l}{h} + i \sin \frac{2\pi l}{h} = \varepsilon^{l-1}, \quad l = 1, 2, \ldots, h,$$

where

$$\varepsilon = \cos \frac{2\pi}{h} + i \sin \frac{2\pi}{h} = e^{i2\pi/h}$$

is the h-order *primitive root* of unity.

Now for the limit (2.21) with $\lambda_1 = r$ we have

$$\lim_{t \to \infty} \frac{L^t \mathbf{x}}{r^t} = \lim_{t \to \infty} P \operatorname{diag}\{1, \varepsilon^t, \varepsilon^{2t}, \ldots, \varepsilon^{(h-1)t}, 0, \ldots, 0\} P^{-1} \mathbf{x} = \lim_{t \to \infty} \mathcal{L}(\mathbf{x}, t), \quad (2.26)$$

the convergence rate being determined, due to (2.18), by

$$\frac{t^{k-1}}{r^{k-1}(k-1)!} \left| \frac{\lambda_{h+1}}{r} \right|^{t-k+1},$$

where λ_{h+1} is an eigenvalue of the second largest modulus and multiplicity k. With a simplifying notation

$$\mathbf{C}(\lambda_l) = \mathbf{C}_l, \qquad \mathbf{R}(\lambda_l) = \mathbf{R}_l, \qquad l = 1, 2, \ldots, h,$$

we rearrange the limit vector function (2.26) into

$$\mathcal{L}(\mathbf{x}, t) = P \begin{bmatrix} \mathbf{R}_1 \\ \varepsilon^t \mathbf{R}_2 \\ \vdots \\ \varepsilon^{(h-1)t} \mathbf{R}_h \\ 0 \\ \vdots \\ 0 \end{bmatrix} \mathbf{x} = [\mathbf{C}_1 \ \mathbf{C}_2 \ \cdots \ \mathbf{C}_h \ \cdots] \begin{bmatrix} \langle \mathbf{R}_1, \mathbf{x} \rangle \\ \varepsilon^t \langle \mathbf{R}_2, \mathbf{x} \rangle \\ \vdots \\ \varepsilon^{(h-1)t} \langle \mathbf{R}_h, \mathbf{x} \rangle \\ 0 \\ \vdots \\ 0 \end{bmatrix}$$

$$= \sum_{l=1}^{n} \varepsilon^{(l-1)t} \langle \mathbf{R}_l, \mathbf{x} \rangle \mathbf{C}_l. \quad (2.27)$$

By (2.23) and (2.24) an i-th component of $\mathcal{L}(\mathbf{x}, t)$ takes on the following form:

$$\mathcal{L}_i(\mathbf{x}, t) = \frac{c_i}{p'(r) r^i} \sum_{k=1}^{n} \frac{q_k(r) r^{k-1} x_k}{c_k} \sum_{l=0}^{h-1} \varepsilon^{l(t+k-i)}.$$

When $t + k - i$ is a multiple of the imprimitivity index h, the last sum in the above expression equals h; otherwise elementary trigonometrical identities show that the sum vanishes. Thus

$$\mathcal{L}_i(\mathbf{x}, t) = \frac{hc_i}{p'(r)r^i} \sum_{\substack{1 \le k \le n \\ k = i - t \pmod{h}}} r^{k-1} q_k(r) \frac{x_k}{c_k}, \qquad i = 1, 2, \ldots, n, \qquad (2.28)$$

where the summation proceeds through all subscripts k that differ from $i - t$ by any multiple of h. In the vector form,

$$\mathcal{L}(\mathbf{x}, t) = \begin{bmatrix} \varphi_1(\mathbf{x}, t) c_1 / r \\ \varphi_2(\mathbf{x}, t) c_2 / r^2 \\ \cdots \cdots \\ \varphi_n(\mathbf{x}, t) c_n / r^n \end{bmatrix}, \qquad (2.29)$$

where

$$\varphi_i(\mathbf{x}, t) = \frac{h}{p'(r)} \sum_{\substack{1 \le k \le n \\ k = i - t \pmod{h}}} r^{k-1} q_k(r) \frac{x_k}{c_k}, \qquad i = 1, 2, \ldots, n. \qquad (2.30)$$

Several statements ensue from (2.29) and (2.30) with regard to the asymptotic behavior of Leslie model trajectories. When matrix L is primitive, i.e., $h = 1$, the summation goes over the total list of subscripts k, 1 through n, and the limit function does not, therefore, depend on t:

$$\lim_{t \to \infty} \frac{L^t \mathbf{x}}{r^t} = \lim_{t \to \infty} \mathcal{L}(\mathbf{x}) = \sum_{k=1}^{n} \frac{r^{k-1} q_k(r) x_k}{p'(r) c_k} \begin{bmatrix} c_1 / r \\ c_2 / r^2 \\ \vdots \\ c_n / r^n \end{bmatrix}. \qquad (2.31)$$

Statement (2.31) is a well-known result of Leslie: for a primitive matrix L the limit distribution is in proportion to the eigenvector corresponding to the dominant eigenvalue, with the proportionality coefficient—denoted by $\varphi(r, \mathbf{x})$—depending linearly on the components of the initial distribution $\mathbf{x}$. It makes sense therefore to call that eigenvector the *dominant* one, too.

This result was obtained by Leslie[1] for the case where the eigenvalues of L are all different; notice that the restriction is not necessary to deduce (2.31).

The result also shows the reason why r is considered as the *total population growth rate*: under the dominant age distribution, the population increases r times per each time step. Parameter r is a matrix-model analog to the well-known *intrinsic rate of increase*, or *Malthusian parameter* μ, of nonstructured models. The latter, often designated by r too, is obviously related to the former by the equality

$$\mu = \ln r.$$

Note that for any growing ($r > 1$) or steady ($r = 1$) population we have $\mathcal{L}_1 \ge \mathcal{L}_2 \ge \cdots \ge \mathcal{L}_n$, that is, the distribution is apparently biased towards younger groups, the farther are survival rates s_i from unity, the stronger is the bias effect.

This type of age distribution is commonly observed, among other types, in natural populations and termed, in some literature on populations, as the "left-sided" age spectrum.[5-6]

D. Periodic Behavior and a Measure of Stability

If matrix L is imprimitive, then the limit function (2.27) is clearly periodic in t since, in any case,

$$\mathcal{L}(\mathbf{x}, t + h) = \mathcal{L}(\mathbf{x}, t). \tag{2.32}$$

However, its average value for h sequential time moments turns out constant, which is attested by the following theorem.

Theorem *(on the cycle average). Let h be the imprimitivity index of matrix L. Then*

$$\lim_{t \to \infty} \frac{1}{h} \sum_{j=1}^{h} \frac{L^{t+j} \mathbf{x}}{r^{t+j}} = \varphi(r, \mathbf{x}) \begin{bmatrix} c_1/r \\ c_2/r^2 \\ \vdots \\ c_n/r^n \end{bmatrix}, \tag{2.33}$$

where

$$\varphi(r, \mathbf{x}) = \sum_{k=1}^{n} \frac{r^{k-1} q_k(r) x_k}{p'(r) c_k}. \tag{2.34}$$

(The proof is given in the Appendix, Theorem A2.3.)

Concerning the true length of the period for function $\mathcal{L}(\mathbf{x}, t)$ in the imprimitive case, there are known more intimate results than equality (2.32). These are given by the following two theorems, whose proofs can be found elsewhere.[9-11]

Theorem *(on periodicity). Let h be the index of imprimitivity of L. Then the limiting vector function $\mathcal{L}(\mathbf{x}, t)$ is periodic in t with a period*

$$T = \text{l.c.m.}\{v_l : \langle \mathbf{R}_l, \mathbf{x} \rangle \neq 0, \quad l - 1 = (u_l/v_l)h\}, \tag{2.35}$$

where l takes on values from $\{1, 2, \ldots, h\}$ and u_l/v_l is an irreducible fraction (for $l = 1$, let $u_1 = 0, v_1 = 1$).

In particular, $h = 1$ for a primitive matrix L, hence there can only be $v_1 = 1$ in (2.35), with the least common multiple $T = 1$. So the primitive limiting function is constant in t as already proved by (2.31).

In general, T is a divisor of the imprimitivity index h in any case and depends on the initial distribution $\mathbf{x}$ in case h does have nontrivial divisors. More rigorously, T depends on whether or not the initial distribution $\mathbf{x}$ belongs to a "plane" in $\mathbb{R}^n$ given by equations $\langle \mathbf{R}_l, \mathbf{x} \rangle = 0$ for some l. Only if it does, then the period is less than h, but normally the period is equal to h. The theorem on periodicity is thus of more theoretical than practical importance, but for the sake of mathematical completeness, we illustrate it with an example (at the end of Section III) where T can indeed be nontrivially less than h.

What is the asymptotic behavior of the total population size $N(t) = \sum_{i=1}^{n} x_i(t)$ when $\mathbf{x}(t)$ is asymptotically periodic? Clearly,

$$N(t) \rightarrow \mathcal{N}(\mathbf{x}, t) = \sum_{i=1}^{n} \mathcal{L}_i(\mathbf{x}, t)$$

and if T is the period of $\mathcal{L}(\mathbf{x}, t)$ then $\mathcal{N}(\mathbf{x}, t + T) = \mathcal{N}(\mathbf{x}, t)$. Hence the true period of $\mathcal{N}(\mathbf{x}, t)$ must at least be a divisor of T. The conditions that make it equal to T are specified by the following

Theorem (*on the total population period*). *The period for the total size $\mathcal{N}(\mathbf{x}, t)$ of the limit distribution $\mathcal{L}(\mathbf{x}, t)$ coincides with the period of $\mathcal{L}(\mathbf{x}, t)$ if and only if*

$$\sum_{i=1}^{n} c_i \lambda_l^{n-i} \neq 0, \qquad l = 1, 2, \ldots, h. \tag{2.36}$$

Simple sufficient conditions for (2.36) to hold are provided by

$$\lambda_1 > \max\{s_1, \ldots, s_{n-1}\} \tag{2.37}$$

or

$$\lambda_1 < \min\{s_1, \ldots, s_{n-1}\}. \tag{2.38}$$

The periodicity theorems indicate that the behavior of trajectories radically depends on the time scale chosen: one can avoid cyclicity in model trajectories, for instance, by extending time intervals so that some age groups with nonzero natality become neighbors. On the other hand, for the organisms whose life span terminates with a single reproductive act (*semelparity* in contrast with *iteroparity*, which designates several reproductions during one life span) the Leslie model can only give asymptotically cyclic trajectories ($h = n$ since $b_1 = 0, \ldots, b_{n-1} = 0$, $b_n > 0$). So, the Leslie formalism implies a highly rigid determination of the asymptotics by the life span structure of the organisms.

As mentioned above, a nontrivial limiting distribution is possible only if the dominant eigenvalue $r = 1$, that is, when

$$p(1) = 1 - \sum_{i=1}^{n} b_i s_1 s_2 \ldots s_{i-1} = 0 \qquad (s_0 = 1). \tag{2.39}$$

Since r is the only positive root of the characteristic polynomial, condition (2.39) is also characteristic for the set of population vital rates. If the condition does not hold, a hypothetical way to redress it would be to change all birth rates proportionally, i.e., to transform

$$\hat{b}_i = kb_i, \qquad k > 0,$$

so that

$$\hat{p}(1) = 0 = 1 - \sum_{i=1}^{n} kb_i s_1 s_2 \ldots s_{i-1}. \tag{2.40}$$

As b_i increases, r also increases and, conversely, r decreases, as b_i decreases. By (2.40) the proportionality coefficient will be

$$k = 1/ \sum_{i=1}^{n} b_i s_1 s_2 \ldots s_{i-1} = \frac{1}{R}. \tag{2.41}$$

Being easily calculable in terms of the vital rates, the value of R serves as a generalized parameter for the whole population reproduction rate, i.e., as a kind of stability measure. If $R = 1$, then $r = 1$ and there is neither exponential growth, nor extinction in the population, instead the population approaches a limit distribution. When $R < 1$ or when $R > 1$, which corresponds respectively to $r < 1$ or $r > 1$, the population, according to (2.26) and (2.29), does decline to extinction or grow infinitely.

III. CLASSIFICATION OF STABLE SOLUTIONS IN THE LESLIE MODEL

The results of the previous section make it possible to describe in terms of stability notions the general picture of asymptotic behavior for trajectories of the linear dynamical system

$$L\mathbf{x}(t + 1) = L\mathbf{x}(t), \qquad t = 0, 1, 2, \ldots, \tag{2.42}$$

given by a Leslie matrix L. The investigation of stability starts traditionally with the simplest type of solution $\mathbf{x}^*(t)$, namely, the equilibrium point, or just equilibrium.

A. Equilibrium Stability

The *equilibrium*, i.e., a solution characterized by the equation

$$\mathbf{x}^* = L\mathbf{x}^*,$$

is nothing else but an eigenvector of L corresponding to the eigenvalue $\lambda = 1$. Since L has only one positive eigenvalue, the unit eigenvalue must be dominant whenever one exists, and possess a positive eigenvector by the Perron–Frobenius theorem.

Linearity of the model guarantees that, under initial perturbation of the equilibrium

$$\delta\mathbf{x}(0) = \mathbf{x}(0) - \mathbf{x}^*,$$

the deviation $\delta\mathbf{x}(t)$ is governed by the same matrix L:

$$\delta\mathbf{x}(t) = \mathbf{x}(t) - \mathbf{x}^* = L^t\delta\mathbf{x}(0).$$

Expanding $\delta\mathbf{x}(0)$ in vectors of the Jordan basis, we can easily see that, after a finite number of steps t, the deviation $\delta\mathbf{x}(t)$ being small can be provided by a sufficiently small initial displacement. Hence the local stability is caused eventually by the long-term behavior of $L^t\delta\mathbf{x}(0)$, which, for $r = 1$, is described by the limit function $\mathcal{L}(\delta\mathbf{x}(0), t)$ (2.27). By choosing $\delta\mathbf{x}(0)$ sufficiently small we can make all components (2.28) of the limit function $\mathcal{L}(\delta\mathbf{x}(0), t)$ arbitrarily close to zero, thereby satisfying the formal definition of the local Lyapunov stability. So, if an equilibrium $\mathbf{x}^*$ exists ($r = 1$), it is locally stable. Notice that there is no asymptotic stability in this case, since for any finite $\delta\mathbf{x}(0)$ the limit function $\mathcal{L}(\delta\mathbf{x}(0), t)$ always differs from zero.

B. Stability of a General Solution

The above statements can be generalized for the case of any trajectory $\mathbf{x}^*(t)$. Indeed, as $t \to \infty$, we have

$$\delta\mathbf{x}(t) = L^t\delta\mathbf{x}(0) \to \begin{cases} \infty, & \text{if } r > 1, \\ 0, & \text{if } r < 1, \\ \mathcal{L}(\delta\mathbf{x}(0), t), & \text{if } r = 1. \end{cases}$$

Thus, any solution is unstable when $r > 1$.

When $r = 1$, all solutions $\mathbf{x}^*(t)$ exhibit local nonasymptotic stability. In more detail, if matrix L is primitive, $\mathbf{x}^*(t)$ has a limit, as $t \to \infty$, proportional to the dominant eigenvector with a coefficient that depends linearly on components of the initial distribution. In a general primitive case, the same is true for the limit of L^t/r^t, i.e., for vector directions only. This means that, whatever the initial distribution may be, all trajectories tend to the unique relative distribution of age group sizes, or the same *age structure*. In other words, the population "forgets" its initial structure in the course of successive generations.

This result is neither surprising nor specific to the Leslie matrix, since it follows from a more general argument. A well-known equivalent definition of matrix A being primitive requires the existence of a positive integer p such that A^p is positive,[12] while a positive matrix (even a primitive nonnegative one) represents a mapping that is contractive in the projective metric[13–14], i.e., in terms of vector directions. The contractive mapping principle then ascertains the existence of a fixed vector, which all directions are successively contracted to in the sequence of A^t as t tends to infinity. What is really surprising is a similar property being valid within a class of Leslie model extensions to time-variable demographic parameters,[15] whose variability is considered to be caused by variable environmental conditions. In other words, variability of the environment alone cannot affect the property of a (model) population to "forget" its initial age structure if the property is valid under fixed environment conditions.

C. Imprimitivity and Cycles

If the Frobenius index $h > 1$, then any solution distinct from an equilibrium tends to a periodic limit function of the period that is a divisor of h and depends on the initial distribution $\mathbf{x}^*(0)$ (see (2.35)). Deviation from this solution, which is always small under sufficiently small $\delta\mathbf{x}(0)$, is also periodic in the limit, with its own period length.

For $r < 1$, all the solutions $\mathbf{x}^*(t)$ are globally asymptotically stable, which reflects just the overall picture of trajectory behavior, i.e., all trajectories approach the zero distribution.

Can there be pure cyclic trajectories, or *cycles*, in the Leslie model? If a vector $\mathbf{x}^{(d)}$ belongs to a cycle of period d, it should satisfy the condition

$$L^d\mathbf{x}^{(d)} = \mathbf{x}^{(d)}. \tag{2.43}$$

Therefore, $\mathbf{x}^{(d)}$ is an eigenvector for matrix L^d with an eigenvalue $\mu = 1$. The same is true for all the other vectors comprising the d-cycle. Since any eigenvalue of L^d equals λ^d, where λ is an eigenvalue of L, the unit eigenvalue for L^d can only occur when there is a d-power root of 1 in the spectrum of L. Then, clearly, $r \geq 1$. Since $r > 1$ corresponds to trajectory instability, we confine to the case of $r = 1$. Then all the roots of modulus 1 satisfy the equation $\lambda^h - 1 = 0$; together with $\lambda^d = 1$, this implies that integers d and h have a divisor in common. Theorem A2.4 in the Appendix shows that the divisor is, in fact, equal to the cycle period d.

Thus, for a Leslie matrix, there only exist cycles of periods d which are divisors of the imprimitivity index h. All these cycles belong to the eigenvector subspace for matrix L^h (denote it by $\Lambda^{(h)}$) that corresponds to its eigenvalue 1. Originated as λ^h, the eigenvalue 1 has apparently the maximum modulus and the multiplicity h. Since the geometric multiplicity of an eigenvalue of any matrix cannot exceed its algebraic multiplicity,[16] we have

$$\dim \Lambda^{(h)} \leq h. \tag{2.44}$$

On the other hand, the relation (2.25) and the theorem on periodicity guarantee that, for any initial vector $\mathbf{x}$,

$$L^h\mathcal{L}(\mathbf{x}, t) = \mathcal{L}(\mathbf{x}, t + h) = \mathcal{L}(\mathbf{x}, t),$$

that is, the space of all possible vectors $\mathcal{L}(\mathbf{x}, t)$ belongs to $\Lambda^{(h)}$:

$$\{\mathcal{L}(\mathbf{x}, t)\} \subset \Lambda^{(h)}. \tag{2.45}$$

But vectors of $\{\mathcal{L}(\mathbf{x}, t)\}$ are, by (2.27), linear combinations of eigenvectors $\mathbf{C}_1, \ldots,$ $\mathbf{C}_h$ of matrix L, which correspond to different eigenvalues 1, $\exp(i2\pi/h), \ldots,$ $\exp(i2\pi(h-1)/h)$, and thereby are linearly independent. Thus

$$\dim\{\mathcal{L}(\mathbf{x}, t)\} = h, \tag{2.46}$$

which, in combination with (2.44) and (2.45), means that both spaces coincide:

$$\{\mathcal{L}(\mathbf{x}, t)\} = \Lambda^{(h)}. \tag{2.47}$$

Call it the *cycle subspace.*

Notice that in the cycle subspace there are also cycles of period 1, i.e., equilibria. These are nothing else than vectors belonging to the characteristic direction for the eigenvalue $r = 1$, i.e., vectors proportional to $\mathbf{C}_1$. Indeed, when $\mathbf{x} = a\mathbf{C}_1$ ($a > 0$), we have by (2.25)

$$\mathcal{L}(a\mathbf{C}_1, t) = \sum_{i=1}^{h} \varepsilon^{(l-1)t} \langle \mathbf{R}_l, a\mathbf{C}_1 \rangle \mathbf{C}_l = a\mathbf{C}_1.$$

Finally, the true cycles in system (2.42) with an imprimitive matrix L are the trajectories constituted by vectors $\mathcal{L}(\mathbf{x}, t)$ with the periods that divide h and depend on $\mathbf{x}$ according to (2.35). In other words, the fate in common for all trajectories is the cycle subspace, which, in contrast to the primitive case, means only "partial loss of memory" of the initial population structure.

In the particular case of $h = n$ there is no loss at all since

$$L^h = I, \tag{2.48}$$

where I is the identity matrix. It means that any vector $\mathbf{x}$ transforms into itself after n steps; in other words, $\mathbf{x}$ belongs to a cycle with a period equal to n or a divisor of n. Matrices L of the property (2.48) correspond to the life cycle in which parents die after a single act of reproduction (semelparity). An example is given by matrix

$$L = \begin{bmatrix} 0 & 0 & 6 \\ 1/2 & 0 & 0 \\ 0 & 1/3 & 0 \end{bmatrix}, \tag{2.49}$$

that was treated by Bernadelli[3] as a model of "population waves."

In a general case of growing or declining population, similar behavior is exhibited now by $\mathbf{x}(t)/r^t$, i.e., all asymptotic statements made for $r = 1$ are again true for the relative population structure.

Concerning *chaotic* behavior, i.e., trajectories $\mathbf{x}^*(t)$ which are neither constant, nor periodic, nor which tend to any constant or cyclic distribution, we can be sure that such a behavior is impossible for system (2.42) with $r = 1$, since for any $\mathbf{x}(t) = L^t\mathbf{x}$ there exists a limit function $\mathcal{L}(\mathbf{x}, t)$ that is either constant or periodic.

When $r < 1$, all trajectories tend to zero. When $r > 1$, $L^t\mathbf{x}$ has no limit, however there is a limit for $L^t\mathbf{x}/r^t$, i.e.,

$$\mathbf{x}(t) \to r^t \mathcal{L}(\mathbf{x}, t)$$

and the "chaos" has quite an ordered structure here: each trajectory approaches its limiting vector function, which grows r times with every time step.

How to use the above results in analysis of a Leslie model can be seen from a particular example of the next section, while in the next chapter we will see how to use them in analysis of more complex models.

D. Hypothetical Example

The index h of imprimitivity of an $n \times n$ matrix is *nontrivial* when $1 < h < n$. If, in addition, we want it to have nontrivial divisors, the least dimension that allows us to do so, is obviously $n = 8$, with $h = 4$. We consider a hypothetical population structured by $n = 8$ age groups in a way such that only the 4th and 8th groups have nonzero birth rates. Let $b_4 = 8$, $b_8 = 12$, and the survival coefficients s_i be such that

$$L = \begin{bmatrix} 0 & 0 & 0 & 8 & 0 & 0 & 0 & 12 \\ 1/3 & & & & & & & \\ & 1/2 & & & & & & \\ & & 1/2 & & & \mathbf{0} & & \\ & & & 1/2 & & & & \\ & & & & 2/3 & & & \\ & \mathbf{0} & & & & 1 & & \\ & & & & & & 1 & 0 \end{bmatrix}. \tag{2.50}$$

Calculated by (2.41), the growth parameter $R = 1$, so that the population size is retained in this model. The characteristic polynomial,

$$p(\lambda) = \lambda^8 - \frac{2}{3}\lambda^4 - \frac{1}{3},$$

has the largest eigenvalue $r = \lambda_1 = 1$ and the index of imprimitivity $h = \text{g.c.d.}\{4, 8\}$ $= 4$. Hence the system can only have equilibria and 2- or 4-cycles. A short calculation yields

$$\begin{aligned} \lambda_2 &= \varepsilon = e^{i\pi/2} = i, \quad \lambda_3 = -1, \quad \lambda_4 = -i, \\ \lambda_5 &= 3^{-1/4}e^{i\pi/4}, \quad \lambda_6 = 3^{-1/4}e^{i3\pi/4}, \quad \lambda_7 = 3^{-1/4}e^{i5\pi/4}, \\ \lambda_8 &= 3^{-1/4}e^{i7\pi/4}. \end{aligned}$$

It follows from (2.23) that the column eigenvectors are:

$$
\mathbf{C}_1 = \frac{\sqrt{3}}{4}
\begin{bmatrix}
1 \\
1/3 \\
1/6 \\
1/12 \\
1/24 \\
1/36 \\
1/36 \\
1/36
\end{bmatrix} ; \quad
\mathbf{C}_2 = \frac{\sqrt{3}}{4}
\begin{bmatrix}
1 \\
-i/3 \\
-1/6 \\
i/12 \\
1/24 \\
-i/36 \\
-1/36 \\
i/36
\end{bmatrix} ;
$$

$$
\mathbf{C}_3 = \frac{\sqrt{3}}{4}
\begin{bmatrix}
1 \\
-1/3 \\
1/6 \\
-1/12 \\
1/24 \\
-1/36 \\
1/36 \\
-1/36
\end{bmatrix} ; \quad
\mathbf{C}_4 = \frac{\sqrt{3}}{4}
\begin{bmatrix}
1 \\
i/3 \\
-1/6 \\
-i/12 \\
1/24 \\
i/36 \\
-1/36 \\
-i/36
\end{bmatrix} ,
$$

while, from (2.24), the row eigenvectors are:

$$
\mathbf{R}_1 = \frac{\sqrt{3}}{4}[1, 3, 6, 12, 8, 12, 12, 12];
$$

$$
\mathbf{R}_2 = \frac{\sqrt{3}}{4}[1, 3i, -6, -12i, 8, 12i, -12, -12i];
$$

$$
\mathbf{R}_3 = \frac{\sqrt{3}}{4}[1, -3, 6, -12, 8, -12, 12, -12];
$$

$$
\mathbf{R}_4 = \frac{\sqrt{3}}{4}[1, -3i, -6, 12i, 8, -12i, -12, 12i].
$$

Every distribution of age-group sizes which is proportional to the vector $\mathbf{C}_1$ remains constant at any step. Any other initial distribution $\mathbf{x}$ approaches, with time, a 4-cycle made up of the vectors $\mathcal{L}(\mathbf{x}, 0)$, $\mathcal{L}(\mathbf{x}, 1)$, $\mathcal{L}(\mathbf{x}, 2)$, and $\mathcal{L}(\mathbf{x}, 3)$, or a 2-cycle under additional constraints. Let, for instance, the initial distribution be uniform and equal, e.g., to

$$
\mathbf{x} = [16, \ldots, 16]^T.
$$

Then

$$
\mathcal{L}(\mathbf{x}, 0) = \sum_{i=1}^{4} \langle \mathbf{R}_l, \mathbf{x} \rangle \mathbf{C}_l = \frac{\sqrt{3}}{4}[66\mathbf{C}_1 - (9 - 9i)\mathbf{C}_2 - 12\mathbf{C}_3 - (9 - 9i)\mathbf{C}_4]
$$

$$
= [108, 96, 36, 15, 9/2, 8, 6, 5]^T;
$$

$$
\mathcal{L}(\mathbf{x}, 1) = L\mathcal{L}(\mathbf{x}, 0) = [180, 36, 48, 18, 15/2, 3, 8, 6]^T;
$$

$$
\mathcal{L}(\mathbf{x}, 2) = L\mathcal{L}(\mathbf{x}, 1) = [216, 60, 18, 24, 9, 5, 3, 8]^T;
$$

$$
\mathcal{L}(\mathbf{x}, 3) = L\mathcal{L}(\mathbf{x}, 2) = [288, 72, 30, 9, 12, 6, 5, 3]^T.
$$

Thus the trajectory initiating at $\mathbf{x} = [16, \ldots, 16]^T$ converges to the 4-cycle of the above vectors. All four are strongly biased to younger age groups.

The rate of convergence is given by $(1/\sqrt[4]{3})^t \approx 0.760^t$; it means that after eight generations the difference between the initial and the limiting distribution decreases approximately 10-fold, after 16 generations approximately 100-fold, and so on.

The criterion (2.36) can be shown to hold, although its sufficient conditions (2.37) and (2.38) appear too coarse for the case. Thus, in any trajectory, the period of the total population size limit coincides with that of the trajectory limit. In particular, $N(t)$ for $\mathbf{x}(0) = [16, \ldots, 16]^T$ also approaches the 4-cycle comprising the values 278.5, 306.5, 343, and 425.

To find 2-cycles we have to apply the theorem on periodicity, which requires the period to be the l.c.m. for a subset of numbers $\{1, 4, 2, 4\}$. The fours can be excluded by the conditions $\langle \mathbf{R}_2, \mathbf{x} \rangle = 0$ and $\langle \mathbf{R}_4, \mathbf{x} \rangle = 0$, which reduce to

$$\begin{cases} x_1 - 6x_3 + 8x_5 - 12x_7 = 0, \\ x_2 - 4x_4 + 4x_6 - 4x_8 = 0. \end{cases}$$

Thus a 2-cycle is a fate of trajectories initiated at the intersection of the above two "planes" in $\mathbb{R}^8$. The intersection contains, for example, vector

$$\mathbf{y} = [96, 16, 16, 4, 0, 0, 0, 0]^T,$$

for which

$$\begin{aligned}
\mathcal{L}(\mathbf{y}, 0) &= \sqrt{3}/4 \, [288\mathbf{C}_1 + 96\mathbf{C}_3] = [72, 12, 12, 3, 3, 1, 2, 1]^T; \\
\mathcal{L}(\mathbf{y}, 1) &= L\mathcal{L}(\mathbf{y}, 0) = [36, 24, 6, 6, 1.5, 2, 1, 2]^T, \\
\mathcal{L}(\mathbf{y}, 2) &= \mathcal{L}(\mathbf{y}, 0), \text{ and so on.}
\end{aligned}$$

The total size converges to the 2-cycle on 106 and 78.5.

Appendix

Theorem A2.1 *A Leslie operator of the form (2.6) has a simple structure if and only if all its eigenvalues are different.*

Proof: A general criterion for a linear operator A to be of simple structure requires all of the so-called *elementary divisors* of A to be linear.[17] By definition, the *elementary divisors* of matrix A are the divisors of its *invariant polynomials*, i.e., of the polynomials

$$i_1(\lambda) = \varphi(\lambda) = \frac{D_n(\lambda)}{D_{n-1}(\lambda)}, \quad i_2(\lambda) = \frac{D_{n-1}(\lambda)}{D_{n-2}(\lambda)}, \ldots, \quad i_n(\lambda) = \frac{D_1(\lambda)}{D_0(\lambda)}, \quad (2.12)$$

where $D_k(\lambda)$ denotes the g.c.d. for k-order minors of the characteristic matrix $\lambda I - L$ (the coefficient at the highest power of λ being taken unitary, $D_0(\lambda) \equiv 1$ by definition). The invariant polynomials are known to be invariant under any similarity transformation of A.[17]

The first of equalities (2.12) yields a formula to calculate the *minimal polynomial* of a matrix A, i.e., its nullifying polynomial of the minimal power (with the leading coefficient equal to unity). The minor of the Leslie matrix that is built on the last $n - 1$ rows and first $n - 1$ columns has the form

$$(\lambda I - L) \begin{pmatrix} 2, 3, \ldots, n \\ 1, 2, \ldots, n-1 \end{pmatrix} = (-1)^{n-1} s_1 s_2 \ldots s_{n-1}$$

and contains no λ. As a result, $D_{n-1}(\lambda) = 1$, hence $D_k(\lambda) = 1$ for $k = n - 2, \ldots, 1$. It follows that

$$i_1(\lambda) = \varphi(\lambda) = D_n(\lambda), \quad i_2(\lambda) = 1, \ldots, \quad i_n(\lambda) = 1, \tag{2.13}$$

and, in particular, the minimal polynomial of the Leslie matrix coincides with its characteristic polynomial. The elementary divisors of the matrix thus coincide with those of the minimal polynomial. Over the field of complex numbers, they are some powers of binomials $(\lambda - \lambda_j)$, where λ_j are different eigenvalues of L. For all the elementary divisors to be of the first power it is necessary and sufficient that all n eigenvalues be different. ∎

Theorem A2.2 *The imprimitivity index for a Leslie operator of the form (2.6) is*

$$h(L) = \text{g.c.d.}\{i_1, i_2, \ldots, i_q\},$$

where $i_1, i_2, \ldots, i_q$ are all the numbers such that $b_{i_k} \neq 0$ in L.

Proof: The imprimitivity index of a matrix is known to equal the greatest common divisor for

$$n - n_1, \ n_1 - n_2, \ \ldots, \ n_{q-1} - n_q,$$

where $n > n_1 > \cdots > n_q$ are the powers of all nonzero terms in the characteristic polynomial.[18] In the polynomial (2.11) the coefficient at λ^m $(0 < m < n)$ is nonzero if $b_{n-m} \neq 0$. Denote by $i_1 < i_2 < \cdots < i_q$ the numbers of all age groups with nonzero birth rates, or in other words, the subscripts of all $b_j \neq 0$. Notice that $i_q = n$ follows from indecomposability. From (2.11) we find that

$$n_{i_s} = n - i_s, \quad s = 1, 2, \ldots, q,$$

whereby

$$
\begin{aligned}
h(L) &= \text{g.c.d}\{n - (n - i_1), (n - i_1) - (n - i_2), \ldots, (n - i_{q-1}) - (n - i_q)\} \\
&= \text{g.c.d.}\{i_1, i_2 - i_1, \ldots, i_q - i_{q-1}\} = \text{g.c.d.}\{i_1, i_2, \ldots, i_q\}. \quad \blacksquare
\end{aligned}
$$

Theorem A2.3 *(on the cycle average). Let h be the imprimitivity index of matrix L.*
Then

$$\lim_{t \to \infty} \frac{1}{h} \sum_{j=1}^{h} \frac{L^{t+j}\mathbf{x}}{r^{t+j}} = \varphi(r, \mathbf{x}) \begin{bmatrix} c_1/r \\ c_2/r^2 \\ \vdots \\ c_n/r^n \end{bmatrix}, \tag{2.33}$$

where

$$\varphi(r, \mathbf{x}) = \sum_{k=1}^{n} \frac{r^{k-1}q_k(r)x_k}{p'(r)c_k}. \tag{2.34}$$

Proof: As follows from (2.28), the i-th component of the limit function for the vector on the left-hand side of (2.33) has the form

$$\frac{1}{h}\sum_{j=1}^{n} \frac{hc_i}{p'(r)r^i} \sum_{\substack{1 \le k \le n \\ k=i-(t+j) \pmod{h}}} \frac{r^{k-1}q_k(r)x_k}{c_k} = \frac{c_i}{r_i} \sum_{k \in \mathcal{K}(i,t)} \frac{r^{k-1}q_k(r)x_k}{p'(r)c_k}, \qquad i = 1, 2, \dots, n.$$

$$\tag{2.51}$$

The set of values the subscript k takes on in the double summation is apparently determined as follows:

$$\mathcal{K}(i, t) = \{k : 1 \le k \le n, \quad k - i + (t+j) = 0 \pmod{h}, \qquad j = 1, 2, \dots, h\}.$$

We show that $\mathcal{K}(i, t)$ does not actually depend on i, nor on t, but coincides merely with the set $k = 1, 2, \dots, n$. Let k_0 be a number from this set. Dividing $k_0 - i + t$ by h results in

$$k_0 - i + t = mh + \rho, \qquad 0 \le \rho < h.$$

If the residue $\rho = 0$, then the integer $k_0 - i + (t + h)$ is also divisible by h with no residue, i.e., $k_0 \in \mathcal{K}(i, t)$. If $0 < \rho < h$ then no residue appears when by h we divide $k_0 - i + (t + h - \rho)$, that is, $k_0 \in \mathcal{K}(i, t)$ again.

We show now that each value of k occurs only once in the double sum of (2.51). If, otherwise, there is a pair of subscripts j_1 and j_2 ($1 \le j_1, j_2 \le h$), such that

$$\begin{aligned} k - i + (t + j_1) &= 0 \pmod{h}, \\ k - i + (t + j_2) &= 0 \pmod{h}, \end{aligned}$$

then

$$j_1 - j_2 = 0 \pmod{h},$$

and since both subscripts j_1 and j_2 themselves cannot exceed h, it follows that

$j_1 = j_2$. Thus $\mathcal{K}(i, t) = \{1, 2, \ldots, n\}$, and the sum in (2.51) is nothing but

$$\frac{c_i}{r_i} \sum_{k=1}^{n} \frac{r^{k-1} q_k(r) x_k}{p'(r) c_k} \; ,$$

which completes the proof.   ∎

Theorem A2.4 (*on the cycle period*). *Let L be a Leslie matrix with the unit dominant eigenvalue and imprimitivity index h. Then there exist cycles among trajectories of the model* $\mathbf{x}(t + 1) = L\mathbf{x}(t)$. *If a cycle has period d, then d is a divisor of h.*

Proof: If matrix L is primitive, its any natural power L^t is indecomposable and primitive too, and has, by the Perron–Frobenius theorem, the dominant eigenvalue $r(L^t) = r(L)^t = 1$ of multiplicity one. Thus the vectors $\mathbf{x}^{(d)}(0), \mathbf{x}^{(d)}(1), \ldots, \mathbf{x}^{(d)}(d-1)$ which comprise a d-cycle, belong to the same positive characteristic direction, that is,

$$\mathbf{x}^{(d)}(j) = \theta_j \mathbf{x}^{(d)}(0), \qquad \theta_j > 0, \qquad j = 1, 2, \ldots, d - 1.$$

By the notion of a cycle

$$L\mathbf{x}^{(d)}(0) = \mathbf{x}^{(d)}(1) = \theta_1 \mathbf{x}^{(d)}(0),$$

therefore $\mathbf{x}^{(d)}(0)$ is an eigenvector of L with the eigenvalue $\theta_1 > 0$. But since the only positive eigenvalue of L is $r = 1$, it follows that $\theta_1 = 1$ and

$$x^{(d)}(1) = x^{(d)}(0).$$

Reasoning along the same line, we come to the chain of equalities

$$\mathbf{x}^{(d)}(0) = \mathbf{x}^{(d)}(1) = \cdots = \mathbf{x}^{(d)}(d - 1),$$

i.e. the d-cycle is an equilibrium in fact. True cycles are impossible for primitive Leslie matrices with $r = 1$.

The same conclusion follows from another argument. Since every initial distribution tends, as $t \to \infty$, to the limiting vector function $\mathcal{L}(\mathbf{x}, t)$, which is constant in t in the primitive case, there is no room for cycles.

If, however, the imprimitivity index $h > 1$, the periodic limiting function $\mathcal{L}(\mathbf{x}, t)$ itself turns into a true cycle. This is true because, referring to (2.27) and $r = 1$, we have

$$L\mathcal{L}(\mathbf{x}, t) = \sum_{l=1}^{h} \varepsilon^{(l-1)t} \langle \mathbf{R}_l, \mathbf{x} \rangle L\mathbf{C}_l = \sum_{l=1}^{h} \varepsilon^{(l-1)t} \langle \mathbf{R}_l, \mathbf{x} \rangle \varepsilon^{l-1} \mathbf{C}_l = \mathcal{L}(\mathbf{x}, t + 1). \quad (2.52)$$

The period of this cycle is obviously equal to the period T of the function $\mathcal{L}(x, t)$, which is a divisor of h by formula (2.35) of the theorem on periodicity.

If system (2.42) had a true cycle of a period k other than a divisor of h, a trajectory initiating from one of the vectors $\mathbf{y}^{(k)}$ of that cycle, would then have to approach $\mathcal{L}(\mathbf{y}^{(k)}, t)$. Then the period of $\mathcal{L}(\mathbf{y}^{(k)}, t)$ would have to be a divisor of h, which contradicts our premise. ■

Additional Notes

To 2.I. The Leslie matrix model owes its name to two fundamental papers by P. H. Leslie[1-2] published in 1945 and 1948, although matrices had been used in earlier works on age-structured population models.[3-4] The "challenge to linear algebra" was readily taken up by mathematicians and modelers, although it was only in the mid-70s when a series of papers by Cull and Vogt[9-11] had left almost nothing to do for a mathematician working within the classical Leslie formulation. The practical need to modify the Leslie model was backed up afterwards by further challenge to mathematical theory. Nowadays the bibliography of works on Leslie-type models is enormous. The survey by Hansen,[19] for instance, which covers English-written works up the the mid 1980s devoted to just mathematical and theoretical aspects, accounts for more than 200 items. An introductory paper by Geramita and Pullman[12] presents a variety of numerical examples of matrix calculations, both in the Leslie framework and in the more general context of nonnegative matrices; the paper contains annotated bibliography of publications on theoretical and applied aspects of matrix modeling in population dynamics from 1973 to 1984. See also the monographs by Pullman,[20] Smith and Keyfitz,[21] and Caswell.[22]

Restricting the main equation (2.5) to the female population only is not crucial and can be relaxed in the sense of, e.g., Williamson,[23] who combined the male and female populations into a single population vector, or Zuber et al.[24], who derived the matrix model equations for them separately, or Pollard,[25] who gave a good exposition of the two-sex model. A more general approach is proposed in Caswell and Weeks.[26]

We have confined ourselves to indecomposable matrices, while almost indecomposable matrices was considered by Geramita and Pullman[12] under the name of *almost irreducible* matrices. They also proved as a theorem the fact that the asymptotic behavior of a model with a complete set of age groups, where b_n may be zero, is the same as that of its indecomposable subset (with a nonzero birthrate in the eldest group).[27]

To 2.II. For the basic concepts related to the notions of the minimal polynomial and the Jordan normal form of a matrix, see, e.g., Gantmacher.[17] Any textbook dealing with nonnegative matrices gives a formulation of the Perron–Frobenius theorem, the formulations varying slightly in form but being equivalent in essence; the cited formulation follows Marcus and Minc.[28] We have used one of Gantmacher's formulas[18] to derive the index of imprimitivity for the Leslie matrix, while Demetrius[29] proved the result by means of graph representation.

The rigid bond between primitivity of the matrix and convergence to the dominant vector direction was stressed by several authors after Leslie[30−31] and some authors treated it as a kind of ergodicity;[32−33] see also Cohen[34] and the references therein.

For the notion of the *intrinsic rate of increase*, or *biotic potential*, see any text on population dynamics, for instance E. Odum[35] and references therein. The *left dominant eigenvector* has also been given a biological interpretation, namely, as a distribution of *reproductive values* among age groups,[36−38] which provided the key to treat evolutionary aspects of the life history in terms of the matrix model.[39−43]

The relationship, like that expressed by equality $\mu = \ln r$, between key parameters of the discrete and continuous models is a traditional theme in population modeling and has its contributors in the field of age-structured models.[24,44−47]

To 2.III. More on application of the projective metric in matrix models can be found in Golubitsky et al.[48] and Caswell and Weeks.[26] A good introduction to the intriguing theme of chaos in population models was given by May[49] and May and Oster[50].

REFERENCES

1. Leslie, P. H., On the use of matrices in certain population mathematics, *Biometrika*, 33, 183–212, 1945.

2. Leslie, P. H., Some further notes on the use of matrices in population mathematics, *Biometrika*, 35, 213–245, 1948.

3. Bernadelli, H., Population waves, *J. Burma Res. Soc.*, 31, 1–18, 1941.

4. Lewis, E. G., On the generation and growth of a population, *Sankhya*, 6, 93–96, 1942.

5. Plant Cenopopulations (Basic Notions and Structures), Uranov, A. A. and Serebryakova, T. I., Eds., *Nauka*, Moscow, 1976, Chap. 2 (in Russian).

6. Plant Cenopopulations (Essays on Population Biology), Serebryakova, T. I. and Sokolova, T. G., Eds., *Nauka*, Moscow, 1988, Chap. 4 (in Russian).

7. Svirezhev, Yu. M. and Logofet, D. O., *Stability of Biological Communities*, Mir Publ., Moscow, 1983 (translated from the 1978 Russian edition), Chap. 2.5.

8. Korn, G. A., and Korn, T. M., *Mathematical Handbook*, 2nd ed., McGrow-Hill, New York et al., 1968, Chap. 1.6-6.

9. Cull, P., and Vogt, A., Mathematical analysis of the asymptotic behavior of the Leslie population matrix model, *Bull. Math. Biol.*, 35, 645–661, 1973.

10. Cull, P., and Vogt, A., The periodic limits for the Leslie model, *Math. Biosci.*, 21, 39–54, 1974.

11. Cull, P., and Vogt A., The period of total population, *Bull. Math. Biol.*, 38, 317–319, 1976.

12. Geramita, J. M., and Pullman, N. J., *An Introduction to the Application of Nonnegative Matrices to Biological Systems*. Queen's Papers in Pure and Applied Mathematics, No 68, Queen's University, Kingston, Ontario, Canada, 1984.

13. Birkhoff, G., Extensions of Jentzsch's theorem, *Trans. Am. Math. Soc.*, 85, 219–227, 1957.

14. Seneta, E., *Non-Negative Matrices and Markov Chains*, 2nd ed., Springer Verlag, New York, 1981, Chap. 3.

15. Gorban, A. N., Non-stationary environment does not affect the population forgetting initial age structure, in *Dynamics of Chemical and Biological Systems*, Bykov, V. I., Ed., Nauka, Novosibirsk, 1989, 206–212 (in Russian).

16. Marcus, M. and Minc, A., *A Survey of Matrix Theory and Matrix Inequalities*, Allyn and Bacon, Boston, 1964, Chap. 3.12.

17. Gantmacher, F. R., *The Theory of Matrices*, Chelsea, New York, 1960, Chaps. 3 to 7.

18. Gantmacher, F. R., *The Theory of Matrices*, Chelsea, New York, 1960, Chap. 13.5.

19. Hansen, P. E., Leslie matrix models: a mathematical survey, in *Papers on Mathematical Ecology, I*, Csetenyi, A. I., Ed., Karl Marx University of Economics, Budapest, Hungary, 1986, 1–139.

20. Pullman, N. J., *Matrix Theory and Its Applications*, Marcel Dekker, Inc., New York, 1976.

21. Smith, D., and Keyfitz, N., *Mathematical Demography*, Springer Verlag, New York, 1977 (Vol. 6 in *Biomathematics series*).

22. Caswell, H., *Matrix Population Models*, Sinauer Associates, Inc., Sunderland, Massachusetts, 1989.

23. Williamson, M. H., Some extensions of the use of matrices in population theory, *Bull. Math. Biophys.*, 21, 13–17, 1959.

24. Zuber, Ye. I., Kolker, Yu. I., and Poluektov, R. A., Control of the population size and age structure, in *Problemy Kibernetiki*, issue 25, Lyapunov, A. A., Ed., Nauka, Moscow, 1972, 129–138 (in Russian).

25. Pollard, J. H., Mathematical Models for the Growth of Human Populations, Cambridge University Press, Cambridge, 1973.

26. Caswell, H., and Weeks, D. E., Two-sex models: chaos, extinction, and other dynamic consequences of sex, *Amer. Natur.*, 128, 707–735, 1986.

27. Geramita, J. M. and Pullman, N. J., Classifying the asymptotic behavior of some linear models, *Math. Biosci.*, 69, 189–198, 1984.

28. Marcus, M. and Minc, A., *A Survey of Matrix Theory and Matrix Inequalities*, Allyn and Bacon, Boston, 1964, Chap. 5.5.

29. Demetrius, L., Primitivity conditions for growth matrices, *Math. Biosci.*, 12, 53–58, 1971.

30. Lopez, A., *Problems in Stable Population Theory*, Office of Population Research, Princeton University, Princeton, NJ, 1961.

31. Sykes, Z. M., On discrete stable population theory, *Biometrics*, 25, 285–293, 1969.

32. Parlett, B., Ergodic properties of population. I. The one-sex model, *Pop. Biol.*, 1, 191–207, 1970.

33. Demetrius, L., Isomorphism of population models, *Kybernetik*, 14, 241–244, 1974.

34. Cohen, J. E., Ergodic theorems in demography, *Bull. Amer. Math. Ass.*, 1, 275–295, 1979.

35. Odum, E. P., *Basic Ecology*, Saunders College Publishing, Philadelphia, 1983, Chaps. 6.3–6.4.

36. Goodman, L. A., On the reconciliation of mathematical theories of population growth, *J. Roy. Stat. Soc.*, Series A, 130, 541–553, 1967.

37. Goodman, L. A., An elementary approach to the population projection matrix, to the population reproductive value, and to related topics in the mathematical theory of population growth, *Demography*, 5, 382–409, 1968.

38. Caswell, H., Stable population structure and reproductive value for populations with complex life cycles, *Ecology*, 63, 1223–1231, 1982.

39. Shaffer, W. M., Selection for optimal life histories: the effects of age structure, *Ecology*, 55, 291–303, 1974.

40. Shaffer, W. M. and Rosenzweig, M. L., Selection for optimal life histories. II. Multiple equilibria and the evolution of alternative reproductive strategies, *Ecology*, 58, 60–72, 1977.

41. Shaffer, W. M., On reproductive value and fitness, *Ecology*, 62, 1683–1685, 1981.

42. Caswell, H., Optimal life histories and the maximization of reproductive value: a general theorem for complex life cycles, *Ecology*, 63, 1218–1222, 1982.

43. Caswell, H., Optimal life histories and the age-specific costs of reproduction, *J. Theor. Biol.*, 98, 519–529, 1982.

44. Lopez, A., Asymptotic properties of a human age distribution under a continuous net maternity function, *Demography*, 4, 680–687, 1967.

45. *The Dynamic Theory of Biological Populations*, Poluektov, R. A., Ed., Nauka, Moscow, 1974, Chap. 8 (in Russian).

46. Anderson, D. H., Estimation and computation of the growth rate in Leslie's and Lotka's population models, *Biometrics*, 31, 701–718, 1975.

47. Sviridov, A. T., Relation between the parameters which characterize the state of a fish population in a matrix model and the continuous fish population model, in *Collection of Works on the Optimal Process Theory*, Vol. 2, Kaliningrad State University, Kaliningrad, 1975, 108–118 (in Russian).

48. Golubitsky, M., Keeler, E. B., and Rotschild, M., Convergence of age structure: applications of the projective metric, *Theor. Popul. Biol.*, 7, 84–93, 1975.

49. May, R. M., Biological populations obeying difference equations: stable points, stable cycles, and chaos, *J. Theor. Biol.*, 51, 511–524, 1975.

50. May, R. M. and Oster, G. F., Bifurcations and dynamical complexity in simple ecological models, *Amer. Natur.*, 110, 573–599, 1976.

CHAPTER 3

Modifications of the Leslie Model

The "challenge to linear algebra" discussed in the previous chapter resulted in an exhaustive description of the asymptotics inherent in the classical Leslie model. This progress has in turn challenged further mathematical efforts to study various modifications of the classical approach. In the present chapter we consider several "vectors" of such efforts, which make use of the previous findings in stability analysis applied to the new models.

I. TACKLING VARIABILITY BY "CONSTANCY" METHODS

Modeling age-structured populations was one of the first steps that mathematical theory took to treat a population more realistically, i.e., as a system of interacting components rather than as a homogeneous entity. But even these steps, when taken within the framework of Leslie matrix formalism, required certain oversimplifying assumptions to be made in terms of dividing the population into age classes of the same duration, and in setting the development of an individual into a constant "schedule" within the life span. In other words, the matrix formalism depended upon certain postulates of constancy in basic features of the population under consideration. Unfortunately, the widely known Leslie matrix model has shown very few examples where we know that its classical form applies to real populations, just because the necessary postulates of constancy could hardly be accepted for those populations.

A. Deviations from a Constant Life Cycle

The constant schedule of the organism's life span, which is one of the principal limitations of the Leslie formalism, often contradicts the variability observed in nature or in experiments. But if we represent the variability in terms of assumptions that can be expressed by means of some constant matrices, then the "constancy" methods will be still effective in analyzing the model.

The limitation stipulated by (2.8) that the age of the last reproductive group is equal to the age of the oldest reproductive individual may be too restrictive in modeling practice. The alternative (2.9) may also be inappropriate. Then one could also consider all postreproductive ages as belonging to the final reproductive group. The final group would thus incorporate only a part from the preceding age, in contrast to the whole as in the classical model. If that part is supposed to be constant, then an element p_n $(0 < p_n < 1)$, the fraction of the n-th group that survives after one time interval, must be added to the Leslie matrix to give

$$L + \Delta(p_n) = \begin{bmatrix} b_1 & b_2 & \cdots & b_{n-1} & b_n \\ s_1 & 0 & & & \\ & s_2 & \ddots & \mathbf{0} & \\ \mathbf{0} & & \ddots & 0 & \\ & & & s_{n-1} & p_n \end{bmatrix}, \tag{3.1}$$

whereby

$$x_n(t+1) = s_{n-1}x_{n-1}(t) + p_n x_n(t).$$

Such a construction brings about a nonzero part of the population living indefinitely long; as a result, a systematic error arises, whose relative value does not however exceed the sum

$$p_n^{M+1} + p_n^{M+2} + \cdots = p_n^{M+1}/(1 - p_n),$$

where M is the real maximum age of the organisms.

An extension of the above case is needed when it is not possible to choose a time scale such that successive moments $t = 1, 2, \ldots$ correspond to the transitions of organisms from one age group to the next, or in other words, when durations of age groups are unequal and/or do not coincide with the time step of the model. Quite common to this case will be the following approach: along with coefficients s_i, we include parameters p_i $(0 < p_i < 1)$ which stand for the relative numbers of organisms aged i that, by the next time $t + 1$, do not mature enough to transfer to the next age group. Then matrix L changes into the following:

$$L + \Delta(p_1, p_2, \ldots, p_n) = \begin{bmatrix} b_1 + p_1 & b_2 & \cdots & b_{n-1} & b_n \\ s_1 & p_2 & & & \\ & s_2 & \ddots & \mathbf{0} & \\ \mathbf{0} & & \ddots & p_{n-1} & \\ & & & s_{n-1} & p_n \end{bmatrix}. \tag{3.2}$$

A particular time scale and age division may be such that one-step transitions not only arise in the next age groups but in more distant groups as well. For example, experiments on the rice weevil *Sitophilus oryzae* revealed that the immature period varied according to a statistical log-normal distribution with a mean duration of 35.3 days.[1] Thus, in a model[1] where 40 daily age groups spanned the immature period (until emergence) and 40 groups spanned that of maturity, there were "irregular" transitions $25 \rightarrow 41, 26 \rightarrow 41, \ldots, 39 \rightarrow 41$ which appeared in addition to the "regular" ones of the classic Leslie model. Transition probabilities were estimated from emergence data and the modified Leslie matrix had the following form:

$$
L+\Delta=
\begin{bmatrix}
0 & 0 & \cdots & 0 & 0 & \cdots & 0 & 0 & b_{41} & b_{42} & & \cdots & b_{80} \\
s_1 & 0 & & & & & & & & & & & \\
 & s_2 & \ddots & & & \mathbf{0} & & & & & & & \\
 & & \ddots & & & & & & & & \mathbf{0} & & \\
 & & s_{25} & 0 & & & & & & & & & \\
 & \mathbf{0} & & s_{26} & \ddots & & & & & & & & \\
 & & & & \ddots & & & & & & & & \\
 & & & & & s_{39} & 0 & & & & & & \\
0 & 0 & \cdots & e_{25} & e_{26} & \cdots & e_{39} & s_{40} & 0 & 0 & & \cdots & 0 \\
 & & & & & & & & s_{41} & 0 & & & \\
 & & \mathbf{0} & & & & & & & s_{42} & & \mathbf{0} & \\
 & & & & & & & & \mathbf{0} & & \ddots & \ddots & \\
 & & & & & & & & & & & s_{79} & 0
\end{bmatrix}. \quad (3.3)
$$

Modifications of the kind found in (3.1) to (3.3) neither alter the Leslie matrix's properties of being nonnegative, nor of being indecomposable. Hence, the Perron–Frobenius theorem still "works," so that, in the primitive case, there exists the limit

$$
\lim_{t \to \infty} \frac{(L+\Delta)^t \mathbf{x}}{r^t} = \varphi(r, \mathbf{x}) \mathbf{C}_1, \quad (3.4)
$$

where $\mathbf{C}_1$ is a positive eigenvector corresponding to the dominant eigenvalue $r(L+\Delta)$ of the modified matrix. The possibility of having an imprimitive case disappears immediately once there appears a positive p_i, generating a 1-cycle in the digraph. However, other modifications may retain the imprimitivity of the matrix, in which case the argument that resulted in expression (2.27) is still valid (although the explicit form of the eigenvectors (2.23)–(2.24) may change): the limit distribution $\mathcal{L}(\mathbf{x}, t)$ is periodic in t, with a period that divides the index of imprimitivity h.

Furthermore, since the entries of the modifying matrix Δ seem nonnegative, the relation

$$
L \leq L + \Delta, \quad (3.5)
$$

should hold element-wise, whereby it should follow that

$$
r(L) < r(L + \Delta), \quad (3.6)
$$

as the dominant eigenvalue of an indecomposable matrix is a strictly monotone function of its entries.[2] Consequently, a modification should increase the rate of population growth as compared with the original model L. However, this does not happen in some practical examples, including the rice weevil model (3.3). In the latter case, the value of r turned out to be practically the same as before, while the value of $|r/\lambda_2|$ was greater, indicating a higher rate at which all trajectories converged to the limit function.

The reason for this paradox is hidden in the estimation of the "modifying" entries for the model matrix. Derived from real data, the estimates logically brought about some compensating decrease in "classical" survival rates s_i. Therefore, the modifying matrix Δ is not necessarily nonnegative, so that the relation (3.6) may generally not hold.

In general, since deviations from a "regular" schedule of the life span may be exhibited only by those organisms which survive until irregular transitions occur, the "compensation" property of matrix Δ must be as follows: its first lower-diagonal consists of zero or negative elements, while the remaining entries are all nonnegative and such that each column of Δ has a zero sum.

B. Small Modifications and the Perturbation Method

Let us now assume, based on the biology of a population we are now studying, that the entries of the modifying matrix Δ are small in comparison with the basic birth and survival rates b_i and s_i. We can then assess the effect of such a modification on the spectrum of the model matrix, or eventually on the pattern of model behavior, by using the method of perturbation theory.

Suppose, furthermore, that a modified (or *perturbed*) Leslie matrix can be represented in the form of

$$L(\varepsilon) = L(0) + \varepsilon\Delta + O(\varepsilon^2), \tag{3.7}$$

where ε is a *small parameter*, $\varepsilon\Delta$ is a linear term with respect to ε, and $O(\varepsilon^2)$ designates the terms of higher order of smallness. Since eigenvalues of a matrix depend on matrix entries continuously, the spectrum of $L(\varepsilon)$ tends to that of $L(0)$ as $\varepsilon \to 0$. It follows[3] from perturbation theory that the (simple) dominant eigenvalue $r(\varepsilon)$ and the corresponding positive eigenvector $C(\varepsilon)$ obey the relations

$$r(\varepsilon) \;=\; r + \varepsilon r^{(1)} + O(\varepsilon^2) \tag{3.8}$$
$$C_1(\varepsilon) \;=\; C_1 + \varepsilon C_1^{(1)} + O(\varepsilon^2) \tag{3.9}$$

The *principal part* $\varepsilon r^{(1)}$ of the eigenvalue perturbation that is induced by perturbation of the matrix, can be found by means of vectors R_1 and C_1 of the two orthogonal sets of eigenvectors (2.23) and (2.24) for the original matrix $L(0)$:

$$r^{(1)} = \langle R_1, \Delta C_1 \rangle, \tag{3.10}$$

where ΔC_1 is the result of matrix operator Δ on vector C_1. Then the perturbation $\varepsilon C_1(1)$ of the steady age structure can be uniquely determined by vectors R_1 and C_1 and the resolvent of $L(0)$.[3] Similar statements are also valid for other (nonmultiple) eigenvalues and their corresponding eigenvectors.

In particular, if $\lambda_2(\varepsilon)$ (or generally $\lambda_{h+1}(\varepsilon)$) is the second in modulus eigenvalue, then simple algebra brings about the following estimation for the rate of decrease in trajectory deviations from the limit function (or the *convergence rate*):

$$\left| \frac{r(\varepsilon)}{\lambda_2(\varepsilon)} \right| = \left| \frac{r}{\lambda_2} \right| \cdot \left| 1 + \varepsilon \left(\frac{r^{(1)}}{r} - \frac{\lambda_2^{(1)}}{\lambda_2} \right) \right| + O(\varepsilon^2).$$

It follows that if

$$\rho(\varepsilon) = |r(\varepsilon)/\lambda_2(\varepsilon)| = \rho(0) + \varepsilon \, \rho^{(1)} + O(\varepsilon^2),$$

then the principal part of variation in $\rho(\varepsilon)$ is equal to

$$\rho^{(1)} = \mathrm{Re} \left(\frac{r^{(1)}}{r} - \frac{\lambda_2^{(1)}}{\lambda_2} \right) . \tag{3.11}$$

Readily calculable in practice, the values of (3.10) and (3.11) show a direction and quantitative level of those variations in the population growth rate and the rate of convergence to the limit (equilibrium or cyclic) distribution which are induced by a (small) modification of the Leslie model. The mysterious "small parameter," which often seems no more than a mathematical trick, acquires quite an obvious meaning in this context.

If, for example, we reformulate matrix (3.3) of the rice weevil population in terms of perturbation theory and with regard to the compensating decrease in survival rates $s_{25}, \ldots, s_{39}$, then the modifying matrix Δ of expression (3.7) will take on the form

$$\varepsilon \Delta = \varepsilon \begin{bmatrix} 0 & 0 & \cdots & 0 & 0 & \cdots & 0 & 0 & \\ 0 & 0 & & & & & & & \\ & 0 & & & & & & & \\ & & \ddots & \ddots & & & & & \mathbf{0} \\ & & & -s_{25} & 0 & & & & \\ & \mathbf{0} & & & \ddots & \ddots & & & \\ & & & & & & -s_{39} & 0 & \\ 0 & 0 & \cdots & s_{25} & s_{26} & \cdots & s_{39} & 0 & \\ & & \mathbf{0} & & & & & \mathbf{0} & \end{bmatrix}, \tag{3.12}$$

under the assumption that irregularity in the maturation period occurs uniformly in the age groups 25 through 39. But if the irregularity obeys a less trivial law of distribution (among the groups), then the proper weight factors β_i ($i = 25, \ldots, 39, \sum_i \beta_i = 1$) should be incorporated into nonzero entries of (3.12), the compensation property being still preserved. The small parameter is then equal to $\varepsilon = \sum_i \beta_i^0$, where each β_i^0 is to be estimated from emergence data as the portion of those survivors of the i-th group ($i = 25, 26, \ldots, 39$) which pass ahead of the regular maturation "schedule." Thus, ε can serve as a measure of variability of this kind, while (3.10) and (3.11) show what parts of the variability convert into the variations in dynamic characteristics of the model.

In the Appendix, Statement A3.1, calculation of $r^{(1)}$ is shown to reduce to

$$r^{(1)} = \frac{r^n}{p'(r)} \left(\frac{r^{15}\beta_{25}}{s_{26}s_{27}\ldots s_{40}} + \frac{r^{14}\beta_{26}}{s_{27}\ldots s_{40}} + \cdots + \frac{r\beta_{39}}{s_{40}} - 1 \right), \qquad (3.13)$$

which is further shown to be greater than or equal to 0 if $r \geq 1$. But if $r < 1$, then expression (3.13) may be of any sign.

Thus, for a steady or growing Leslie model population, the maturation irregularities of the above type may only accelerate the growth, whereas for a declining population, whether $r(\varepsilon)$ will be less or greater than $r(0)$ depends on fairly delicate relations among particular values of survival probabilities s_i ($i = 26, \ldots, 40$) for the immature ages that exhibit irregularity in the emergence schedule and the distribution, $\{\beta_i\}$, of the irregularities themselves.

With regard to the principal part of the variation in $|r(\varepsilon)/\lambda_2(\varepsilon)|$ relative to $|r/\lambda_2|$, it is similarly shown to be equal to

$$\mathrm{Re}\left\{ \frac{r^{(1)}}{r} - \frac{\lambda_2^{(1)}}{\lambda_2} \right\} = \mathrm{Re}\left\{ \beta_{25}\frac{\hat{r}r^{15} - \hat{\lambda}_2\lambda_2^{15}}{s_{26}s_{27}\ldots s_{40}} + \beta_{26}\frac{\hat{r}r^{14} - \hat{\lambda}_2\lambda_2^{14}}{s_{27}\ldots s_{40}} + \cdots \right.$$
$$\left. + \beta_{39}\frac{\hat{r}r - \hat{\lambda}_2\lambda_2}{s_{40}} - (\hat{r} - \hat{\lambda}_2) \right\}, \qquad (3.14)$$

where

$$\hat{r} = \frac{r^{n-1}}{p'(r)}, \qquad \hat{\lambda}_2 = \frac{\lambda_2^{n-1}}{p'(\lambda_2)}.$$

Expression (3.14) is proved to be positive at least for any nondeclining population with positive $\hat{r} - \mathrm{Re}\,\hat{\lambda}_2$ (see Appendix, Statement A3.1).

So, the constancy in r and increase in $|r/\lambda_2|$ that were observed in the rice weevil model as a result of matrix modification,[1] are nothing else than particular consequences of equations (3.13) and (3.14). In general, however, the effect of variability in the pattern of life span on important dynamic characteristics, such as r and $|r/\lambda_2|$, is quite far from being unidirectional, although it is quite calculable in each particular case.

However, a general feature to note is that modification normally disrupts imprimitivity or decreases its quantitative index. Indeed, an equivalent definition of matrix A's *index of imprimitivity* determines it to be the g.c.d. of lengths of all directed cycles in the digraph associated with matrix A.[4] Since the modifying entries, no matter how small they may be, bring new links, and hence new cycles, into the graph of age transitions, the g.c.d. of the cycle lengths would remain unchanged only in a very special case, i.e., when none of the new links generates a cycle of new length or when the cycles they do generate have the length multiple of the existing ones. But the most probable case is where a cycle will emerge whose length is mutually prime to the existing ones, thus reducing the index of imprimitivity immediately to unity. Modification (3.2) with any positive p_i (generating a

1-cycle) is the nearest example. So, imprimitivity represents a structurally unstable property of matrix models.

For particular modified matrices, one could, of course, also calculate the dominant eigenvalue r and the convergence rate $|r/\lambda_2|$ directly, by means of a computer. But, in contrast to the Leslie matrix, whose characteristic polynomial is known explicitly, hence the eigenvalues could be found iteratively up to any desired level of accuracy, the computation of eigenvalues for a matrix of a more general form by a computer routine appears often to be unreliable when the matrix is ill-conditioned or the algorithm is numerically unstable. Unavoidable round-off errors may also violate the logic of the Perron–Frobenius theorem in the imprimitive case. So, using equations (3.13) and (3.14) may contribute to avoiding erroneous conclusions from computation alone. If, in addition, the modeler wants to keep the population growth rate unchanged after modification, then equating expression (3.13) (and, perhaps, coefficients at higher powers of the small parameter) to zero brings about a reduction in the degree of uncertainty in numerical values of model parameters, which is so desirable in modeling practice.

C. Switching Matrices

Another kind of limitation of the Leslie model, when used to simulate dynamics of real populations, is related to the pattern of trajectory behavior in general and the period of cycles in particular. Typical for many populations, the cycles can somehow be mimicked in a Leslie model only if their period does not exceed the life span of the organisms; the matrix should then be constructed in such a way that its index of imprimitivity would either be divisible by or equal to the cycle period. Together with the absence of chaotic behavior, this shows linearity of birth and survival mechanisms to constrict the qualitative diversity of trajectories known for nonlinear models of (even nonstructured) population dynamics.

One of the attempts to reconcile the analytical simplicity of the linear Leslie model with complicated dynamics of natural populations resulted in the so-called "matrix jump" model.[5] Here, cyclic or almost cyclic population dynamics is modeled by means of two Leslie matrices, one differing from another by the set of survival coefficients s_i in a way such that one matrix (L_1) has the largest eigenvalue $r_1 > 1$, while the other one (L_2) has $r_2 < 1$ ($r_1 r_2 = 1$). When the total size $N(t)$ of the model population is less than some average (fixed) value N, say $N = 1000$, the population is governed by matrix L_1, which yields an increase in $N(t)$. As soon as $N(t)$ outgrows N, the matrix switches to L_2 that provides a decrease in $N(t)$. Thus the concept of cyclicity is imposed on the model structure itself, which now ceases to be globally linear.

Unfortunately, almost nothing is known about strict analytical results on cycles in the "matrix jump" model. Model trajectories can easily be obtained in a computer. They show a wide variety of "quasi-cycles," i.e., trajectories formed by group sizes rounded off to integers. These "quasi-cycles" successfully mimic natural population dynamics, for instance, in mammals, with oscillation periods of several years.

While the discontinuity of L as a function of $N(t)$ prevents straightforward linearization, a traditional method of stability analysis for nonlinear systems, the constancy of L outside of its break points helps to avoid this obstacle. When looking for an "equilibrium" in this kind of models, we have to stipulate that either we deal with a true equilibrium in the interior of a region where L is constant (i.e., L_1 or L_2 alone), or we mean an invariable mode of switching between L_1 and L_2 in the vicinity of a critical solution, e.g., L_1, L_2, L_1, L_2, and so on, irrespective of what is the current state of the population. The former case obviously admits all the routine described above, under certain constraints on the domain of stability in the state space. The latter case reduces to the search of a true cycle for the fixed switching mode, which may be caused, for instance, by strict seasonality in demographic parameters.

In the simplest case there should be a pair of vectors, say $\mathbf{a}$ and $\mathbf{b}$, for which

$$\mathbf{b} = L_1\mathbf{a},$$
$$\mathbf{a} = L_2\mathbf{b},$$

whereby

$$\mathbf{a} = L_2 L_1 \mathbf{a}.$$

Thus $\mathbf{a}$ can be found as an eigenvector of the matrix product $L_2 L_1$, associated with the unit eigenvalue if such exists. And if it does, then

$$\mathbf{b} = L_1\mathbf{a} = L_1 L_2 \mathbf{b}$$

must be an eigenvector of $L_1 L_2$ also associated with the unit eigenvalue. (The fact that nonsingular matrices AB and BA have the same spectra follows immediately from the observation that the matrices are *similar* in the sense that $AB = P^{-1}BAP$ with a nonsingular matrix P: here we have $P = B$.)

Stability of this cycle can be investigated as the stability of an equilibrium in the model with a double time step, i.e., with $\tau = 2t$ and

$$\mathbf{x}(\tau + 1) = L_2 L_1 \mathbf{x}(\tau),$$

whose trajectories

$$\mathbf{x}(\tau) = (L_2 L_1)^{\tau}\mathbf{x}(0)$$

are obvious subsequences of the original sequence. The product $L_1 L_2$ thus becomes the hero of spectral analysis in the sense of the previous sections, generating a seasonal cycle that matches an equilibrium in the product, or longer cycles if the product itself has a cyclic structure. It would certainly be a local analysis, valid unless deviations from the cycle vectors $\mathbf{a}$ and $\mathbf{b}$ disturb the switching mode itself.

These arguments can be easily generalized for any fixed mode of switching that comprises a finite mixture of L_1 and L_2 or even of more "ingredients," provided only that the corresponding eigenvalue problem has a nontrivial solution. Unfortunately, multiplication of distinct matrices does not generally result in the same multiplication of their eigenvalues. Thus a general solution to the problem of stable cycles in the "matrix jump" model can hardly be obtained in this way, excepting the case where a fixed switching mode is due to an external reason, e.g., a prescribed seasonality in the birth and survival rates.

II. BLOCK STRUCTURE: TACKLING VARIABILITY WITHIN AGE GROUPS

In the previous section we have seen how a "bank" of mathematical results on the classic Leslie matrix can be used in cases where variability in individual life cycles brings about modification of the classic model by some additional entries to the main, "Leslian," pattern. Now we concentrate on variability within age groups, which were formerly supposed to be homogeneous.

A. Age-Status Set-Up and Block Matrices

Practice in modeling age-structured populations often requires a modification of the classical Leslie model such that tackling the variability within each age class results in a model where, in addition to the age structure, one has to consider a subdivision of each class into a finite number of groups with respect to another character, e.g., genotype, size, physiological status, etc., the transitions among groups of this "additional" structure occurring in parallel with ontogenesis or/and the reproductive process.[6-10]

Suppose, for instance, that we consider an optimal control problem for an animal population subject to harvesting by age-specific strategies, and we also have to account for the following rule, often acting as an axiom for harvesters: the weakest animals are harvested first. The model then must inevitably incorporate an additional structure that should distinguish the physical status of animals, while the status may both vary as a function of living conditions and modify age-specific parameters of the population.

For the convenience of terminology, we shall hereafter refer to the additional structure as a generalized *status structure*; let there also be a finite number m of *status groups* defined within each of n age classes. The concrete meaning of "status" is certainly dependent on the biology of the species and formulation of the problem under consideration. For example, in a model by Csetenyi[10] for a reindeer population under the pressure of harvesting, there are $n = 17$ yearly age classes and there is a rank classification of individuals with respect to their physiological status ("weak," "normal," "strong"), i.e., $m = 3$; in a model by Law[8] for a population of monocarpic "biennial" plants which reproduce only after reaching a certain critical size, specified are $m = 4$ size groups and $n = 5$ yearly age classes (see Figure 4).

Let furthermore the components x_{is} of the population state vector $\mathbf{x}(t)$ be ordered lexicographically with respect to (i, s), with i indicating an age group number and

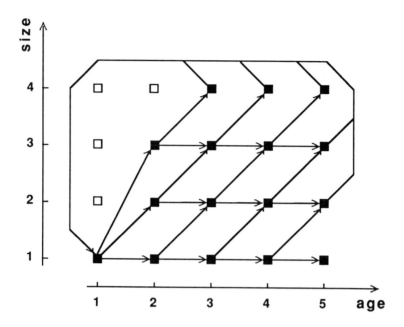

Figure 4.   The associated graph for a block matrix of the model by Law[8].

s indicating a status group number, i.e.,

$$\mathbf{x} = [x_{11}, x_{12}, \ldots, x_{1m}, x_{21}, x_{22}, \ldots, x_{2m}, \ldots, x_{n1}, x_{n2}, \ldots, x_{nm}]^T . \qquad (3.15)$$

Splitting the size of an age class into those of its status groups calls logically for distributing each of the age-specific vital rates among its status-specific components. So, let $a_{kl}^{(ij)}$ denote the portion of those individuals of age j and status l who survive, in one time step, to age i and acquire status k ($i \neq 1$), and let $a_{kl}^{(1j)}$ denote the average number of status k progeny produced per one individual of age j and status l ($i, j = 1, \ldots, n$; $k, l = 1, \ldots, m$). Then the projection matrix of the population growth equation (2.5) can be easily seen to take the block form

$$\mathcal{A} = [A_{ij}], \qquad (3.16)$$

where the $m \times m$-blocks $A_{ij} = [a_{kl}^{(ij)}]$ of the age-status parameters replace the corresponding entries of an $n \times n$ Leslie matrix of the form (2.6) (or of a more general form), the *nonblock prototype* of the block structure.

   The question that springs up is whether the additional population structure can generate any new features in the dynamic behavior of the model. In particular, can the convergence to equilibrium change into limit cycles, or vice versa. In view of the previous results, these questions reduce immediately to whether the block structure of the projection matrix retains the indecomposability and (im)primitivity properties of its nonblock prototype. These issues are dealt with in the present section, mostly by means of standard representation of the matrix structure by its associated directed graph (digraph). We deduce a general criterion of indecomposability for block models, as well as conditions for a block matrix to be primitive or imprimitive; these help to analyze some more specific cases where the additional subdivision structure introduces actually no new features into the main properties of the projection matrix.

Let us recall that indecomposability of a matrix A is equivalent to strong connectedness of its digraph $D(A)$, while matrix A's *index of imprimitivity* is equal to the g.c.d. of lengths of all dicycles in $D(A)$.[4] In what follows this matrix term will be used, in the above sense, for the digraph D as well.

Since the age-status groups are labeled with two subscripts, the vertices of the digraph can consequently be represented by nodes of a (finite) two-dimensional lattice, one axis corresponding to age, another one to status (see Figures 4 and 5,a), while the arcs show all possible ways to change status in the course of survival and reproduction processes. (In the model by Law,[8] the groups of great sizes were not defined for younger classes, so that the projection matrix, strictly speaking, loses its block structure. It can nevertheless be restored by formally introducing the lacking age-status groups, designated by empty dots in Figure 4. These groups would then correspond to isolated vertices in the digraph, though depriving it of connectedness but retaining the former dominant eigenvalue.[11])

If in the digraph of Figure 5,a we reject for a while the status structure of the population, then we would come to a digraph (Figure 5,b) which is the graph of a classical Leslie model for the same number of age classes (cf. Figure 3). Obviously, that Leslie digraph is strongly connected (or just *strong*) and the corresponding Leslie matrix is indecomposable, as there is a nonzero birth rate in the oldest age class. This example shows that introducing the additional structure may quite probably deprive the digraph of its strong connectedness (no path, e.g., from vertex (1,1) to (3,2)) hence the projection matrix of its indecomposability, so that a problem of developing a constructive criterion of indecomposability arises naturally in such models.

B. Factorgraphs and Factor-closure

The criterion proposed below relies upon some ideas that follow from the block structure of the matrices under study, and one of such ideas is the notion of the age factorgraph. For a digraph defined on a two-dimensional lattice of vertices, the age factorgraph is essentially an outcome of the procedure used above to revert to the digraph of Figure 5,b from that of Figure 5,a.

In formal terms (see Definitions A3.1 and A3.2 in Appendix), the *age factorgraph*, $L(\mathcal{A})$, is the factorgraph, or the condensation, of the digraph $D(\mathcal{A})$ with respect to partitioning the set of its vertices into groups of the same ages: there is an arc $i \rightarrow j$ between vertices i and j in the age factorgraph $L(\mathcal{A})$ if in $D(\mathcal{A})$ there is a one-step transition from a status of the i-th age into a status of the j-th age.

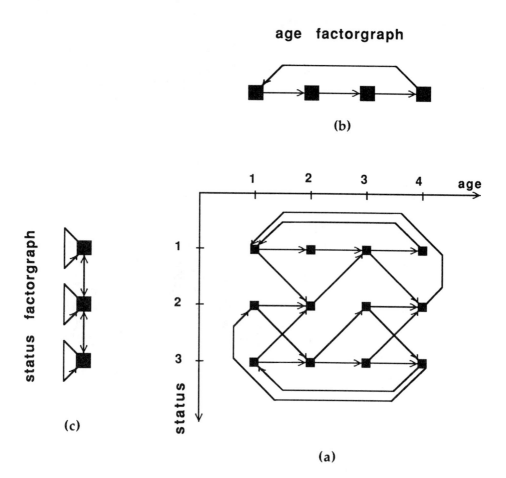

Figure 5.   Digraphs for a hypothetical population with $n = 4$ age classes and $m = 3$ status classes (from Csetenyi and Logofet,[11] used with permission).

The *status factorgraph*, $S(\mathcal{A})$, is defined similarly as the factorgraph with respect to partitioning into groups of the same statuses.

For example, the digraph shown in Figure 5,a has its age and status factorgraphs as shown in Figures 5,b and 5,c respectively. They are the same for the digraph in Figure 6.

Both age and status factorgraphs can be imagined as "projections" of digraph $D(\mathcal{A})$ onto the age and status axes respectively. This may be useful when a digraph D is defined on a two-dimensional lattice $V_1 \times V_2$ with no reference to any block matrix $\mathcal{A}$ (see Definitions A3.1' and A3.2' in the Appendix).

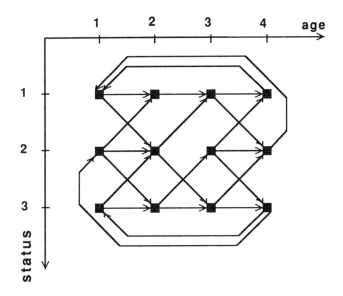

Figure 6.   An example of a strongly connected digraph (from Csetenyi and Logofet,[11] used with permission).

It is clear intuitively and provable formally that for a digraph $D(A)$ to be strong it is necessary that both its age and status factorgraphs be strongly connected too. For the age factorgraph this requirement is quite natural, signifying merely the continuity of the life cycle; it is less obvious for the status factorgraph, which is interpreted as the possibility of attaining, even during the course of successive generations, any one status from any other in a finite number of time steps. But these necessary conditions are generally not yet sufficient for the digraph itself to be strong. The digraph of Figure 5,a, for example, is not strong, even though it has strong age and status factorgraphs.

Besides partitioning the set of all vertices of a digraph $D(A)$ by age or by status, one may consider other partitions too, that would have their own senses. For example, one may combine into a single group all but the initial ages of a given status. Repeated for each status, this combination will result in a so-called *bi-partite* factorgraph to display a pattern of reproduction among various status groups.

Such partitionings make sense in the problem of strong connectedness, since the following theorem is true (the proof is given in the Appendix):

Theorem 3.1 *A digraph D is strongly connected if and only if any partitioning of its vertex set results in a strongly connected factorgraph.*[12]

Theorem 3.1 can be used to establish the lack of strong connectedness in some particular cases (if one succeeds in finding a partition whose factorgraph would apparently not be strong). But to ascertain strong connectedness in $D(A)$ the theorem is even less constructive than the definition of strong connectedness itself.

Let us consider a particular case, though rare in practice but important in methodology, where the picture of status transitions is common to all age classes. This means that the pattern by which zero elements are allocated within a matrix $A = [A_{ij}]$ is the same within each of its $(m \times m)$-blocks (such a block structure will be referred to as *regular*). Then the digraph of the block matrix can be represented as the so-called *Kronecker product*, or *conjunction*, of its age and status factorgraphs, the properties of connectedness and imprimitivity in the product being completely described by those of the "co-factors."

A formal definition of the Kronecker product (Definition A3.3 in the Appendix) requires that an arc occurs between each particular pair of age-status groups in the product once it occurs between the proper ages in the age graph and the proper statuses in the status graph.

By a theorem of McAndrew[13] the number of strongly connected components (or *strong components*) of a digraph $D = D_1 \wedge D_2$ equals g.c.d. $\{h(D_1), h(D_2)\}$, where $h(D_i)$, the imprimitivity index of digraph D_i, is equal to the g.c.d. for the length of all directed loops in D_i ($i = 1, 2$); for any strong component C of the digraph D its imprimitivity index $h(C) = $ l.c.m. $\{h(D_1), h(D_2)\}$. In particular, $D = D_1 \wedge D_2$ is strongly connected if and only if both D_1 and D_2 are strong, while their indices $h(D_1)$ and $h(D_2)$ are mutually prime.

In order to make use of these results in investigating strong connectedness of the digraph $D(A)$ for a matrix A whose block structure is not necessarily regular, we introduce another notion as follows.

Definition 3.1 *If D is the digraph of an arbitrary block matrix A, then the conjunction of its age and status factorgraphs will be called the* factor-closure *of D and denoted by $F[D(A)]$, i.e., $F[D(A)] = L(A) \wedge S(A)$.*

Any digraph $D(A)$ is obviously a *skeleton* subgraph of its factor-closure (i.e., a subgraph that includes all the vertices of the original graph). It coincides with its factor-closure only in the special case where the matrix A has a regular block pattern, i.e. where matrix Sgn A itself (which is equivalent to the *adjacency matrix* of the digraph $D(A)$) represents a Kronecker product $L \otimes S$ of some $(n \times n)$ matrix L and $(m \times m)$ matrix S. Clearly, a category of connectedness of a digraph $D(A)$ can never be higher than that of its factor-closure $F[D(A)]$.

Application of McAndrew's theorem[13] to the factor-closure of a digraph $D(A)$ brings about a criterion of strong connectedness for a digraph in the form of $F_1 \wedge F_2$, formulated below as Theorem 3.2, or an equivalent criterion of indecomposability for the matrix A of a regular block pattern, formulated further as Theorem 3.2'.

Theorem 3.2 *For a digraph $D(A) = L(A) \wedge S(A)$ of a block matrix A to be strongly connected, it is necessary and sufficient that its age and status factorgraphs, $L(A)$ and $S(A)$, have the same property and the imprimitivity indices of the factorgraphs be mutually prime; their product then gives the imprimitivity index of the strong digraph $L(A) \wedge S(A)$.*

Theorem 3.2' *For a matrix $\mathcal{A}$ of a regular block pattern* Sgn $\mathcal{A} = L \otimes S$ *to be indecomposable, it is necessary and sufficient that the co-factor matrices L and S have the same property and the indices of imprimitivity of matrices L and S be mutually prime; their product then gives the imprimitivity index of indecomposable matrix $\mathcal{A}$.*

If, for example, matrix L is primitive (let there be, for instance, nonzero birth rates in at least two consecutive age classes), then matrix $L \otimes S$ will be indecomposable and primitive in all cases where matrix S is indecomposable. But if L is imprimitive, then, assuming the existence of a nontrivial common divisor for the imprimitivity indices of L and S, the block matrix $L \otimes S$ turns out to be decomposable. It is thus conceivable to have a situation in which introducing even the same, strongly connected pattern of status transitions into each age class brings about the block extension of the initial Leslie matrix (or any of its modifications).

C. Basic Arcs and the Criterion of Connectedness

Note that strong connectedness of the factor-closure of a digraph $D(\mathcal{A})$ is not yet sufficient, in general, for the $D(\mathcal{A})$ itself to be strongly connected. For example, the imprimitivity indices for the factorgraphs in Figure 5, $h(L(\mathcal{A})) = 4$ and $h(S(\mathcal{A})) = 1$, are mutually prime, hence $F[D(\mathcal{A})]$ is strongly connected by Theorem 3.2, whereas $D(\mathcal{A})$ is not strong as noted above.

If a digraph $D(\mathcal{A})$, which is not identical to its factor-closure $F[D(\mathcal{A})] = L \wedge S$, is nevertheless strongly connected, this means that the set of arcs $\{F[D] \setminus D\}$, which complements D with respect to its factor-closure, is not of critical importance for $F[D]$ being strongly connected: deleting any arcs of this set from the graph does not change the category of its connectedness. Such a set of arcs is called *neutral* in graph theory.[14] Hence, in order that a digraph D with a strong factor-closure $F[D]$ be strongly connected it is necessary and sufficient that the set of arcs $\{F[D] \setminus D\}$ be neutral in $F[D]$. Given below in Theorem 3.3, the constructive criterion for strong connectedness of $D(\mathcal{A})$ reduces the problem to treating just this set of arcs and uses the notion of a *basic arc*.

Definition 3.2 *An arc x of a digraph D is called* basic *if there is no other directed path in D that would go from the initial vertex of x to its terminating vertex.*[14]

For example, arcs $(2, 2) \rightarrow (3, 3)$ and $(4, 2) \rightarrow (1, 1)$ are basic in the digraph of Figure 6, thus being of critical importance for the structural integrity of the model.

Theorem 3.3 *A digraph $D(\mathcal{A})$ is strongly connected if and only if its factor-closure is strongly connected and neither arc y of the set $\{F[D] \setminus D\}$ is basic in the digraph $D + y$. (The proof is given in the Appendix).*

Theorem 3.3 is especially useful to verify strong connectedness in the case where the digraph $D(\mathcal{A})$ lacks only a few arcs to be identical with its factor-closure. If, for example, we add some arcs to the nonstrong digraph of Figure 5, it can turn into a strongly connected one. Such a digraph is shown in Figure 6, where its age and status factorgraphs appear to be the same as in Figures 5,b and 5,c, respectively. Thus, its factor-closure $F[D]$ meets all conditions of Theorem 3.2, hence being strongly connected. The set $Y = \{F[D] \setminus D\}$ consists of five arcs, viz. $(2,1) \to (3,2); (4,1) \to (1,2); (2,2) \to (3,2); (4,2) \to (1,2)$, and $(4,2) \to (1,3)$, which are not shown in Figure 6. Arc $y_1 : (2,1) \to (3,2)$, for instance, is not basic in the digraph $D + y$ since in D there is a directed path

$$
\begin{aligned}
(2,1) \quad &\to \quad (3,1) \to (4,1) \to (1,1) \to (2,2) \to (3,3) \to (4,3) \\
&\to \quad (1,3) \to (2,3) \to (3,2).
\end{aligned}
$$

Similarly, all the other arcs of set Y are nonbasic too, so that the digraph $D(\mathcal{A})$ is strong by Theorem 3.3 and the corresponding block matrices are indecomposable. Direct verification of the definition would require a combinatorial number of pairs to be checked for being connected, whereas the theorem reduces this number to a moderately small one.

D. Sufficient Conditions of Indecomposability

Besides Theorem 3.2, some sufficient conditions of strong connectedness can be proposed, which are useful in checking whether the projection matrix of a model for an age- and status-structured population is indecomposable in some particular cases amenable to ready interpretation.

It is clear, for example, that in a strongly connected digraph there must be neither *impasse* nor *anti-impasse*, i.e., vertices with no outgoing or ingoing arcs respectively. (It is this requirement that is apparently violated by the digraph in Figure 4.) In other words, for any age-status group of a model population a transition must be possible into at least one other group and each group must be recruited from some other group(s) by aging or reproduction. For a population projection matrix the former condition means that individuals of any age class proceed into the next age or/and produce a progeny, while the latter represents the natural continuity and recurrence of ontogenesis for individuals of any status.

Recall also that for a digraph to be strongly connected it is necessary that any partitioning of its vertex set generates a strong factorgraph (Theorem 3.1), in particular, the age and status factorgraphs must be strong.

As regards sufficient conditions of strong connectedness, one group reduces to a situation where the graph of status transitions is *age-invariant* (i.e., the same for all ages) and *status-complete* (i.e., any status is attainable from any other one in one step), though obvious in theory but rare in practice. For a block matrix of the form (3.16) this means that there are no zero entries in either of the nontrivial blocks $A_{ij} \neq 0$ of the matrix, i.e., both preservation and any change of status are possible in each age class while it is surviving to the next one or/and producing a progeny.

If we attenuate the condition and replace it with the possibility to retain the current status or of transiting into "adjacent" statuses only (i.e., of upgrading or downgrading the status by one unit in one step), then nontrivial blocks A_{ij} of matrix $\mathcal{A}$ take on the three-diagonal form, the diagonals having no zero elements. This is the case, for instance, in a matrix model by Csetenyi[10] for a population of reindeer classified by their physiological status into three ranks ("weak," "normal," "strong"), i.e., into three status classes. Based on real observations, it was assumed that the status rank of an individual may preserve or may change by no more than one unit in one time step as a function of wintering conditions. If the digraph of a block matrix has strong age and status factorgraphs and has no impasses, then we can prove it to be strongly connected (Theorem A3.4 in Appendix), hence the matrix to be indecomposable. So, indecomposability of the projection matrix (of dimension 51×51) for Csetenyi's model can now be ascertained at an early stage of formulating the model.

Serving as a theoretical prerequisite to the analysis of model dynamics, the indecomposability of a projection matrix, when treated as strong connectedness of the associated graph of transitions, also signifies a certain structural integrity among the components that have been specified in the population. But if a projection matrix turns out decomposable, then the problem naturally arises that its indecomposable diagonal blocks (i.e., the components of strong connectedness in its digraph) have to be identified and given a proper interpretation in terms of the population modeled.

For an indecomposable projection matrix, the asymptotic properties of model trajectories (such as convergence to a steady state or cyclicity) can be determined then from the primitivity/imprimitivity properties of the matrix. The block pattern of matrices under consideration has allowed for a constructive theorem on the index of imprimitivity to be proven only in the special case where the block pattern is regular, i.e., where the matrix digraph coincides with its factor-closure. To find the the index of imprimitivity in a more general case, still under strong connectedness of the digraph, we have to use its definition and a general combinatorial method for calculating lengths of all possible directed cycles in the given digraph.

Finally, to argue the role in tackling variability that is played by the "constancy" methods, i.e., the matrix and graph methods that are traditionally used in analysis of linear deterministic models for population dynamics, we see them to have a nonempty although fairly restricted area where the methods could give certain theoretical conclusions on the effects of variability in a population pattern. Therefore, treating such effects in modeling practice should be recognized as an aspect of ecological modeling which still continues to be more of an art than a science.

III. TIME-DEPENDENT MODIFICATIONS: LIMITED UNCERTAINTY IN ASYMPTOTIC BEHAVIOR

The obvious observation that vital rates of a population may generally vary in time also brings about modification of the classical Leslie model. In the particular simple case considered earlier, the variation may be reduced just to switching between two particular sets of parameter values. But if, in other cases, the art of tackling variability by methods of constancy fails to meet the requirements of a specific problem, the failure has to be marked by the notation

$$\mathbf{x}(t + 1) = L(t)\mathbf{x}(t),$$

instead of (2.5), and resorting either to general theory of nonautonomous difference equations, or to mere computer simulation, or both.

A. Lyapunov Exponents and Exponential Separation

While systems with constant matrices are investigated by means of their eigenvalues, those with time-variable matrices, i.e., *nonautonomous* systems, invoke a generalization of the eigenvalue concept into the so-called Lyapunov exponent. The theory of Lyapunov exponents[15-16] was developed from the idea of comparing the rates of growth in solutions to a system of differential (or difference) equations with the rates of growth of some exponential curves. The numbers that characterize those rates are called the *Lyapunov exponents* (or *Lyapunov characteristic numbers*). For any solution $\mathbf{x}(t)$, its (upper) exponent, is defined as

$$\Lambda(\mathbf{x}(t)) = \overline{\lim_{t \to \infty}} \ln |\mathbf{x}(t)|/t,$$

so that for a constant-matrix linear system of differential equations the exponents are merely the real parts of the matrix eigenvalues (or their logarithms in the case of difference equations), hence being calculable in each particular case. Different solutions may generate different values of the exponent, the *highest* one, $\Lambda(A)$, being of particular importance.

Unfortunately, the exponents of nonlinear systems and those of nonautonomous linear systems can only be found in some special cases. Therefore, qualitative theories of differential and difference equations deal with a wide spectrum of problems on matching the exponents of a perturbed system with those of the unperturbed (usually linear) one and on asymptotic similarity in solutions of both systems.

We consider now a nonautonomous Leslie-type model

$$\mathbf{x}(t + 1) = A(t)\mathbf{x}(t), \tag{3.17}$$

where $A(t) = [a_{ij}(t)] \geq 0$ is a nonnegative nonsingular matrix for any $t = 0, 1, \ldots$ with a finite norm such that

$$\sup_t \|A(t)^{\pm 1}\| < \infty \tag{3.18}$$

The solution to equation (3.17) corresponding to any nonnegative initial state $\mathbf{x}(0)$ is obviously given by the formula

$$\mathbf{x}(t) = A(t - 1)A(t - 2)\ldots A(0)\mathbf{x}(0)$$

and the task is to estimate its asymptotic growth rate as t tends to infinity. The notion that arises from the idea of comparing the asymptotic behavior with the growth of exponential functions is that of *exponential separateness* in solutions.

Definition 3.3 *System (3.17) is called* exponentially separated under index $n - 1$ *if the state space* $\mathbb{R}^n$ *can be represented as a direct sum* $\mathbb{R}^1 \oplus \mathbb{R}^{n-1}$ *such that any two nontrivial solutions* $\mathbf{x}_1(t)$ *and* $\mathbf{x}_2(t)$ *initiating at different subspaces* $(\mathbf{0} \neq \mathbf{x}_1(0) \in \mathbb{R}^1,$ $\mathbf{0} \neq \mathbf{x}_2(0) \in \mathbb{R}^{n-1})$ *satisfy the following inequality*

$$\frac{\|\mathbf{x}_1(t)\|}{\|\mathbf{x}_1(s)\|} : \frac{\|\mathbf{x}_2(t)\|}{\|\mathbf{x}_2(s)\|} \geq d \exp\{\theta(t - s)\}, \tag{3.19}$$

where $t \geq s \geq 0$, *and* $d, \theta > 0$. *The subdivision* $\mathbb{R}^n = \mathbb{R}^1 \oplus \mathbb{R}^{n-1}$ *is called* the separation-defining subdivision.[17-18]

In other words, the component of any solution that belongs to $\mathbb{R}^1$ grows faster with time than its complement, the difference also growing exponentially.

If we recall the behavior of any trajectory in a model (3.17) with constant primitive projection matrix A, we can see that

$$\frac{\|\mathbf{x}_1(t)\|}{\|\mathbf{x}_1(s)\|} \sim C_1 \lambda_1^{(t-s)}, \qquad \frac{\|\mathbf{x}_2(t)\|}{\|\mathbf{x}_2(s)\|} \sim C_2 \lambda_2^{(t-s)},$$

where $\lambda_1 > 0$ and λ_2 $(|\lambda_2| < \lambda_1)$ are the dominant and the second in modulus eigenvalues of A. Hence,

$$\frac{\|\mathbf{x}_1(t)\|}{\|\mathbf{x}_1(s)\|} : \frac{\|\mathbf{x}_s(t)\|}{\|\mathbf{x}_s(s)\|} \sim C|\lambda_1/\lambda_2|^{(t-s)} = C \exp\{\ln|\lambda_1/\lambda_2|(t - s)\},$$

demonstrating the exponential separateness with $\theta = \ln|\lambda_1/\lambda_2|$. In other words, any solution behaves asymptotically as its part that initially belongs to $\mathbb{R}^1$. In the primitive case $\mathbb{R}^1$ is given by the ray of positive dominant eigenvectors. Nonautonomous generalization implies that there is a one-dimensional subspace of the solutions which have the highest value of the Lyapunov exponent, whereas the exponent of any orthogonal solution is lower. Since any positive initial vector has a nonzero component belonging to $\mathbb{R}^1$, any solution initiating in the interior of the positive orthant has the highest value of the Lyapunov exponent and is asymptotically close to the direction of $\mathbb{R}^1$.

Imprimitive matrices can no longer satisfy the definition of exponential separateness under index $n - 1$ since the limits of trajectories constitute now a subspace of dimension $h > 1$. The generalization would thus require the notion of separateness under an index lower than $n - 1$.

B. Reducibility to a Strongly Positive System

Such a standard operation as a linear change of variables in a system of equations (which may also have a biological sense in the search for pertinent aggregation and the interpretation of model variables) should certainly not affect the qualitative pattern of trajectory behavior. In the autonomous case, this is mathematically confirmed by the fact that the eigenvalues of a matrix are invariant under any of its (nonsingular) similarity transformations (a linear change of variables induces a similarity transformation of the projection matrix). In the nonautonomous case, a linear change of variables

$$\mathbf{x}(t) = L(t)\mathbf{y}(t), \tag{3.20}$$

with a time-dependent $n \times n$ matrix $L(t)$, can be easily shown to transform system (3.17) into

$$\mathbf{y}(t + 1) = L(t + 1)^{-1}A(t)L(t) \tag{3.21}$$

It is called a *Lyapunov transformation* if

$$\sup_t \|L(t)^{\pm 1}\| < \infty, \qquad \sup_t |\det L(t)| > 0,$$

i.e., $L(t)$ has finite norms together with its inverse and does not degenerate for any t.[16,18]

Lyapunov transformations are relevant to asymptotic analysis of system (3.17), since both the Lyapunov exponents and the property of system (3.17) to be exponentially separated under index $n-1$ turn out invariant with respect to any Lyapunov transformation.[16,18] If, therefore, one manages to find a Lyapunov transformation that reduces the system to that of known asymptotic properties, the problem would be advantageously solved.

The following theorem of A. I. Ivanov[18] specifies an important class of nonautonomous systems whose asymptotic properties are known in the sense that they are determined by the highest Lyapunov exponent.

Theorem 3.4 *System (3.17) is exponentially separated under index $n - 1$ if and only if it is* reducible *(by a Lyapunov transformation) to a* strongly positive *system* $\mathbf{y}(t + 1) = B(t)\mathbf{y}(t)$, *i.e. to a system with a matrix* $B(t) = [b_{ij}(t)]$ *of the property that*

$$\inf_t \min_{i,j}\{b_{ij}(t)\} > 0. \tag{3.22}$$

Thus, any reducible system exhibits well-defined asymptotics. Also, for any solution of a reduced system that takes on a nonnegative vector at a moment t_0, there exists a region (a cone in $\mathbb{R}_+^n$) which the solution never leaves, the region being determined eventually by the entries of matrix $A(t_0)$.[19]

To illustrate the idea that time-dependent strongly positive systems represent a nonautonomous generalization of constant primitive matrices, we consider below a general model

$$\mathbf{x}(t + 1) = A\mathbf{x}(t) \tag{3.23}$$

with a (nonsingular) primitive nonnegative matrix A ($A \geq 0$). For any primitive matrix A there exists a positive integer number p, the *index of primitivity*, such that A^p becomes (strictly) positive ($A^p > 0$).[2] Let $A_\rho = A/\rho$, where $\rho > 0$ is the dominant eigenvalue of A, and consider

$$L(t) = (A_\rho)^{(1-p)t} \tag{3.24}$$

as a Lyapunov transformation of (3.23). Then we have

$$\mathbf{y}(t) = (A_\rho)^{(p-1)t}\mathbf{x}(t),$$

while

$$\mathbf{y}(t+1) = \rho(A_\rho)^p \mathbf{y}(t). \tag{3.25}$$

It is a strongly positive system in the sense of definition (3.22) as $\rho(A_\rho)^p$ is just a positive matrix.

Note that the highest Lyapunov exponent for system (3.25) is equal to $\ln \rho$, which is the same as that for (3.24). Theorem 3.4 now states that model (3.23) is exponentially separated under index $n-1$, i.e. all its solutions grow asymptotically in the same way as the component belonging to $\mathbb{R}^1$. Clearly, $\mathbb{R}^1$ is the direction of the dominant eigenvector, so that the maximum growth rate is associated with this steady-state age structure.

In a similar way one can show that the "seasonal matrix jump" model

$$\mathbf{x}(2t+1) = A_1\mathbf{x}(2t), \quad \mathbf{x}(2t+2) = A_2\mathbf{x}(2t+1), \qquad t = 0, 1, \ldots$$

with a primitive product $A_1 A_2$ is also reducible to a strongly positive system.

These statements add, of course, nothing new to the picture drawn from direct analysis of the linear autonomous models by the "constancy" methods of the previous sections. The real use of the theory mentioned above begins with models (3.17) in which the projection matrix does depend nontrivially on t. In particular, if the matrix has a classical Leslie form (2.6) which is primitive for any t, then the following conditions

$$\min_{1 \leq j \leq n-1} \inf_t \{s_j(t)\} > 0,$$

$$\inf_t \min_j \{b_{k+j}(t)\} > 0 \text{ for some } k, \quad 1 \leq k \leq n-1,$$

$$\inf_t b_n(t) > 0 \tag{3.26}$$

can be proved sufficient for the reducibility to a strongly positive system.[20] The conditions are fairly nonrestrictive since the first one just means that neither survival rate ever drops to zero, the second one requires nonzero birthrates in a group of $n - k + 1$ adjacent age classes at each moment of time, and only the third one establishes that the oldest age group must always be reproductive (i.e., the matrix never degenerates).

C. Stability under Permanently Acting Perturbations

Another advantage of the theory is related to models whose matrix can be represented as

$$A(t) = A_0(t) + \Delta(t),\tag{3.27}$$

where $A_0(t)$ has known asymptotic properties (e.g., primitive or reducible to a strongly positive system), while its *permanently acting perturbations* (PAP) $\Delta(t)$ are assessed from some additional conjectures. Any change in model structure or coefficients logically evokes the question whether and how the modification of matrix $A_0(t)$ affects the behavior of trajectories. The system is stable if sufficiently small PAPs generate a trajectory which is close enough to the unperturbed one. In the classical Leslie case and its autonomous modifications, a clear answer follows from the standard perturbation theory (see Section 3.I.B), but it is not so clear in the general case.

The problem is that the Lyapunov exponents are generally not stable under PAP: i.e., any arbitrarily small perturbation may generate a finite change in the exponent.[21−23] Therefore, to restrict a class of PAPs under which any small change in coefficients will induce a respectively small change in the exponent, is a task of practical importance. The following theorem of Rakhimberdiev and Ivanov[24] gives a sufficient condition for stability of the highest Lyapunov exponent.

Theorem 3.5 *The exponent $\Lambda(A)$ of system (3.17) is stable under perturbations $\Delta(t) = [\delta_{ij}(t)]$ which meet the conditions*

$$|\delta_{ij}(t)| \leq \delta a_{ij}(t), \qquad i, j = 1, 2, \dots, n\tag{3.28}$$

with some $\delta > 0$.

The theorem means that one may consistently perturb all nonzero entries of matrix $A(t)$, whereas any small variation of a zero entry will lead the system out of the class (3.28). Within that class, all perturbed models retain asymptotic stability (or instability) of their trajectories, the decline (or growth) rates being sufficiently close to those of the unperturbed model $A_0(t)$. If, in particular, we perturb a reducible system, e.g., a nonautonomous Leslie model, then its remarkable property of convergence to a single vector direction will remain.

Perturbation of the nonzero entries can not bring about new transitions in the life-cycle scheme represented by the digraph associated with $A_0(t)$, while the new transitions, due to the zero entries being perturbed, may easily decrease the index of imprimitivity of the matrix. From this viewpoint, it is quite understandable that the perturbation class should not affect zero entries. On the other hand, there is the need to modify exactly zero entries, which often arises in attempts to improve the model, as was shown in Section 3.I. Thus, to make use of Theorem 3.5, one should apply the "art of ecological modeling" to constructing a matrix $A_0(t)$ that would already include all possible pathways of the life cycle, while reducing the perturbations just to temporal variations in the corresponding transition rates. Using the same "art" one should apply the task of finding a Lyapunov transformation (hence the subspace $\mathbb{R}^1$) appropriate to the particular problem at hand.

All these theoretical results still leave room for uncertainty in the behavior of time-dependent Leslie-type models, although they impose limits to the uncertainty, both in the state space of reducible systems and their asymptotic growth rates under perturbations.

IV. DENSITY-DEPENDENT MODIFICATIONS: NONLINEARITY VS. ASYMPTOTIC DIVERSITY

The classical Leslie model presupposes the birth and survival rates to be constant, irrespective of what is the current population density. But if we take into account any factor of population dynamics that depends on the sizes of age groups or on the total population size, then the model takes on the following general form:

$$\mathbf{x}(t + 1) = L_{\mathbf{x}(t)}\mathbf{x}(t), \tag{3.29}$$

where entries of matrix L_x are now functions of the population vector $\mathbf{x}$, thus making the model no longer linear.

A. Linearization at Equilibrium

An equilibrium $\mathbf{x}^*$ has to be a solution to the equation

$$\mathbf{x}^* = L_{\mathbf{x}^*}\mathbf{x}^*, \quad \mathbf{x}^* > 0, \tag{3.30}$$

which indicates that there exists a set of group sizes $\mathbf{x}^*$ such that matrix $L_{\mathbf{x}^*}$ has a unitary eigenvalue; then $\mathbf{x}^*$ is the eigenvector belonging to this value. Equation (3.30) now represents a nonlinear eigenvalue problem. In some classes of these problems, there are theorems generalizing the results of the classic Perron–Frobenius theorem,[25] but how to find the eigenvectors depends entirely on the particular form of functions chosen in $L_{\mathbf{x}}$.

Note that the "matrix jump" model described above is a special case of system (3.29) with a piecewise dependence

$$L_{\mathbf{x}} = \begin{cases} L_1 & \text{if } \sum_{i=1}^{n} x_i \leq N, \\ L_2 & \text{if } \sum_{i=1}^{n} x_i > N. \end{cases}$$

In this case discontinuity of the functions prevents linearization, a traditional method of stability analysis. But once the functions comprising $L_{\mathbf{x}}$ are sufficiently smooth, the linearization of (3.29) at $\mathbf{x}^*$ yields, up to the second-order terms,

$$\mathbf{x}(t + 1) = L_{\mathbf{x}^*}\mathbf{x}^* + \left(L_{\mathbf{x}^*} + \sum_{i=1}^{n} \left(\frac{\partial L}{\partial x_i} \right)_* X_i^* \right)(\mathbf{x}(t) - \mathbf{x}^*), \tag{3.31}$$

where $(\partial L/\partial x_i)_*$ is a matrix with entries derived from those of L by differentiation with respect to x_i at $\mathbf{x}^*$, and matrix X_i^* has a single nonzero i-th column equal to vector $\mathbf{x}^*$.

Denoting the deviation from equilibrium by $\delta\mathbf{x} = \mathbf{x} - \mathbf{x}^*$ and using (3.31), we have

$$\delta\mathbf{x}(t + 1) = L_\delta \delta\mathbf{x}(t),$$

where

$$L_\delta = \sum_{i=1}^{n} \left(\frac{\partial L}{\partial x_i}\right)_* X_i^* + L_{\mathbf{x}^*}. \tag{3.32}$$

Matrix L_δ is no longer dependent on t, so that

$$\delta\mathbf{x}(t) = (L_\delta)^t \delta\mathbf{x}(0),$$

and one can infer stability of equilibrium $\mathbf{x}^*$ from the spectrum of L_δ. In general there is no reason to regard matrix L_δ of (3.32) as nonnegative. But if its largest in modulus eigenvalue ξ—call it *dominant* as before—exceeds 1, then deviations $\delta\mathbf{x}(t)$ from equilibrium increase with time, hence the equilibrium is unstable. If $|\xi| < 1$, the deviations vanish with time and $\mathbf{x}^*$ is asymptotically stable. When $|\xi| = 1$, we need to examine higher-order terms in the expansion (3.31). In all cases except when $\xi > 0$, one should expect oscillatory behavior of trajectories around equilibrium.

For practical purposes, it often appears sufficient to use a more special case of system (3.29) ensuing from the assumption that all birth and survival rates depend only upon the total population size $N(\mathbf{x})$, i.e.,

$$\mathbf{x}(t + 1) = L_{N[\mathbf{x}(t)]}\mathbf{x}(t) \tag{3.33}$$

The problem (3.30) is then reduced to searching for a positive number N^* such that matrix L_{N^*} has an eigenvalue $\lambda = 1$. In other words, N^* is a positive solution to the equation

$$\det(L_{N^*} - I) = 0. \tag{3.34}$$

For each $N^* > 0$ found, the equilibrium $\mathbf{x}^* > 0$ is determined as an eigenvector of the numerical matrix $L_{N^*} \geq 0$ corresponding to the eigenvalue $\lambda = 1$ (which exists by the Perron–Frobenius theorem) and being normalized so as all its components totals N^*.

The linearization matrix (3.32) is then simplified to

$$L_\delta = L_{N^*} + \left(\frac{\partial L}{\partial N}\right)_* X^*, \tag{3.35}$$

where matrix X^* consists of n identical columns $\mathbf{x}^*$.

B. Leslie's Nonlinear Model: Linearization and Perturbation Findings

A simplest example of the model (3.33) was suggested by P. Leslie[26] himself, who assumed that all the elements of a constant matrix L of the dominant eigenvalue $\lambda_1 > 0$ vary with N in a way such that the dominant value for matrix L_N equals $\lambda_1/q(N)$, where $q(N)$ is a linear function of the total size N. Actually it means that all the elements of L are multiplied by the same quantity $1/q(N)$, or that all the nonzero birth and survival rates have a single dependence on N. This reduces (3.33) further to

$$\mathbf{x}(t+1) = L\mathbf{x}(t)/q(N), \qquad (3.36)$$

where L is a Leslie matrix with constant entries and dominant eigenvalue λ_1.

An equilibrium $\mathbf{x}^*$ here exists if

$$\lambda_1/q(N^*) = 1,$$

and is represented by the dominant eigenvector of matrix L. Since the sum of $\mathbf{x}^*$'s components should satisfy the above relation, the equilibrium $\mathbf{x}^*$ is now determined uniquely rather than multiply, to within multiplying by a constant, as in the classical Leslie model. When the population size N is so small as to have no regulation effect, the population is governed by matrix L, i.e., $q(0) = 1$. For these reasons the linear function $q(N)$ takes on the following form:

$$q(N) = 1 + \frac{\lambda_1 - 1}{K}N, \qquad (3.37)$$

where $K = N^*$ is the population size in the state of equilibrium $\mathbf{x}^*$.

So, given a model of type (3.36)–(3.37), the equilibrium can be calculated as

$$\mathbf{x}^* = \frac{K}{\|\mathbf{C}(\lambda_1)\|_\Sigma}\mathbf{C}(\lambda_1), \qquad (3.38)$$

where $\mathbf{C}(\lambda_1)$ is defined in (2.23) and $\|\mathbf{C}\|_\Sigma$ designates the norm which is the sum of vector components in modulus.

If $\mathbf{x}(t)$ is an age distribution proportional to $\mathbf{x}^*$, then

$$\mathbf{x}(t+1) = \frac{L\mathbf{x}(t)}{q[N(t)]} = \frac{\lambda_1\mathbf{x}(t)}{q[N(t)]}$$

and summing the components of $\mathbf{x}(t+1)$ results, by (3.37), in

$$N(t+1) = \frac{\lambda_1 N(t)}{1 + (\lambda_1 - 1)N(t)/K}. \qquad (3.39)$$

The general solution to the nonlinear difference equation (3.39) is

$$N(t) = \frac{K}{1 + Ce^{-rt}}, \qquad (3.40)$$

where the constant C is defined by the initial condition as

$$C = \frac{K - N(0)}{N(0)},$$

and $r = \ln \lambda_1$. Formula (3.40) represents also the general solution to the well-known (differential) logistic equation of growth for a nonstructured population with the parameters r (the *intrinsic growth rate*) and K (the *carrying capacity* of the environment).[27] Thus, if a trajectory of model (3.36) originates from a vector which corresponds to the steady (relative) age structure for matrix L, then the total population grows according to the logistic equation, while keeping the age structure invariable.

But how the total population size varies in the case of a nonsteady initial age structure has remained with no definite answer in the Leslie formulation. Some statements to clarify this can now be obtained from stability investigation of equilibrium $\mathbf{x}^*$ in the nonlinear problem (3.36)–(3.37). Linearization results here in the following matrix:

$$L_\delta = \frac{L}{q(N^*)} + [q^{-1}(N)]'_* L X^* = \frac{1}{\lambda_1} \left(L + \frac{1 - \lambda_1}{K} X^* \right), \tag{3.41}$$

whose dominant eigenvalue, if found, could indicate the conditions of stability. Since

$$L_\delta \mathbf{x}^* = \frac{1}{\lambda_1} \left(L\mathbf{x}^* + \frac{1 - \lambda_1}{K} X^* \mathbf{x}^* \right) = \frac{1}{\lambda_1} \left(\lambda_1 \mathbf{x}^* + \frac{1 - \lambda_1}{K} N^* \mathbf{x}^* \right) = \frac{1}{\lambda_1} \mathbf{x}^*,$$

the vector $\mathbf{x}^*$, being an eigenvector for L, is an eigenvector for L_δ too, the corresponding eigenvalue being $1/\lambda_1 > 0$. When $\lambda_1 < 1$, we can show that $\xi = 1/\lambda_1$ is the dominant value. In this case, indeed, it follows from (3.41) that matrix L_δ is positive, and consequently, by the Perron theorem, it has the dominant value $\mu > 0$ with the corresponding eigenvector $\mathbf{y} > 0$. Assuming that $\mu \neq \xi$ would make vectors $\mathbf{x}^*$ and $\mathbf{y}$ linearly independent, and would thereby lead to an indecomposable positive matrix, L_δ, having two linearly independent positive eigenvectors, which is impossible.[2]

So, when $\lambda_1 < 1$, the positive number $\xi = 1/\lambda_1 > 1$ is the dominant eigenvalue for matrix L_δ and, hence, the equilibrium $\mathbf{x}^*$ is (monotonically) unstable. Any deviation from equilibrium $\mathbf{x}^*$ would exponentially increase, along with the deviation from the equilibrium population size $N^* = K$, until the linear approximation is valid.

When $\lambda_1 > 1$, the investigation becomes much more complicated as L_δ is no longer nonnegative and the Perron–Frobenius theorem can no longer apply. But it is this case which is worthy of consideration in view of density regulation of the otherwise exponential growth. If the original matrix L is primitive and $|\lambda_2| < 1$, it follows from the continuity argument that the positive number $\xi = 1/\lambda_1 < 1$, when close enough to 1, continues to be the dominant eigenvalue for L_δ; hence equilibrium $\mathbf{x}^*$ is asymptotically stable, the behavior of $N(t)$ being close to the logistic one given by (3.40).

The same conclusion follows from the perturbation method with the quantity $\lambda_1 - 1 > 0$ interpreted as a small parameter ε. The perturbed matrix is then represented as

$$L_\delta(\varepsilon) = \frac{1}{\lambda_1}\left(L - \varepsilon\frac{X^*}{K}\right), \tag{3.42}$$

and the perturbation method for nonmultiple eigenvalues[3] yields

$$\xi(\varepsilon) = 1 - \frac{\varepsilon}{\lambda_1}\xi^{(1)} + O(\varepsilon^2), \tag{3.43}$$

where

$$\xi^{(1)} = \left\langle \mathbf{R}(\lambda_1), \frac{1}{K}X^*\mathbf{C}(\lambda_1)\right\rangle, \tag{3.44}$$

and vectors $\mathbf{C}(\lambda_1), \mathbf{R}(\lambda_1)$ are defined in (2.23)–(2.24). Some algebra with (3.37) and (2.25) then shows that $\xi^{(1)} = 1$, so that

$$\xi(\varepsilon) = 1 - \frac{\varepsilon}{\lambda_1} + O(\varepsilon^2). \tag{3.45}$$

Thus, under small increase in λ_1 beyond 1, the dominant eigenvalue of L_δ becomes slightly less than 1, providing for the above-mentioned stability of $\mathbf{x}^*$.

But if λ_1 is not close enough to 1, the dominant eigenvalue $\xi = 1/\lambda_1$ of L_δ may now lack its primitivity or even dominance, whereby the monotone pattern of convergence to $\mathbf{x}^*$ may change to an oscillatory behavior. This follows readily from a lemma proved in the Appendix: the spectrum of matrix L_δ consists of the simple positive eigenvalue $\xi = 1/\lambda_1$ and values $\lambda(L)/\lambda_1$ with the same multiplicities that the eigenvalues $\lambda(L)$ have in the spectrum of L. The statement certainly is compatible with what has been stated above and makes it particularly evident that, when the original Leslie matrix is imprimitive, the dominant eigenvalues of L_δ are numbers $\lambda_2/\lambda_1, \ldots, \lambda_h/\lambda_1$, of modulus 1. Hence the linearization gives no certain answer on stability of $\mathbf{x}^*$ and the standard analysis would therefore require an account of higher-order terms in the expansion (3.31), again with no a priori certainty that we will get a certain answer.

C. Leslie's Nonlinear Model: Direct Analysis Findings

A certain answer does however ensue from the simple form (3.36)–(3.37) of the nonlinear problem itself. Since any trajectory of the nonlinear model advances along the same vector directions as does that of the linear model with matrix L originating from the same initial state, it follows that the asymptotic patterns of both models are identical to within vector directions. Nonlinearity then provides convergence to a limit distribution, constant or periodic, but unique for any initial state of any fixed vector direction.

To be more precise let, by definition,

$$\pi(t; \mathbf{x}(0)) = \prod_{\tau=0}^{t-1} q(\mathbf{x}(\tau))/\rho^t, \qquad t = 1, 2, \ldots, \tag{3.46}$$

Then the general form of a solution to equation (3.36)–(3.37) is

$$\mathbf{x}(t) = L_\rho^t \mathbf{x}(0)/\pi(t; \mathbf{x}(0)), \qquad t = 0, 1, 2, \ldots, \tag{3.47}$$

which can be proved, for instance, by induction in t.

Two theorems proved in the Appendix (Theorems A3.6 and A3.7) specify the asymptotics of $\mathbf{x}(t)$ for any $\mathbf{x}(0) = \mathbf{x} \geq 0$. If L is primitive, then $\mathbf{x}(t)$ tends to the unique solution of the nonlinear eigenvalue problem (3.36)–(3.37), which thereby is asymptotically stable globally in the positive orthant. But if L is imprimitive, then $\mathbf{x}(t)$ tends to a T-cycle which belongs to the directions of the limit cycle $\mathcal{L}(\mathbf{x}, 0), \mathcal{L}(\mathbf{x}, 1), \ldots, \mathcal{L}(\mathbf{x}, T-1)$ for the linear model with matrix L/ρ. The nonlinear cycle is unique for the whole beam of initial states $\mathbf{x}(0) = c\mathbf{x}$, $c > 0$, and is locally nonasymptotically stable, similar to the limit cycles in the linear model. Strictly speaking, if a perturbation of the initial state $\mathbf{x}$ means simple extension (or contraction) of $\mathbf{x}$, then the perturbed trajectory tends to the same limit cycle, whereas any other perturbation will result in another limit cycle, close to the original one if the perturbation is sufficiently small.

The above statements are also valid for a nonnegative matrix L of a more general form than the classic Leslie matrix. The constructive feature of the statements however becomes somewhat weaker, because no explicit expressions are known for vectors $\mathcal{L}(\mathbf{x}, 0), \mathcal{L}(\mathbf{x}, 1), \ldots, \mathcal{L}(\mathbf{x}, T-1)$ in the general case, although the index h, that determines all possible values of T, can always be calculated as the g.c.d. for lengths of all closed paths in the digraph associated with matrix L.

Thus, even a highly artificial construction which is given by model (3.36)–(3.37) has asymptotic properties that may be classified as a kind of stabilization in the population dynamics and interpreted as a model illustration to the effect of density regulation: instead of the whole beam of limit cycles (of equilibria in the primitive case) in the nonregulated growth we have only one cycle in the model with regulation. Nonlinearity thus reduces the variety of asymptotic behavior in the classic Leslie model and its constant-matrix modifications.

D. Beddington's Nonlinear Model

The next step in making the model (3.36) more realistic, is to assume that the dependence of birth rates on N is different from that of survival rates, but both functions are again the same for all age groups. In some cases this already gives an adequate description for dynamics of a real population. An example of this kind of nonlinear equation (3.33) was successfully applied by Beddington[28] to dynamics of a collembolan population (*Folsomia candida L.*) in laboratory culture.

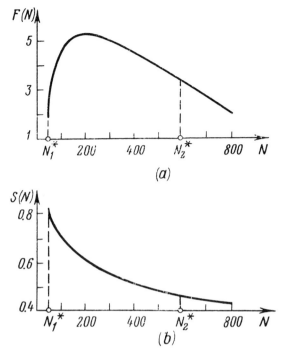

Figure 7.   Fecundity and survival relationships for a culture of *Folsomia candida* L. grown at 10 C°: (a) $F(N) = 18.53 \ln N - 1.74(\ln N)^2 - 44.04$: (b) $S(N) = 1.35 - 0.14 \ln N$ (from Beddington[28], used with permission).

The population was subdivided into 4 weekly age groups, of which the 3rd and the 4th ones produced progeny, so that the model took on the following matrix form:

$$
\begin{bmatrix} x_1(t+1) \\ x_2(t+1) \\ x_3(t+1) \\ x_4(t+1) \end{bmatrix} = \begin{bmatrix} 0 & 0 & F(N) & F(N) \\ S(N) & 0 & 0 & 0 \\ 0 & S(N) & 0 & 0 \\ 0 & 0 & S(N) & 0 \end{bmatrix} \begin{bmatrix} x_1(t) \\ x_2(t) \\ x_3(t) \\ x_4(t) \end{bmatrix}. \tag{3.48}
$$

Empirical relationships of fecundity (F) and survival (S) rates to the total population size (N) were statistically reproduced, within a certain range of the total population size, by two curves of the following form:

$$
\begin{aligned}
F(N) &= a \ln N - b(\ln N)^2 - c, \\
S(N) &= d - e \ln N
\end{aligned} \tag{3.49}
$$

(with positive $a, b, c, d,$ and e) shown in Figures 7a,b, with an obvious optimum of

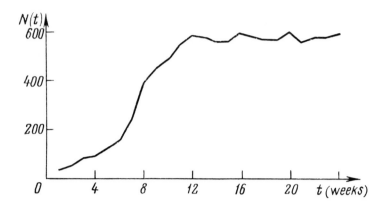

Figure 8.   Total population trajectory for a culture of *Folsomia candida* L. grown at 10 C° (from Beddington[28], used with permission.

N in $F(N)$ and monotone decrease in $S(N)$. Searching for equilibrium values N^* was thus reduced to the solution of the following transcendental equation for N^*:

$$\det \begin{bmatrix} -1 & 0 & F(N^*) & F(N^*) \\ S(N^*) & -1 & 0 & 0 \\ 0 & S(N^*) & -1 & 0 \\ 0 & 0 & S(N*) & -1 \end{bmatrix} = 0. \tag{3.50}$$

For particular values of the parameters (see Figures 7a,b) this equation revealed two positive roots $N_1^* \cong 41.13$ and $N_2^* \cong 596.45$ with the corresponding equilibrium distributions x_1^* and x_2^*. Calculating the spectrum of matrix L_6 (3.35) for system (3.48) linearized at the points x_1^* and x_2^*, one can see that, in the former case, the dominant value $\xi(x_1^*) > 1$, i.e., equilibrium x_1^* is unstable, whereas in the latter case, the pair of complex conjugate numbers $\xi_{1,2}$, $|\xi_i| < 1$, is dominant in modulus, i.e., equilibrium x_2^* is asymptotically stable, the local deviations vanishing in the form of damped oscillations. While the theory guarantees the local stability of x_2^*, the computer simulation shows the property of global convergence to x_2^*, within the whole range where the regulating functions $F(N)$ and $S(N)$ are positive.

These conclusions are verified by an experimental curve of $N(t)$ (Figure 8), which exhibits a sigmoid type of growth from an initial size up to a value close to N_2^*, the upper value predicted by the model. The model thus gives one more illustration of the concept of density regulation. Moreover, the idea of movement from an unstable low population level to a stable high level was accepted by some ecologists as a model mechanism of population outbreaks.[28-30]

So, even a fairly simple form (3.48) to incorporate the self-regulation effect into a Leslie model, can furnish an adequate simulation of population dynamics in proper cases. This is probably the main reason for the wide-spread use of the model modifications not only to represent the age structure dynamics, but also to study issues of optimal population management, of optimal life-cycle strategies, and others in population theory.

However, it should be noted that the search for steady states (to say nothing of cycles) and their stability analysis has become much more complicated in density-dependent generalizations of the Leslie model, since the very first step calls for solution of a transcendental equation. Therefore, one can hardly suggest a regular general method of stability analysis that would bring about a definite conclusion about the dynamic behavior immediately from the form of matrix L, as it was in the classical Leslie model and its linear-regulation extension (3.36).

E. Stability of the Total Population

In examples (3.36) and (3.48) we have a certain kind of stability in the age distribution $\mathbf{x}(t)$ despite the fact that the projection matrix is a function of the total population $N(t)$ alone. The latter is stable too as a clear consequence of the former. In formal terms, it follows from the fact that convergence in the Euclidean norm is equivalent to convergence in the norm

$$\|\mathbf{x}\|_{\Sigma} = \sum_{i=1}^{n} |x_i|.$$

This, in turn, means that the expression

$$\|\mathbf{x}(t) - \mathbf{x}^*\|_{\Sigma} = \sum_{t} |x_i(t) - x_i^*| \geq \left|\sum_{i}(x_i(t) - x_i^*)\right| = |N(t) - N^*|$$

converges to zero (as $t \to \infty$).

From this estimation it is also clear that the convergence of $N(t)$ does not, in general, provide the convergence of distribution $\mathbf{x}(t)$, i.e., the total population stability is a weaker property than stability of the age distribution. Thus, one can imagine, at least theoretically, a situation where stability of $N(t)$ takes place under instability of $\mathbf{x}(t)$. While, in a classical Leslie model, this can be realized only for highly specific values of demographic parameters (Example A3.1 in the Appendix), the study of this situation in more realistic models—e.g., with density dependence incorporated—is of a certain interest from both mathematical and ecological points of view.

Coming back to the general picture of asymptotic behavior in Beddington's case of density regulation, we should recognize that nonlinearity again reduces the variety of asymptotic behavior observed in the classic Leslie model and its constant-matrix modifications.

Fortunately, this observation yields no vast generalization: another, exponential, form of the regulating functions may generate even a bifurcation and chaotic behavior similar to that of one-dimensional nonlinear mappings.[31-33] Unfortunately, the structural methods of matrix and graph representations can scarcely add any-thing new to the fundamental results of functional analysis in that area, and we restrict our consideration here.

Appendix

Statement A3.1 *If a Leslie model population does not decline, i.e., if its dominant eigenvalue $r \geq 1$, then any small perturbation of type (3.12) or similar will not decrease the dominant value. If, in addition,*

$$\hat{r} - Re\ \hat{\lambda}_2 = r^{n-1}/p'(r) - Re\ \{\lambda_2^{n-1}/p'(\lambda_2)\} > 0,$$

then the perturbation will increase the rate of convergence to the limit distribution.

Proof: The perturbation matrix has the following form:

$$\varepsilon\Delta = \varepsilon \begin{bmatrix} 0 & 0 & \cdots & 0 & 0 & \cdots & 0 & 0 \\ 0 & 0 & & & & & & \\ & 0 & & & & & & \\ & & \ddots & \ddots & & & & \mathbf{0} \\ & & & -s_{25}\beta_{25} & 0 & & & \\ & \mathbf{0} & & & \ddots & \ddots & & \\ & & & & & -s_{39}\beta_{39} & 0 & \\ 0 & 0 & \cdots & s_{25}\beta_{25} & s_{26}\beta_{26} & \cdots & s_{39}\beta_{39} & 0 \\ & & & \mathbf{0} & & & & \mathbf{0} \end{bmatrix}, \qquad (3.51)$$

which reduces calculation of $r^{(1)}$, the principal part of the variation in r, to

$$\langle \mathbf{R}_1, \Delta\mathbf{C}_1 \rangle = \frac{r^{n-1}}{p'(r)}[R_{41}(s_{25}\beta_{25}C_{25} + \cdots + s_{39}\beta_{39}C_{39}) \\ - (R_{26}s_{25}\beta_{25}C_{25} + R_{27}s_{26}\beta_{26}C_{26} + R_{40}s_{39}\beta_{39}C_{39})],$$

where R_j and C_j are, respectively, the components of the row and column dominant eigenvectors of L. Formulae (2.23) and (2.24) transform the expression into

$$r^{(1)} = \frac{q_{41}(r)}{r^{n-1}} \left(\frac{r^{15}\beta_{25}}{s_{26}s_{27}\dots s_{40}} + \frac{r^{14}\beta_{26}}{s_{27}\dots s_{40}} + \cdots + \frac{r\beta_{39}}{s_{40}} \right) \\ - \frac{1}{r^{n-1}}(\beta_{25}q_{26}(r) + \beta_{26}q_{27}(r) + \cdots + \beta_{39}q_{40}(r)).$$

Since $b_1 = b_2 = \cdots = b_{40} = 0$, the pertinent "tails" of the characteristic polynomial $p(\lambda)$ are all equal:

$$
\begin{aligned}
q_{26}(r) &= q_{27}(r) = \ldots = q_{40}(r) = q_{41}(r) \\
&= b_{41}c_{41}r^{n-41} + b_{42}c_{42}r^{n-42} + \cdots + b_n c_n = r^n,
\end{aligned}
$$

whereby

$$
\begin{aligned}
r^{(1)} &= \frac{r^n}{p'(r)}\left[\left(\frac{r^{15}\beta_{25}}{s_{26}s_{27}\ldots s_{40}} + \frac{r^{14}\beta_{26}}{s_{27}\ldots s_{40}} + \cdots + \frac{r\beta_{39}}{s_{40}}\right) - (\beta_{25} + \cdots + \beta_{39})\right] \\
&= \frac{r^n}{p'(r)}\left(\frac{r^{15}\beta_{25}}{s_{26}s_{27}\ldots s_{40}} + \frac{r^{14}\beta_{26}}{s_{27}\ldots s_{40}} + \cdots + \frac{r\beta_{39}}{s_{40}} - 1\right). \quad (3.13)
\end{aligned}
$$

Note that $p'(r)$ is always positive for a Leslie matrix (otherwise r would not be the only positive root of $p(r)$). If $r \geq 1$, then each β_i in (3.13) has a coefficient not less than one, so that

$$
r^{(1)} \geq \frac{r^n}{p'(r)}(\beta_{25} + \cdots + \beta_{39} - 1) = 0
$$

($r^{(1)} = 0$, for instance, when $r = 1$ and all $s_i = 1$, $i = 25, \ldots, 39$). But if $r < 1$, then (3.13) may have any sign.

In a way similar to derivation of (3.13), we calculate the principal part of variation in the convergence rate as follows:

$$
\begin{aligned}
\mathrm{Re}\left\{\frac{r^{(1)}}{r} - \frac{\lambda_2^{(1)}}{\lambda_2}\right\} &= \mathrm{Re}\left\{\frac{r^n}{p'(r)}\left[\frac{r^{15}\beta_{25}}{s_{26}s_{27}\ldots s_{40}} + \frac{r^{14}\beta_{26}}{s_{27}\ldots s_{40}} + \cdots + \frac{r\beta_{39}}{s_{40}} - 1\right]\right. \\
&\quad \left. - \frac{\lambda_2^{n-1}}{p'(\lambda_2)}\left[\frac{\lambda_2^{15}\beta_{25}}{s_{26}s_{27}\ldots s_{40}} + \frac{\lambda_2^{14}\beta_{26}}{s_{27}\ldots s_{40}} + \cdots + \frac{\lambda_2\beta_{39}}{s_{40}} - 1\right]\right\} \\
&= \mathrm{Re}\left\{\beta_{25}\frac{\hat{r}r^{15} - \hat{\lambda}_2\lambda_2^{15}}{s_{26}s_{27}\ldots s_{40}} + \beta_{26}\frac{\hat{r}r^{14} - \hat{\lambda}_2\lambda_2^{14}}{s_{27}\ldots s_{40}} + \cdots\right. \\
&\quad \left. + \beta_{39}\frac{\hat{r}r - \hat{\lambda}_2\lambda_2}{s_{40}} - (\hat{r} - \hat{\lambda}_2)\right\}, \quad (3.14)
\end{aligned}
$$

where

$$
\hat{r} = \frac{r^{n-1}}{p'(r)}, \qquad \hat{\lambda}_2 = \frac{\lambda_2^{n-1}}{p'(\lambda_2)}.
$$

Since $\hat{r} > \mathrm{Re}\,\hat{\lambda}_2$ and since $r \geq 1$ is dominating, the nominator of any term in (3.14) is greater, in the real part, than the last term, thereafter the fraction itself is greater too. Hence, we have

$$
\mathrm{Re}\left\{\frac{r^{(1)}}{r} - \frac{\lambda_2^{(1)}}{\lambda_2}\right\} > \beta_{25}(\hat{r} - \mathrm{Re}\,\hat{\lambda}_2) + \beta_{26}(\hat{r} - \mathrm{Re}\,\hat{\lambda}_2) + \cdots + \beta_{39}(\hat{r} - \mathrm{Re}\,\hat{\lambda}_2) - (\hat{r} - \mathrm{Re}\,\hat{\lambda}_2) = 0.
$$

∎

Definition A3.1 *Let the set V of all vertices of a digraph D be subdivided into disjoint subsets $S_1, \ldots, S_r$, the components of V. Then the digraph $F_S(D)$ is called the* factorgraph,[34] *or* condensation,[14] *of digraph D with respect to subdivision $S_1, \ldots, S_r$ if it is constructed by the following rule: its r vertices are identified with the components $S_1, \ldots, S_r$ and there is an arc $S_i \to S_j$ in $F_S(D)$ if and only if there exists at least one arc in D that goes from a vertex of subset S_i to a vertex of subset S_j.*

Theorem 3.1 *A digraph D is strongly connected if and only if any partitioning of its vertex set results in a strongly connected factorgraph.*[12]

Necessity can be proved by contradiction: if there exists a subdivision S of the vertex set V such that the factorgraph $F_S(D)$ is not strong, one can also find a pair of components in that subdivision, say $S_i \neq S_j$, which has no path in $F_S(D)$ from S_i to S_j; this obviously means that for neither vertex $v_i \in S_i$ is there a path in D to a vertex $v_j \in S_j$, which contradicts the strong connectedness of D.

Sufficiency is evident since the factorgraph $F_V(D)$ of a digraph D with respect to subdividing the vertex set V into all its individual elements coincides with D itself. ■

Definition A3.2 *Condensation of a digraph $D(A)$ on the subdivision $V = \cup_{i \in V_1} \times (i, V_2)$ will be called the* age condensation *(or* age factorgraph*) and denoted by $L(A)$; condensation of $D(A)$ on the subdivision $V = \cup_{s \in V_2}(V_1, s)$ will be called the* status condensation *(or* status factorgraph*) and denoted by $S(A)$.*

When a digraph D on a two-dimensional lattice $V_1 \times V_2$ is considered with no reference to any block matrix A, it makes sense to use the following definition.

Definition A3.2′ *The factorgraph of a digraph D with respect to the subdivision $V_1 \times V_2 = \cup_{i \in V_1}(i, V_2)$ is called the* projection *of D to the axis V_1 and denoted by $F_1(D/V_2)$. The projection to the axis V_2, i.e., $F_2(D/V_1)$, is defined in a similar way.*

Clearly, the age factorgraph $L(A)$ is the projection of $D(A)$ to the age axis, i.e., $L(A) = F_1(D/V_2)$, and the status factorgraph $S(A)$ is the projection to the status axis, i.e., $S(A) = F_2(D/V_1)$.

Definition A3.3 *Let digraphs D_1 and D_2 have the vertex sets V_1 and V_2 respectively. The digraph $D = D_1 \wedge D_2$ is called the* conjunction,[35] *or* (Kronecker) product,[13] *of D_1 and D_2 if it is constructed on the vertex set $V_1 \times V_2$ by the following rule: there is an arc $(u_1, u_2) \to (v_1, v_2)$ in D, if and only if there exist the arcs $u_1 \to v_1$ in D_1 and $u_2 \to v_2$ in D_2.*

By McAndrew's theorem[13] the number of strongly connected components in a digraph $D = D_1 \wedge D_2$ is equal to g.c.d. $\{h(D_1), h(D_2)\}$, where $h(D_j)$ is the index of imprimitivity of digraph D_j. For any strong component C of the digraph D we have $h(C) = $ l.c.m. $\{h(D_1), h(D_2)\}$. In particular, $D = D_1 \wedge D_2$ is strongly connected if and only if both D_1 and D_2 are strong while $h(D_1)$ and $h(D_2)$ are mutually prime; their product $h(D_1)h(D_2)$ then gives the index of imprimitivity of D.

Definition A3.4 *If $D(\mathcal{A})$ is the digraph of an arbitrary block matrix $\mathcal{A}$, then its factor-closure[12] is the digraph $F[D(\mathcal{A})]$, which is the conjunction of $D(\mathcal{A})$'s age and status factorgraphs, i.e.,*

$$F[D(\mathcal{A})] = L(\mathcal{A}) \wedge S(\mathcal{A}).$$

In a more general case, namely, for a digraph defined on a two-dimensional lattice $V_1 \times V_2$ with no reference to any block matrix $\mathcal{A}$, the following is appropriate.

Definition A3.4′ *The factor-closure $F[D]$ of a digraph D is the conjunction of its projections to each of the axes, i.e.,*

$$F[D] = F_1(D/V_2) \wedge F_2(D/V_1).$$

Application of McAndrew's theorem[13] to the factor-closure of a digraph $D(\mathcal{A})$ results in the criterion of strong connectedness for a digraph of the form $F_1 \wedge F_2$ (or the equivalent criterion of indecomposability for a matrix $\mathcal{A}$ of a regular block structure), which are formulated as Theorem 3.2 and Theorem 3.2′ in the main text.

Definiton A3.5 *An arc x of a digraph D is called a* basic *arc[14] if in D there is no other directed path from the beginning vertex of x to its terminal vertex.*

Theorem 3.3 *A digraph $D(\mathcal{A})$ is strongly connected if and only if its factor-closure $F[D]$ is strong and neither arc y of the set $\{F[D]\backslash D\}$ is basic in the digraph $D+y$.*[12]

Necessity. If $D(\mathcal{A})$ is strongly connected, the same is true for any of its extensions $D + X$, where X is an arbitrary set of arcs between vertices of D. If the arc set $Y = \{F[D] \setminus D\}$ is nonempty, then any arc $y \in Y$ is neutral in $D + y$ because removing y results in the strong digraph D. By Theorem 7.17 of Harary et al.[14] the neutral arc y of the strong digraph $D + y$ is not basic in $D + y$.

Sufficiency. If an arc $y_1 \in Y$ is not basic in digraph $D + y_1$, then it is a fortiori nonbasic in the extension of $D + y_1$ up to the strongly connected $F[D]$. It follows that arc y_1 is neutral in $F[D]$ and removing it from $F[D]$ keeps $F[D] - y_1$ strongly connected. Another nonbasic arc $y_2 \in Y$, $y_2 \neq y_1$, is also not basic in the extension of $D + y_2$ up to the strongly connected $F[D] - y_1$. Hence, arc y_2 is neutral in $F[D] - y_1$, so that the digraph $F[D] - y_1 - y_2$ is strong. By repeating the argument as many times as the number of arcs in set Y we come to the strong connectedness of the digraph $F[D] \setminus Y = D$.   ∎

Theorem A3.4 *If the digraph $D(\mathcal{A})$ of a block matrix $\mathcal{A}$ has no impasses, its age factorgraph is strongly connected, and all nontrivial blocks $A_{ij} \neq 0$ of $\mathcal{A}$ are of a three-diagonal form:*

$$
A_{ij} =
\begin{bmatrix}
a_1 & b_2 & & & 0 \\
c_2 & a_2 & b_2 & & \\
& c_3 & a_3 & b_3 & \\
& & \ddots & \ddots & \ddots \\
& & c_{m-1} & a_{m-1} & b_{m-1} \\
& & & c_m & a_m
\end{bmatrix}
\tag{3.52}
$$

with nonzero entries in the diagonals, then matrix $\mathcal{A}$ is indecomposable.[12]

Proof: Let an arbitrary pair of vertices, say (i,j) and (k,l), be chosen in the digraph of a matrix $\mathcal{A}$. We then prove that there exists a path from (i,j) into (k,l). If $i \neq k$, then let $i < k$ for definiteness. Since there are no impasses in the digraph and its age factorgraph is strongly connected, we can start from vertex (i,j) in the direction of l (let, for instance, $j > l$), downgrading the status by one unit with each transition. In case status l is reached before age k, the rest of the path ought to (and can) be passed at the constant status l. But if status l is not reached upon the first passage through k, then downgrading should be continued in passing through older-than-k ages and even further, if needed, by repeating the life cycle a proper number of times through an arc of reproduction until the status becomes equal to l; the rest of the path to age k then ought to (and can) be passed at the constant status l.

When $j < l$, the argument is similar, the only difference being that transitions must be chosen which upgrade the status, and if $i > k$, then the path to vertex (k,l) has to pass through the whole life cycle once more. ■

Lemma A3.5 *Let a square matrix B be represented in the form of*

$$
B = L - (\rho - 1)\hat{X},
$$

where L is a nonsingular Leslie matrix, $\rho \neq 1$ is its spectral radius, and the columns of matrix $\hat{X}$ are all identical and equal to the positive eigenvector $\hat{\mathbf{x}}$ of matrix L corresponding to the eigenvalue ρ and being such that the sum of its components $N(\hat{\mathbf{x}}) = 1$. Then the spectrum of B consists of the number 1 with multiplicity 1 and the eigenvalues λ of matrix L, $\lambda \neq \rho$, with the same multiplicities as they have in the spectrum of L.

Proof: For vector $\hat{\mathbf{x}}$ we have

$$
B\hat{\mathbf{x}} = L\hat{\mathbf{x}} - (\rho - 1)\hat{X}\hat{\mathbf{x}} = \rho\hat{\mathbf{x}} - (\rho - 1)\hat{\mathbf{x}} = \hat{\mathbf{x}},
$$

whereby 1 is an eigenvalue of B.

Searching for an eigenvector in the form of $\mathbf{e} + \beta\hat{\mathbf{x}}$, where $\mathbf{e}$ is an eigenvector of L corresponding to an eigenvalue $\lambda \neq \rho$ and β is a scalar, results in the equality $B(\mathbf{e} + \beta\hat{\mathbf{x}}) = \lambda(\mathbf{e} + \beta\hat{\mathbf{x}})$ if

$$\beta = \frac{\rho - 1}{1 - \lambda}N(\mathbf{e}). \tag{3.53}$$

Here $N(\mathbf{e})$ designates the sum of vector $\mathbf{e}$'s coordinates, and the denominator does not vanish since the dominant eigenvalue $\rho > 0$ is also the sole positive eigenvalue of a Leslie matrix.

If the eigenvalues λ are all different, then we have found all n eigenvalues of B and the statement is true. If the multiplicity of some $\lambda = \mu$ is $m > 1$, then, in the Jordan normal form of the Leslie matrix, all entries μ concentrate in a single Jordan cell,[36] hence there are m linearly independent vectors $\mathbf{e}_1 = \mathbf{e}(\mu), \mathbf{e}_2, \ldots, \mathbf{e}_m$ in the corresponding cyclic subspace of operator L:[37]

$$L\mathbf{e}_1 = \mu\mathbf{e}_1, L\mathbf{e}_2 = \mu\mathbf{e}_2 + \mathbf{e}_1, \ldots, L\mathbf{e}_m = \mu\mathbf{e}_m + \mathbf{e}_{m-1}. \tag{3.54}$$

Let us seek a similar set of vectors for matrix B in the form of $\mathbf{e}_j + \xi_j\hat{\mathbf{x}}, j = 1, \ldots, m$. Then the equalities which are similar to (3.54) result in the following:

$$\xi_j = [\xi_{j-1} + (\rho - 1)N(\mathbf{e}_j)]/(1 - \mu), \qquad j = 1, \ldots, m, \qquad \xi_0 = 0, \tag{3.55}$$

whereby the values of ξ_j can be successively found. Belonging to different cyclic subspaces of operator L, the vectors $\hat{\mathbf{x}}$ and $\mathbf{e}_1, \mathbf{e}_2, \ldots, \mathbf{e}_m$ are linearly independent, hence the vectors $\mathbf{e}_j + \xi_j\hat{\mathbf{x}}, j = 1, \ldots, m$, are linearly independent too. Thus, matrix B generates the same structure of splitting into cyclic subspaces as does matrix L, so that the multiplicity of eigenvalues λ in the spectrum of B is the same as in the spectrum of L.   ∎

Below are two theorems, whose proofs[38] rely upon the following formula for the general solution to (3.36)–(3.37):

$$\mathbf{x}(t) = L^t\mathbf{x}(0)/\prod_{\tau=0}^{t-1} q(\mathbf{x}(\tau)), \qquad t = 1, 2, \ldots \tag{3.56}$$

(easily provable by induction in t) and the following identity (ensuing from the formula):

$$\prod_{\tau=0}^{t} q(\mathbf{x}(\tau)) = \prod_{\tau=0}^{t-1} q(\mathbf{x}(\tau)) + (\rho - 1)N(L^t\mathbf{x}(0)). \tag{3.57}$$

If we denote by L_ρ the matrix L/ρ(of the spectral radius 1) and let, by definition,

$$\pi(t; \mathbf{x}(0)) = \prod_{\tau=0}^{t-1} q(\mathbf{x}(\tau))/\rho^t, \qquad t = 1, 2, \ldots, \tag{3.58}$$

then equalities (3.56) and (3.57) will transform into the following:

$$\mathbf{x}(t) = L_\rho^t \mathbf{x}(0)/\pi(t; \mathbf{x}(0)), \qquad t = 0, 1, 2, \ldots, \qquad (3.59)$$

$$\pi(t+1; \mathbf{x}(0)) = \left[\pi(t; \mathbf{x}(0)) + (\rho - 1)N(L_\rho^t \mathbf{x}(0))\right]/\rho, \qquad \pi(0) = 1. \quad (3.60)$$

Theorem A3.6 *(on the nonlinear T-cycle problem). For any one-parameter family of proportional T-cycles for the linear Leslie model (i.e. for the family of vectors of the form (2.27), with $t = 0, 1, \ldots, T - 1$, $\mathbf{x} = c\mathbf{x}_0$, and c running through all positive values), there exists, and is unique for the family, a set of vectors that constitute a T-cycle for the nonlinear equation (3.36)–(3.37).*

Proof: For the sake of simplicity, is given in the case of $K = 1$ and $T = 3$, the generalization being evident. Let $\mathbf{l}_0, \mathbf{l}_1, \mathbf{l}_2$ be vectors of the form (2.27) representing a 3-cycle of the operator L/ρ. Then the proportional vectors

$$\mathbf{u} = u\mathbf{l}_0, \mathbf{v} = v\mathbf{l}_1, \mathbf{w} = w\mathbf{l}_2 \quad (u, v, w > 0) \qquad (3.61)$$

constitute a 3-cycle of the nonlinear operator in the right-hand side of (3.36)—denote it by L_N—if and only if the following equalities hold:

$$\begin{cases} \mathbf{v} = L_N\mathbf{u}, \\ \mathbf{w} = L_N\mathbf{v}, \\ \mathbf{u} = L_N\mathbf{w}. \end{cases} \qquad (3.62)$$

By (3.36) these equalities are reduced to the condition

$$q(\mathbf{u})q(\mathbf{v})q(\mathbf{w}) = \rho^3, \qquad (3.63)$$

for the cycle to exist, or, after the first two equalities (3.62) substituted and with regard to (3.37) and (3.61),

$$1 + u(\rho - 1)N(\mathbf{l}_0) + u(\rho - 1)\rho N(\mathbf{l}_1) + u(\rho - 1)\rho^2 N(\mathbf{l}_2) = \rho^3. \qquad (3.64)$$

Treated as an equation for $u > 0$, the equality (3.64) has a unique solution u_0, since the left-hand side is continuous in u and increases with u monotonically from 1 to infinity. (Note that solution u_0 depends continuously on $\mathbf{l}_0, \mathbf{l}_1, \mathbf{l}_2$.) The quantity u_0 and equalities (3.62) then determine uniquely the vectors $\mathbf{u}, \mathbf{v}$, and $\mathbf{w}$ of the cycle under study.

If one considers a proportional cycle $c\mathbf{l}_0, c\mathbf{l}_1, c\mathbf{l}_2$ $(c > 0)$ of the operator L/ρ instead of the original one, then equation (3.62) will have the unique solution $u_c = u_0/c$, that will obviously bring about the previous set of vector $\mathbf{u}, \mathbf{v}, \mathbf{w}$ for the cycle of the nonlinear model. ∎

Theorem A3.7 *(on convergence to a T-cycle). A trajectory $\mathbf{x}(t; \mathbf{x})(t = 0, 1, \ldots)$, of the nonlinear equation (3.36)–(3.37) matching the initial condition $\mathbf{x}(0) = \mathbf{x} \geq 0$*

($\mathbf{x} \neq 0$), *converges as $t \to \infty$ to the solution of the T-cycle problem (Theorem A3.6) for vectors $\mathcal{L}(\mathbf{x}, 0), \mathcal{L}(\mathbf{x}, 1), \ldots, \mathcal{L}(\mathbf{x}, T - 1)$ of the limit cycle in the corresponding linear model, i.e. the equation (3.36) at $q(N) \equiv \rho$. There will converge to the very same solution any trajectory of the family $\{\mathbf{x}(t; c\mathbf{x})\}$ matching initial conditions of the form $\mathbf{x}(0) = c\mathbf{x}$, $c > 0$.*

Proof: Convergence of $\mathbf{x}(t)$ to a T-cycle is equivalent to each of T subsequences $\{\mathbf{x}(Tt)\}, \{\mathbf{x}(Tt + 1)\}, \ldots, \{\mathbf{x}(Tt + (T - 1))\}$ $(t = 0, 1, \ldots)$ converging to its own limit, the limits satisfying the cycle equations with operator L_N. According to (3.47) we have

$$\lim_{t \to \infty} \mathbf{x}(Tt) = \lim_{t \to \infty} (L_\rho^T)^t \mathbf{x}(0) \Big/ \lim_{t \to \infty} \pi(Tt; \mathbf{x}(0)), \tag{3.65}$$

if there exist the limits standing on the right-hand side. The nominator limit does exist in accordance with the asymptotics of the linear model, while equation (3.60) gives the following for the denominator:

$$\pi(Tt; \mathbf{x}) = 1/\rho^{Tt} + (\rho - 1)\left[N(\mathbf{x})/\rho^{Tt} + N(L_\rho^{Tt-1}\mathbf{x})/\rho^{Tt-1} + \cdots \right.$$
$$\left. + N(L_\rho^{Tt-(T-1)}\mathbf{x})/\rho\right]. \tag{3.66}$$

Since all subsequences of the form $\{L_\rho^{Tt-j}\mathbf{x}, t = 1, 2, \ldots\}$, $j = 0, 1, \ldots, T - 1$, converge to their finite limits, all of them are bounded above (by the same constant M) in any of the proper norms, for instance, in the sum of absolute values of vector coordinates. In the cone of nonnegative vectors, this means that all $N(L_\rho^{Tt-j}\mathbf{x})$ are bounded, i.e., all terms of the sum in square brackets are majorized by the respective terms of the converging series $\sum_{t=1}^{\infty} M/\rho^{Tt}$, hence, there exists a finite limit of expression (3.66). Therefore, a finite limit (3.65) does exist and, in a similar way, all of the rest subsequences $\{\mathbf{x}(Tt + 1)\}, \ldots, \{\mathbf{x}(Tt + (T - 1))\}$ can be proved to converge too.

Once the limits exist, they should have to meet the cycle equations of the type (3.62). This is guaranteed by Theorem A3.6 and, since a proportional initial vector tends to the proportional limit cycle in the linear model, the same cycle equations will hold for any initial state $c\mathbf{x}$, $c > 0$. ∎

Example A3.1 To illustrate the idea that stability of the total population size $N(t)$ may exist under instability of the population vector $\mathbf{x}(t)$, we consider a Bernadelli-type matrix of a more general form than (2.49):

$$B = \begin{bmatrix} 0 & 0 & b \\ s_1 & 0 & 0 \\ 0 & s_2 & 0 \end{bmatrix}, \quad b \geq 1, \quad s_{1,2} \leq 1, \tag{3.67}$$

with the cycle condition $B^3 = I$ that reduces to

$$bs_1 s_2 = 1.$$

Searching for a triple of alternating successive vectors $\mathbf{x} \in \mathbb{R}^3_+$, $B\mathbf{x}$, and $B^2\mathbf{x}$ that retain the sum of their components at each time step is equivalent to solving the system of linear algebraic equations

$$\begin{cases} x_1 + x_2 + x_3 = N, \\ s_1 x_1 + s_2 x_2 + b x_3 = N, \\ s_1 s_2 x_1 + b s_2 x_2 + b s_1 x_3 = N, \end{cases}$$

in the positive octant (a particular value of N does not matter due to the linearity of the model; let, for instance, $N = 100$). Each of the equations represents a plane in $\mathbb{R}^3$ and the existence of solution(s) means geometrically that there is a point of intersection or a straight line of intersection points of those three planes. The former case corresponds to the single-solution case of linear systems (when the determinant is nonzero) and is obviously inappropriate since it has been already occupied by the model equilibrium $\mathbf{x}^*$. In the latter case, it is an algebraic exercise to investigate whether the straight line goes through $\mathbb{R}^3_+$ for the general case of matrix (3.67). At least, its extremal particular case given by the permutation matrix

$$P_3 = \begin{bmatrix} 0 & 0 & 1 \\ 1 & 0 & 0 \\ 0 & 1 & 0 \end{bmatrix},$$

furnishes an apparent plentitude of solutions as P_3 reduces the Leslie operator to mere cyclic permutation of the vector components, hence keeps their sum invariable.

Additional Notes

To 3.I. "Few examples" of the classical Leslie matrix application can be found in Williamson.[39] The need to modify Leslie's set-up was realized soon after Leslie and stressed in later publications.[6,40–42] Thus modifications (3.1) or (3.2) are more abundant in applications: see, e.g., Usher,[41] Jeffers,[43] Longstaff,[44] and references therein. Several authors[45–47] commented on the relation between a classical Leslie model and its version (3.2).

The analysis of perturbation effects is a particular case of sensitivity analysis (how sensitive are model characteristics to changes in model parameters?), which is a favorite stream for modeling efforts (see, e.g., Cullen[48] or Caswell[49]).

Models similar to that of "matrix jump" were also considered in a different population context: see, e.g., Loreau,[50] where strict seasonality induced a unique, and the simplest, mode of switching between two model matrices.

To 3.II. Hansen[51] surveyed more examples of block-structured projection matrices, which sometimes appeared in the population literature under the names of *Goodman* or *Lefkovitch* matrices.[8] Other than (3.15) ordering of group sizes into the population vector would generate another block structure in the projection matrix, the matrix though being the same to within a similarity transformation (see, e.g., Smirnov,[52] who proposes some auxiliary matrix to be constructed to verify the indecomposability and primitivity properties).

To 3.III. The theory of characteristic exponents (or just exponents) was developed by A. M. Lyapunov[15] (who called them *characteristic numbers*) for solutions of linear differential equations. Nowadays the main properties of exponents can be deduced from the general theory of exponents defined on a linear vector space since the solutions constitute a pertinent linear space (see Bylov et al.[16] for the theory and its application to systems of differential equations). The modification for nonautonomous difference equations like system (3.17), though falling into general lines of analogy between these two major classes of equations, is of interest in its own right.[23] A survey of results for the subclass of difference systems with nonnegative matrices can be found in Ivanov.[53] In more general terms, the Lyapunov exponents and their continuity properties are considered by Millionshchikov[54−55], who also commented on the Perron example[21] illustrating instability of the Lyapunov exponent in a nonautonomous case.

The definition of exponential separateness follows Millionshchikov[17] and is common to both differential and difference classes of equations, while the definition of Lyapunov transformation, following Ivanov,[18] has a specific form (3.21).

Borrowed from the qualitative theory of differential equations,[16] the idea of exponential separation looks formally different from, though consonant with, the concept of weak ergodicity pioneered in the Leslie model context by Lopez[56−57] and developed further by several authors (see a survey in Hansen[51]). Each of both notions implies a kind of convergence in the population age structure and future mathematical efforts will probably reveal a logical relation between them. The convergence is commonly interpreted as population "forgetting its past," which, though having some evidence in demography,[58−59] may be just a general implication of linearity in the mathematical description of birth and survival processes.

Another approach to asymptotics in model (3.17), which seems quite natural from the mathematical point of view, is to find conditions on the sequence of matrices $A(t)$ that guarantee existence of the $\lim_{t \to \infty} \mathbf{x}(t)/\lambda^t = c\mathbf{e}$ for some $\lambda_1 > 0$ and vector $\mathbf{e}$. It appears that convergence $A(t) \to A$ alone is insufficient and some additional constraints are in order, to provide for a behavior similar to that of the constant primitive case.[60−61] Both the convergence itself and the additional constraints seem too artificial to be readily interpreted in population terms. Rather the formulations are practical which admit permanent perturbations of demographic parameters in the form of "white" or correlated noise (see, e.g., Ginzburg et al.[62]). Stochasic convergence established for this kind of models (Cohen;[63] Tuljapurkar and Orzack[64]) may have a fundamental reason in the deterministic property of exponential separateness.

To 3.IV. A synopsis of results that generalize the Perron–Frobenius theorem for some classes of nonlinear eigenvector problems, is presented in Morisima.[25] Unfortunately, these results require the regulation functions to be homogeneous of a fixed order, which looks an artificial constraint from the standpoint of biological populations (see also Smirnov[65]).

The general solution to Leslie's nonlinear model was obtained independently by Allen[66] and Logofet[38] (the latter gives a more precise formulation of periodicity properties); Ziebur[67] produced a formulation for a slightly different modification.

In spite of mathematical difficulties, density-dependent modifications were applied both in modeling practice[45,49,68] and further development of the population dynamical theory in certain of its branches such as optimal harvesting[49,68–73] and optimal life-cycle evolution.[49,74–75] For the "chaos" in matrix models see Caswell,[49] Efremova,[76] and references therein.

REFERENCES

1. Longstaff, B., An extension of the Leslie matrix model to include a variable immature period, *Austral. J. Ecol.*, 9, 289–293, 1984.

2. Gantmacher, F. R., *The Theory of Matrices*, Chelsea, New York, 1960, Chap. 13.

3. Lancaster, P., *Theory of Matrices*, Academic Press, New York-London, 1969, Chap. 7.

4. Voyevodin, V. V. and Kuznetsov, Yu. A., *Matrices and Calculations*, Nauka, Moscow, 1984, Chap. 18 (in Russian).

5. Williamson, M. H., The analysis of discrete time cycles, in *Ecological Stability*, Usher, M. B. and Williamson, M. H., Eds., Chapman & Hall, London, 1974, 17–33.

6. Williamson, M. H., Some extensions of the use of matrices in population theory, *Bull. Math. Biophys.*, 21, 13–17, 1959.

7. Goodman, L. A., The analysis of population growth when the birth and death rates depend upon several factors, *Biometrics*, 25, 659–681, 1969.

8. Law, R., A model for the dynamics of a plant population containing individuals classified by age and size, *Ecology*, 64, 224–230, 1983.

9. Logofet, D. O., Tackling variability by "constancy" methods: some examples from modelling structured populations, Paper C5, ESF Workshop on Theoretical Ecology: Tackling Variability, Helsinki, 1988 (European Science Foundation).

10. Csetenyi, A. I., A Matrix Model for a Reindeer Population with Regard to Ecological and Physiological Indices, Computer Center of the USSR Acad. Sci., Moscow, 1988 (in Russian).

11. Csetenyi, A. I. and Logofet, D. O., Leslie model revisited: some generalizations to block structures, *Ecological Modelling*, 48, 277–290, 1989.

12. Logofet, D. O., On indecomposability and primitivity of nonnegative block-structured matrices, *Soviet Math. Dokl.*, 40, 306–310, 1990.

13. McAndrew, M. H., On the product of directed graphs, *Proc. Amer. Math. Soc.*, 14, 602–606, 1963.

14. Harary, F., Norman, R. Z., and Cartwright, D., *Structural Models: an Introduction to the Theory of Directed Graphs*, John Wiley, New York, 1965, Chap. 7.

15. Lyapunov, A. M., *General Problem of Motion Stability*, Gostekhizdat, Moscow, Leningrad, 1950, Chap. 1 (in Russian).

16. Bylov, B. F., Vinograd, R. E., Grobman, D. M., and Nemytsky, V. V., *Theory of Lyapunov Exponents and Its Application to Issues of Stability*, Nauka, Moscow, 1966, Chap. 1 (in Russian).

17. Millionshchikov, V. M., On indices of exponential separation, *Math. USSR Sbornik*, 52, 439–470, 1985.

18. Ivanov, A. I., On exponential separateness in solutions of linear strongly positive systems of difference equations. Deposited paper 4773-B88, VINITI (All-Union Institute of Scientific and Technical Information), Moscow, U.S.S.R., 1988 (in Russian).

19. Ivanov, A. I., On non-stationary matrix models of biological systems, in *Ecological Problems of Wild Nature Conservation. Theses of The All-Union Conference*, Part 1. Krasilov, V. A., Ed., U.S.S.R. State Committee on Nature Conservation, Moscow, 1990, 55–56 (in Russian).

20. Ivanov, A. I., personal communication, 1991.

21. Perron, O., Über Stabilität und asymptotisches Verhalten der Integrale von Differential-gleichungssystemen, *Math. Z.*, 29, 129–160, 1928/29 (in German).

22. Millionshchikov, V. M., Robust properties of linear systems of differential equations, *Differentsialnye Uravneniya (Differential Equations)*, 5, 1775–1784, 1969 (in Russian).

23. Shirobokov, N. V., Stability of central indices in discrete difference equations, *Differentsialnye Uravneniya (Differential Equations)*, 19, 722–723, 1983 (in Russian).

24. Rakhimberdiev, M. I. and Ivanov, A. I., On stability of the major exponent of linear discrete systems with non-negative coefficients, *Izvestiya AN Kazakh SSR (Transactions of the Kazakh SSR Acad. Sci.)*, Physical-Mathematical Series, 1986, No. 1, 55–57 (in Russian).

25. Morishima, M., *Equilibrium, Stability, and Growth*, Clarendon Press, Oxford, 1964, Appendix.

26. Leslie, P. H., Some further notes on the use of matrices in population mathematics, *Biometrika*, 35, 213–245, 1948.

27. Odum, E. P., *Basic Ecology*, Saunders College Publishing, Philadelphia, 1983, Chap. 6.

28. Beddington, J. R., Age distribution and the stability of simple discrete time population models, *J. Theor. Biol.*, 47, 65–74, 1974.

29. Takanashi, F., Reproduction curve with two equilibrium points: a consideration on the fluctuation of insect population, *Res. Popul. Ecol.*, 6, 28–36, 1964.

30. Isaev, A. S. and Khlebopros, R. G., Stability principle in population dynamics of forest insects, *Doklady AN SSSR (Proc. U.S.S.R. Acad. Sci.)*, 208, 225–227, 1973 (in Russian).

31. Sharkovsky, A. N., Coexistence of cycles in a continuous mapping of the axis into itself, *Ukranian Math. Journal*, 16, 61–71, 1964 (in Russian).

32. Li, T.-Y. and Yorke, J. A., Period three implies chaos, *Amer. Math. Monthly*, 82, 982–985, 1975.

33. Jacobson, M. V., On properties of the dynamical system generated by mapping $x \to Ae^{-\beta x}$, in *Modelling of Biological Communities*, Shapiro, A. P., Ed., Far-Eastern Scientific Center of the U.S.S.R. Acad. Sci., Vladiviostok, 1975, 141–162 (in Russian).

34. Zykov, A. A., *Theory of Finite Graphs, I*, Nauka, Novosibirsk, 1969, Chap. 8 (in Russian).

35. Harary, F., *Graph Theory*, Addison-Wesley, Reading, Mass., 1969, Chap. 16.

36. Svirezhev, Yu. M. and Logofet, D. O., *Stability of Biological Communities*, Mir Publ., Moscow, 1983 (translated from the 1978 Russian edition), Chap. 2.6.

37. Gantmacher, F. R., *The Theory of Matrices*, Chelsea, New York, 1960, Chap. 6.

38. Logofet, D. O., Non-linear Leslie model revisited: asymptotics of trajectories in the primitive and imprimitive cases, *Soviet Math. Dokl.*, 43, 1992 (in press).

39. Williamson, M. H., Introducing students to the concepts of population dynamics, in *The Teaching of Ecology*, Lambert, J. M., Ed., Blackwells, Oxford, 1967, 169–175.

40. Goodman, L. A., Population growth of the sexes, *Biometrics*, 9, 212–225, 1953.

41. Usher, M. B., Developments in the Leslie matrix model, in *Mathematical Models in Ecology. The 12th Symposium of the British Ecological Society*, Jeffers, J. N. R., Ed., Blackwell, Oxford, 1972, 29–60.

42. Pip, E. and Stewart, J. M., A method for fitting population data to a matrix model when the growth rate is unknown, *Int. Res. des Hydrobiol.*, 60, 669–773, 1975.

43. Jeffers, J. N. R., *An Introduction to Systems Analysis: with ecological applications*, Edward Arnold, London, 1978, Chap. 4.

44. Longstaff, B. C., The dynamics of collembolan populations: a matrix model of single species population growth, *Can. J. Zool.*, 55, 314–324, 1977.

45. Cook, D., and Leon, J. A., Stability of population growth determined by 2×2 Leslie matrix with density-dependent elements, *Biometrics*, 32, 435–442, 1976.

46. Horwood, J. W. and Shepherd, J. G., The sensitivity of age-structured populations to environmental variability, *Math. Biosci.*, 57, 59–82, 1981.

47. Fisher, M. E. and Goh, B. S., Stability results for delayed-recruitment models in population dynamics, *J. Math. Biol.*, 19, 147–156, 1984.

48. Cullen, M. R., *Linear Models in Biology*, Ellis Horwood, Chichester, 1985, Chap. 12.

49. Caswell, H., *Matrix Population Models*, Sinauer Associates, Suderland, Massachusetts, 1989, Chaps. 6, 9.

50. Loreau, M., Coexistence of temporally segregated competitors in a cyclic environment, *Theor. Popul. Biol.*, 36, 181–201, 1989.

51. Hansen, P. E., Leslie matrix models: a mathematical survey, in *Papers on Mathematical Ecology, I*, Csetenyi, A. I., Ed., Karl Marx University of Economics, Budapest, Hungary, 1986, 54–106.

52. Smirnov, A. I., Properties of generalized Leslie matrices and their use in the analysis of model ecosystems, in *Mathematical Modelling of Medical and Biological Systems*, Mazurov, V. D. and Smirnov, A. I., Eds., The Urals Branch of the U.S.S.R. Acad. Sci., Sverdlovsk, 1988, 87–94 (in Russian).

53. Ivanov, A. I., Asymptotic Properties of Linear Difference Systems with Non-negative Coefficients. Candidate of Sciences Thesis, Institute of Mathematics and Mechanics, Kazakh SSR Academy of Sciences, Alma-Ata, 1989, Chap. 1 (in Russian).

54. Millionshchikov, V. M., Lyapunov exponents of a family of endomorphisms of a metrized vector bundle. *Matematicheskiye Zametki*, 38, 92–109, English transl. in *Math. Notes*, 38, 1985.

55. Millionshchikov, V. M., Lyapunov exponents as functions of a parameter, *Math. USSR Sbornik*, 65, 369–384, 1990.

56. Lopez, A., *Problems in Stable Population Theory*, Office of Population Research, Princeton University, Princeton, NJ, 1961, Chap. 2.

57. Lopez, A., Asymptotic properties of a human age distribution under a continuous net maternity function, *Demography*, 4, 680–687, 1967.

58. Coale, A. J., How the age distribution of a human population is determined, in *Cold Spring Harbor Symposia on Quantitative Biology*, 22, Warren, K. B., Ed., Long Island Biol. Assoc., New York, 1957, 83–89.

59. Kim, Y. J. and Sykes, Z. M., An experimental study of weak ergodicity in human populations, *Theor. Popul. Biol.*, 10, 150–172, 1976.

60. Thompson, M., Asymptotic growth and stability in populations with time dependent vital rates, *Math. Biosci.*, 42, 267–278, 1978.

61. Artzrouni, M., Generalized stable population theory, *J. Math. Biol.*, 21, 363–381, 1985.

62. Ginzburg, L. R., Johnson, K., Pugliese, A., and Gladden, J., Ecological risk assessment based on stochastic age-structured models of population growth, in *Statistics in Environmental Sciences,* Gertz, S. M. and M. D. London, Eds., STTP 845, American Soc. for Testing and Materials, Philadelphia, PA, 1984, 31–45.

63. Cohen, J. E., Ergodicity of age structure in populations with Markovian vital rates, I: Countable states, *J. Amer. Statist. Assoc.*, 71, 335–339, 1976.

64. Tuljapurkar, S. D. and Orzack, S. H., Population dynamics in variable environments. I. Long-run growth rates and extinction, *Theoretical Population Biology*, 18, 314–342, 19??.

65. Smirnov, A. I., Solvability conditions for the problem of driving a population to a given structure, in *Classification and Optimization in Control Problems*, Mazurov, V. D. and Kislyak, V. M., Eds., The Urals Scientific Center of the U.S.S.R. Acad. Sci., Sverdlovsk, 1981, 90–102 (in Russian).

66. Allen, L. J. S., A density-dependent Leslie matrix model, *Math. Biosci.*, 95, 179–187, 1989.

67. Ziebur, A. D., Age-dependent models of population growth, *Theor. Popul. Biol.*, 26, 315–319, 1984.

68. Abakumov, A. I., Optimal harvesting problems in age-structured populations, in *Mathematical Studies in Population Ecology*, Frisman, E. Ya., Ed., Far-Eastern Scientific Center of the U.S.S.R. Acad. Sci., Vladiviostok, 1988, 19–32 (in Russian).

69. Beddington, J. R. and Taylor, D. B., Optimum age specific harvesting of a population, *Biometrics*, 29, 801–809, 1973.

70. Doubleday, W. G., Harvesting in matrix population models, *Biometrics*, 31, 189–200, 1975.

71. Mendelsohn, R., Optimization problems associated with a Leslie matrix, *Amer. Natur.*, 110, 339–349, 1976.

72. Reed, W. J., Optimum age-specific harvesting in a non-linear population model, *Biometrics*, 36, 579–593, 1980.

73. Getz, W. M., The ultimate-sustainable-yield problem in nonlinear age-structured populations, *Math. Biosci.*, 48, 279–292, 1980.

74. Schaffer, W. M., Selection for optimal life histories: the effects of age structure, *Ecology*, 55, 291–303, 1974.

75. Schaffer, W. M. and Rosenzweig, M. L., Selection for optimal life histories. II. Multiple equilibria and the evolution of alternative reproductive strategies, *Ecology*, 58, 60–72, 1977.

76. Efremova, S. S., Stability and Bifurcations in Discrete Models of Age-structured Population Dynamics. Candidate of Sciences Thesis, Leningrad Polytechnic Institute, Leningrad, 1990 (in Russian).

CHAPTER 4

Lotka–Volterra Models of *n*-Species Communities

Presented in this chapter are the results of stability analysis applied to models of multispecies community dynamics, which were originated in the works of A. Lotka[1] and V. Volterra.[2] The further development of the technique, and extended interpretation and generalization of the findings have revealed a close tie between Lyapunov stability of equilibrium in a model system and ecological stability in model trajectories. Relying upon the Lyapunov concept, we establish a logical hierarchy for a number of matrix stability concepts which are used in the contemporary literature and proposed anew, and interpret them in terms of corresponding models for population dynamics.

I. A GRAPH OF SPECIES INTERACTIONS AND THE COMMUNITY MATRIX

The first logical step in studying a community of several interacting species, so common in ecology, is a description of the structure of interactions among (and within) species by means of a graph. Vertices of the graph represent individual species or whole groups of species if they play the same role in the structure. Arcs (or links) indicate the existence of ties, i.e., interactions, between individual (groups of) species. Depending on the rule adopted in drawing the arcs, the graph may be one of several formally defined types: nondirected, directed, or even signed directed graphs.

If, for example, the graph has to display a structure of only trophic relations in the community, i.e., relations of the "prey-predator" (or "resource-consumer") type, then the links might be nondirected. But arrows can be apparently useful in indicating "who eats whom," so that directed graphs are quite naturally appropriate in representing trophic webs. Figure 9 illustrates a part of the trophic web (*subweb*) observed in a marine rocky intertidal zone.[3] Two species of the upper level are shown to share resources in the lower level.

In a directed graph used to model a trophic web, the direction of the arc between a pair of vertices is essentially the direction of the matter or biomass flow due to the trophics. However, another type of species interaction, such as resource competition or symbiosis, may not be bound uniquely with the matter or energy movement

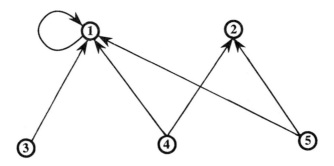

Figure 9.   Feeding relations in a community of an intertidal rocky zone. Two species of small muricid gastropods, *Thais biserialis* (1) and *Acanthina brevidentata* (2), feed on groups of species: (3) carrion, 2 spp.; (4) bivalves, 2 spp., and (5) barnacles, 3 spp.; also, species 1 is cannibalistic. After Paine.[3]

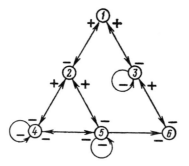

Figure 10.   Signed directed graph of a 6-species community: species 1 feeds on 2 and 3, which prey on 4, 5, and 6; species 4 and 5 are linked by competition, while species 6 is an amensal of 5; species 3, 4, and 5 are density self-regulated.

between the species. Therefore, to display the structure of various interactions, directed graphs are often used whose arcs correspond merely to directed influences of species. An arc between, for instance, vertices i and j, directed $j \rightarrow i$, indicates that species j has a certain influence on species i. If, furthermore, the sign of the influence ("+" if species j stimulates an increase in the population of species i, and "−" if species j suppresses the population of species i) is ascribed to any arc $j \rightarrow i$, then a construction arises which is called a *signed directed graph* (*SDG*).[4–7] An example is shown in Figure 10.

Being a simple and apparent means of displaying a community structure, the graphs may also give certain information on stability properties in the corresponding models of community dynamics. For example, in n-species communities where the predator-prey relation is the sole type of species interactions, even a nondirected graph is sufficient to display the relation structure, while its properties turn out to be sufficient to reveal the existence of a stable equilibrium in the model (see Chapter 5.I). A wider class of model ecosystems is amenable to the so-called "qualitative stability" analysis by means of SDGs, again relying only on the graph properties and making no use of any quantitative information on particular species interactions (see Chapter 5.II).

The quantitative information, in a special form, is contained in the so-called *community matrix*, which also represents the structure of intraspecies and (pairwise) interspecies relations and plays an important role in studying multispecies communities.

Let, for example, a model of n-species community dynamics is given in the form of a system of ordinary differential equations for the sizes of n populations composing an n-dimensional vector $\mathbf{N}(t) = [N_1(t), \dots, N_n(t)]^T \in \mathbb{R}_+^n$, namely,

$$d\mathbf{N}/dt = \mathbf{F}(\mathbf{N}; t), \tag{4.1}$$

or, in the particularly important and widely used *autonomous* case,

$$d\mathbf{N}/dt = \mathbf{F}(\mathbf{N}). \tag{4.1'}$$

Suppose that function $\mathbf{F}(\mathbf{N}; t)$ or $\mathbf{F}(\mathbf{N})$ meets all the conditions under which the initial-value problem is *correctly posed* for system (4.1) or (4.1') (i.e., there exists a unique solution $\mathbf{N}(t)$ starting at any given initial point $\mathbf{N}(0) \in \mathbb{R}_+^n$ and continuously depending on it), the conditions being not in fact too restrictive. Suppose also that there exists an *equilibrium* solution $\mathbf{N}(t) = \mathbf{N}^*(t)$, that is one satisfying the equation

$$\mathbf{F}(\mathbf{N}^*(t)) \equiv \mathbf{0}.$$

In the autonomous case, the equilibrium is to be identically constant, $\mathbf{N}^*(t) \equiv \mathbf{N}^*$, and is also said to be a *rest point*, or a *steady state*, of system (4.1').

The *community matrix* is then nothing but the matrix,

$$A = [a_{ij}] = \left[\frac{\partial F_i}{\partial N_j} \bigg|_{\mathbf{N}=\mathbf{N}^*} \right], \tag{4.2}$$

of the first-order partial derivatives calculated at an equilibrium point, or the *Jacobian matrix* of the system of equations linearized at the equilibrium.

According to definition (4.2) any entry a_{ij} of A is generally a function of the equilibrium sizes of all species in the model community, and we will see later in this chapter how this function operates in the Lotka–Volterra models. But once the equilibrium point is determined, the function yields a particular number, a_{ij}^*, which, according to the meaning of the partial derivative, indicates the change in the i-th population rate of increase in response to a small change in the j-th population size, assuming that all of the remaining populations are kept unchanged at their equilibrium values.

In other words, the sign of an entry a_{ij} in the community matrix shows the qualitative nature of the effect that species j exerts upon species i (where "+" means stimulating, "−" means suppressing, and "0" means neutral relations), while the absolute value of a_{ij} measures the quantitative level of that effect. The pair of signs of two symmetric entries, a_{ij} and a_{ji}, in the community matrix provides the basis for classifying pair-wise interactions between species and generates the following six major types:[7–9]

+	+	*mutualism* or *symbiosis*;
+	−	*prey-predator* (or *resource-consumer*, or *host-parasite*);
+	0	*commensalism*;
−	−	*competition*;
−	0	*amensalism*;
0	0	*neutralism*.

This approach to classification is attractive for a mathematician, as it gives a key to relate the structure of interactions with the stability properties in the corresponding community model. If we define the *sign matrix*

$$S = \text{Sign } A = [\text{Sign } a_{ij}], \quad \text{Sign } x = \begin{cases} - \\ 0 \\ + \end{cases}, \tag{4.3}$$

as the *sign pattern* of matrix A, then the sign pattern of a community matrix can easily be seen to represent, in analogy with the SDG, the qualitative structure of pair species interactions in the community. Moreover, the SDG is nothing but the digraph associated with matrix A and signed by its sign pattern. Clearly, there is a natural one-to-one correspondence between the set of all $n \times n$ sign matrices and the set of all SDGs having n vertices, both the matrix and the associated SDG representing the interaction structure uniquely, to within renumbering of graph vertices, or simultaneous permutation of the matrix lines and columns. Thus, the enumeration of vertices shown in Figure 10 corresponds to the following sign pattern of the community matrix:

$$S = \begin{bmatrix} 0 & + & + & 0 & 0 & 0 \\ - & 0 & 0 & + & + & 0 \\ - & 0 & - & 0 & 0 & + \\ 0 & - & 0 & - & - & 0 \\ 0 & - & 0 & - & - & 0 \\ 0 & 0 & - & 0 & - & 0 \end{bmatrix}.$$

Examining certain properties of matrix A (4.2), in particular, its spectrum, gives information on stability properties of solutions to the model system and on how community structures can be classified with respect to stability in the corresponding models.

II. LOTKA–VOLTERRA MODELS AND ECOLOGICAL BALANCE EQUATIONS

In the literature on mathematical theory of population and community dynamics the following system is commonly referred to as a *Lotka–Volterra* model:[7,9−11]

$$\frac{dN_i}{dt} = N_i \left(\varepsilon_i - \sum_{j=1}^{n} \gamma_{ij} N_j \right), \qquad j = 1, \ldots, n. \tag{4.4}$$

Here $N_i(t)$ is the population size (or population density) of species i, ε_i is an intrinsic rate of increase or decrease in the size of species i population in the absence of all others, while the sign and the absolute value of γ_{ij} $(i \neq j)$ indicate, respectively, the nature and intensity of the effect species j has on species i, and γ_{ii} is a parameter of intraspecies regulation for species i.

Representing the structure of species relations, matrix $\Gamma = [\gamma_{ij}]$ is often termed the *community matrix*, as well as matrix (4.2); it will be easily seen below that to sign matrix S introduced above in (4.3) matrix Γ is related by $S = -\text{sign } \Gamma$. Therefore, in terms of major qualitative types of species interactions, Lotka–Volterra models are able to describe any kind of community structure.

A. The Hypothesis of "Encounters and Equivalents"

V. Volterra[2] derived system (4.4) from the following argument: the dynamics of species i in the absences of all others can be described by the well-known logistic equation (normally $\gamma_{ii} > 0$ indicating intraspecies competition, or density self-limitation) and the effect of species j on the growth rate of species i can be expressed as a term proportional to the product $N_i N_j$ according to the so-called "encounters and equivalents" hypothesis. The hypothesis assumes the change in population size of species i (prey) due to species j (predator) preying on it to be determined by the frequency of encounters between individuals of both species, whereby the product $N_i N_j$ arises, while the increase in the predator population is assumed to be such as if there were immediate transformation of prey eaten by predators into the predator population size according to proportionality coefficient, the "equivalent."

Originated apparently from the "encounters" principle of statistical physics, the "encounters and equivalents" hypothesis can explain the form of equations (4.4) for prey-predator interactions, but is apparently unable to justify it for other types of biological interactions. Moreover, its nonbiological origin makes the hypothesis open to criticism on the part of ecologists. It turns out, however, that the Lotka–Volterra equations can be deduced from more natural assumptions.

B. Conservation Principle and Balance Equations

Figure 11 illustrates the generally known concept of an ecosystem as a system of mass and energy flows between its basic components. In accordance with these thoughts, a simplified version of a general dynamic model may consist of three following groups of variables:

Producers with biomasses (or concentrations) x_i ($i = 1, 2, \ldots, m$). These are primarily green plants capable of fixing radiant energy and assimilating simple substances.

Consumers with concentrations y_j ($j = 1, 2, \ldots, n$). This group refers to animals that prey upon other organisms, as well as decomposers which break the dead organics down into simple substances to be taken up by the producers.

Substrates with concentrations s_k ($k = 1, 2, \ldots, p$). These are the abiotic substances (mainly the products of the consumers' vital activity), taken up by the producers.

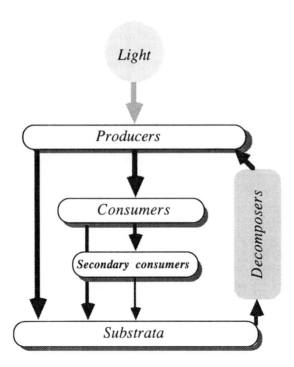

Figure 11.   General diagram of mass and energy flows among aggregated components of terrestrial ecosystems.

Equations to describe the mass balance for each of these components are nothing but formal expressions for the mass conservation principle, which can be written in the following form:

$$\frac{dx_i}{dt} = (F^i_x - D^i_x)x_i - \sum_{j=1}^{n} V_{ij}y_j + R_x, \qquad i = 1, \ldots, m, \tag{4.5}$$

$$\frac{dy_j}{dt} = (F^j_y - D^j_y)y_i - \sum_{r=1}^{n} v_{jr}y_r + R_y, \qquad j = 1, \ldots, n, \tag{4.6}$$

$$\frac{ds_k}{dt} = \sum_{j=1}^{n} U_{kj}y_j - \sum_{i=1}^{m} W_{ki}x_i + R_s, \qquad k = 1, \ldots, p. \tag{4.7}$$

Here $F^i_{x(y)}$ and $D^i_{x(y)}$ are the birth and mortality functions for producers (consumers); V_{ij} is a *grazing function* describing the rate at which the biomass of producer species i is consumed by the unit biomass of consumer species j; v_{jr} is a trophic function for species j uptake by species r (among consumers); W_{kj} is a function of substrate k uptake by producer species j; U_{kj} is the intensity of substrate k production by consumer j; R_x, R_y, and R_s, are the resulting sums of inflows and outflows of respective components. Naturally, all these functions depend upon parameters of the environment (seasonal variations of humidity, temperature, etc.).

The birth rate of the ith producer species, F^i_x may be assumed to depend only upon the intensity of the radiation flux I and the amount of substrates taken up, $\sum_{k=1}^{n} W_{ki}x_i$. From simple arguments which rely on the conservation principle, we can assume that

$$F^i_x = \sum_{k=1}^{p} K^{(i)}_k(I)W_{ki}, \tag{4.8}$$

i.e., stoichiometrics takes place with coefficients $K^{(i)}_k$, which can be interpreted as the fraction of substrate k used to reproduce a unit of producer i biomass. Notice that if we implement consistently the stoichiometrics hypothesis, then

$$K^{(i)}_k = K_k, \qquad F^i_k = \sum_{k=1}^{p} K_k(I)W_{ki}. \tag{4.9}$$

Similarly to (4.8), we have the growth rate of consumer j equal to

$$F^j_y = \sum_{i=1}^{m} K^{(j)}_i V_{ij} + \sum_{r=1}^{n} P^{(j)}_r v_{rj}. \tag{4.10}$$

where $K^{(j)}_i$ and $P^{(j)}_r$ are stoichiometric coefficients for the producer–consumer interactions and for preying upon consumers.

Let, as a special case of the general model, the birth rates of producers be limited neither in light nor in mineral nutrients, and be restricted only by mere physiological limits. Then F_x^i = const. Since the substrate components in this case have no effect on the dynamics of other components, the model can be reduced to producer-consumer interactions only. A similar situation arises under the light limitation since light is an external factor with respect to this system, and when the radiation flux is constant, F_x^i can be considered constant as well. Considering the system to be closed, we should set $R_x = R_y = R_s = 0$.

The most simple assumption on the form of mortality functions is the hypothesis that they are constant. In real communities the situation is more complicated: both the intraspecies and interspecies competition for a common resource (food) bring about increased mortality in response to an increase in the population sizes. This mechanism can hardly be described by any kind of balance relationships. The simplest assumption is therefore a linear relationship between the mortality function and the population sizes of the species involved:

$$D_x^i = m_x^i + \sum_{s=1}^{m} \mu_{is} x_s, \qquad D_y^j = m_y^j + \sum_{r=1}^{n} \nu_{jr} y_r. \qquad (4.11)$$

Here m_x^i and m_y^j are intrinsic (physiological) death rates, μ_{ii} and ν_{jj} describe the intensity of intraspecies competition (should there be no competition for a common resource, the corresponding ε and ν values will be zero).

The simplest assumption about the form of grazing functions is the hypothesis of a linear relationship between these functions and the concentration (size, biomass) of the species preyed on:

$$V_{ij} = \alpha_{ij} x_i, \qquad \nu_{jr} = \beta_{jr} y_j, \qquad r = 1, \ldots, n. \qquad (4.12)$$

Of course, this assumption of a linear relationship between the grazing function and the diet is valid only for a fairly narrow range of concentrations of food to be grazed (e.g., for low concentrations) and for a consumer's trophic strategy that makes no preference among species in the diet. But these assumptions, though being restrictive, do not contradict the biology. Moreover, they are true for quite a nonempty class of ecological situations.

Generally, there may exist both upper physiological bounds to the amount of food consumed (thus the grazing functions should be presented by saturation-type curves), and more complicated forms of the trophic strategy (for example, the trophic preference results in a sequential rather than a simultaneous uptake of species; in this case the grazing function of a species has to be of a more complicated form depending on concentrations of the other species too).

Under the above assumptions, model (4.5)–(4.7) can be rewritten in the following form:

$$\frac{dz_k}{dt} = \varepsilon_k z_k - \frac{1}{b_k} \sum_{l=1}^{m+n} a_{kl} z_k z_l, \qquad k = 1, \ldots, m+n, \qquad (4.13)$$

where

$$z_k = \begin{cases} x_i, & k = i = 1, \ldots, m, \\ y_j, & k - m = j = 1, \ldots, n; \end{cases}$$

$$\varepsilon_k = \begin{cases} F_x^i - m_x^i, & k = i = 1, \ldots, m, \\ -m_y^j, & k - m = j = 1, \ldots, n; \end{cases}$$

$$\frac{1}{b_k} a_{kl} = \begin{cases} \mu_{is}, & k, l = i, s = 1, \ldots, m, \\ \alpha_{ij}, & k = i = 1, \ldots, m, l - m = j = 1, \ldots, n, \\ -K_i^{(j)} \alpha_{ij}, & k - m = j = 1, \ldots, n, \; l = i = 1, \ldots, m, \\ \beta_{jr} & -P_r^{(j)} \beta_{rj} + \nu_{jr}, k - m, l - m = j, r = 1, \ldots, n. \end{cases}$$

System (4.13) (which can be written in the form of (4.4) by means of scaling transformation of variables $N_k = b_k z_k$ and by introducing a matrix of entries $\gamma_{kl} = a_{kl}/(b_k b_l)$) represents the most general form of Lotka–Volterra models. Volterra[2] performed a detailed analysis for more particular cases—with some restrictions on the matrix $[a_{kl}]$. Considering the example of an "n predators–m prey" system we shall try to clear up the biological meaning of those restrictions. The model has the following form:

$$\frac{dz_k}{dt} = z_k \left(\varepsilon_k - \frac{1}{b_k} \sum_{l=1}^{m+n} a_{kl} z_l \right), \qquad k = 1, \ldots, m + n,$$

$$a_{kk} = 0, \quad a_{kl} = -a_{kl}, \quad k \neq l. \quad (4.14)$$

Let all the predators belong to one trophic level, and all the prey to another, with no intra- or interspecies competition among species of the same level. If $z_1, \ldots, z_m$ denote the prey populations and $z_{m+1}, \ldots, z_{m+n}$ denote the predator populations, then

$$\frac{1}{b_k} a_{kl} = \begin{cases} 0, & k, l = 1, \ldots, m \text{ or } k, l = m + 1, \ldots, m + n, \\ a_{kl}, & k = 1, \ldots, m, \quad l = m + 1, \ldots, m + n, \\ -K_l^{(k)} a_{lk}, & k = m + 1, \ldots, m + n, \quad l = 1, \ldots, m. \end{cases}$$

It follows that a restriction of the type $a_{kl} = -a_{lk}$ will hold if, for instance,

$$K_l^{(k)} = K^{(k)}, \quad b_k = \begin{cases} 1, & k = 1, \ldots, m, \\ 1/K^{(k)}, & k = n + 1, \ldots, m + n, \end{cases} \quad (4.15)$$

or if

$$K_l^{(k)} = K_l, \quad b_k = \begin{cases} K_k, & k = 1, \ldots, m, \\ 1, & k = m + 1, \ldots, m + n. \end{cases} \quad (4.15')$$

To interpret conditions (4.15) and (4.15$'$) we turn again to the conservation principles. To this end, consider the case where m nonrenewable substrates (with concentrations x_i) provide food for n species (with biomass densities y_j). If only a fraction, $K_i^{(j)}$, of the consumed resource i is included in species j biomass, then the "consumption-reproduction" balance should satisfy the equations

$$\frac{dx_i}{dt} = -\sum_{j=1}^{n} V_{ij} y_j;$$

$$(i = 1, \ldots, m)$$

$$\frac{dy_j}{dt} = \sum_{i=1}^{m} K_i^{(j)} V_{ij} y_j$$

$$(j = 1, \ldots, n). \tag{4.16}$$

If, in analogy with (4.15′), we have $K_i^{(j)} = K_i$ for any j (strict stoichiometrics), then it follows from (4.16) that

$$\sum_{i=1}^{m} K_i x_i + \sum_{j=1}^{n} y_j = \text{const.} \tag{4.17′}$$

This form of the mass conservation principle implies that only a certain fraction of each resource mass contributes to biomass production of the consuming species (irrespective of what the species is). Everything is predetermined by the energetic and biochemical value of the uptake resource, while species specificity is characterized by the choice of a grazing function.

There can yet be another approach. Let a balanced diet be determined by the set of grazing functions, while species specificity is characterized be the fact that different species spend different fractions of their diets ($K^{(j)}$) for their biomass reproduction. Then the conservation principle takes on the form of

$$\sum_{i=1}^{m} x_i + \sum_{j=1}^{n} y_j / K^{(j)} = \text{const.} \tag{4.17}$$

Hence, both of the cases described above, which may be regarded as the opposite "poles" of idealization, fall into the formalism of Lotka–Volterra models with antisymmetric interactions. It is not very important to judge which of the "poles" is closer to reality; the truth perhaps lies somewhere in between. But what is important is the fact that the Lotka–Volterra description is quite meaningful in biological terms.

III. STABILITY IN LOTKA–VOLTERRA COMMUNITY MODELS

A. Lyapunov Stability of a Feasible Equilibrium

Stability in (4.4.)-type models is usually taken to mean the existence and Lyapunov stability of a positive steady-state solution, $\mathbf{N}^*$, to the system of model equations.

In community models, a *positive*, or *feasible*[12] equilibrium is any equilibrium that belongs to the interior of the positive orthant of the phase space, i.e.,

$$\mathbf{N}^* \in \text{int } \mathbb{R}^n_+ = \{\mathbf{x} \in \mathbb{R}^n : x_i > 0, \quad i = 1, \ldots, n\}, \text{ or } \mathbf{N}^* > \mathbf{0}.$$

The biological sense is obvious: there are all those species conserved in the equilibrium which were initially present in the community as described by the model.

Stability of $\mathbf{N}^*$ is clearly the most stringent of all formal requirements to model trajectories that can be given a proper ecological sense. In other words, the Lyapunov stability of equilibrium $\mathbf{N}^*$ is sufficient for any other, "ecological," kind of stability to take place in the model. On the other hand, there exists a whole class of models, namely, those conservative or dissipative in the Volterra sense, for which the Lyapunov stability of $\mathbf{N}^*$ turns out to be equivalent to the "ecological" stability understood as boundedness from above and below the trajectories $N_i(t)$ for all $i = 1, \ldots, n$.

A positive equilibrium, or steady-state, solution to system (4.4) must satisfy the system of linear algebraic equations whose matrix form is

$$\Gamma \mathbf{N}^* = \varepsilon, \tag{4.18}$$

where $\Gamma = [\gamma_{ij}]$, $\varepsilon = [\varepsilon_1, \ldots, \varepsilon_n]^T$, n is the number of species. Let vector ε be such that all components of the solution to system (4.18) are positive, or in other words, $\mathbf{N}^*$ lies within the positive orthant $\mathbb{R}^n_+$ of the n-dimensional space. In system (4.14), for instance, this implies that at least $\varepsilon_i > 0$ for all the prey species and $\varepsilon_j < 0$ for all the predators. Then Lyapunov stability of $\mathbf{N}^*$ can generally be examined by means of linearization at $\mathbf{N}^*$ (local stability analysis) or searching (not always successful) for a pertinent Lyapunov function (global stability analysis). Both methods are illustrated below by application to systems which were thoroughly investigated by Volterra,[2] with no use of the Lyapunov terminology.

B. Volterra Conservative and Dissipative Systems

When analyzing systems of type (4.4), Volterra[2] distinguished two particular classes, namely, *conservative* and *dissipative* systems.

Definition 4.1 *System (4.4) is termed* conservative *if there exists a set of positive numbers* $\alpha_1, \ldots, \alpha_n$ *such that the following quadratic form:*

$$F(N_1, \ldots, N_n) = \sum_{i,j=1}^n \alpha_i \gamma_{ij} N_i N_j \tag{4.19}$$

is identically zero; if the form $F(N_1, \ldots, N_n)$ *is positive definite, the system is called* dissipative.

Biological implications of these definitions are as follows. If the quantities α_i are identified with the average biomasses of an individual in each species, then the expression

$$V(N_1, \ldots, N_n) = \alpha_1 N_1 + \cdots + \alpha_n N_n$$

will give the total community biomass. From equation (4.4) it follows immediately that the total biomass growth rate is

$$\frac{dV}{dt} = \sum_{i=1}^{n} \alpha_i \varepsilon_i N_i - F(N_1, \ldots, N_n).$$

The first term on the right-hand side represents the effect of species growth or decline patterns (which are constant in this case), and the second term represents the effect of species interactions. Thus, conservative communities are characterized by species interactions exerting no influence on the dynamics of the total community biomass, while, in dissipative systems, these interactions slow down the total biomass growth.

C. Dynamical Properties of Conservative Systems

Since a conservative system should obviously meet the conditions

$$\gamma_{ii} = 0, \quad \alpha_i \gamma_{ij} + \alpha_j \gamma_{ji} = 0, \quad i, j = 1, \ldots, n, \tag{4.20}$$

i.e., γ_{ij} and γ_{ji} must be of opposite signs, it is clear that only those communities can be conservative which have predator-prey interactions as their only type of interspecies relations except neutralism, and lack species self-limiting. Volterra showed[2] that the necessary and sufficient condition for system (4.4) to be conservative is provided by the following combination of requirements on the entries of matrix Γ:

1. all $\gamma_{ii} = 0$,

2. for any $i \neq j$ either $\gamma_{ij} = \gamma_{ji} = 0$, or $\gamma_{ij}\gamma_{ji} < 0$;

3. for any sample of $m \geq 3$ different integers $p, q, r, \ldots, y, z$ from the set $\{1, 2, \ldots, n\}$ the equality holds $\gamma_{pq}\gamma_{qr} \cdots \gamma_{yz}\gamma_{zp} = (-1)^m \gamma_{qp}\gamma_{rq} \cdots \gamma_{zy}\gamma_{pz}$.

Therefore, excluding the case $n = 2$, conservativeness calls for the system parameters to satisfy some relationships of the equality type, which real-life systems are unlikely to meet to because there always exists an uncertainty related to measurement procedures. Conservative communities in mathematical ecology is an idealization similar to systems without friction in mechanics, and just as we often neglect friction in mechanics, the conservative hypothesis may also be accepted for some hypothetical communities.

The basic property of conservative systems (4.4) is the existence of the first integral along trajectories belonging to the positive orthant in case matrix Γ is nonsingular (for even n):

$$\left(\frac{e^{N_1}}{N_1^{q_1}}\right)^{\alpha_1} \left(\frac{e^{N_2}}{N_2^{q_2}}\right)^{\alpha_2} \cdots \left(\frac{e^{N_n}}{N_n^{q_n}}\right)^{\alpha_n} = \text{const}, \tag{4.21}$$

where $\mathbf{q} = [q_1, \ldots, q_n]^T$ is a solution to system (4.18). At $\mathbf{q} = \mathbf{N}^* > \mathbf{0}$, the change of variables $n_i = N_i/N_i^*$ reduces (4.21) to

$$\left(\frac{e^{n_1}}{n_1}\right)^{\alpha_1 N_1^*} \left(\frac{e^{n_1}}{n_2}\right)^{\alpha_2 N_2^*} \cdots \left(\frac{e^{n_n}}{n_n}\right)^{\alpha_n N_n^*} = \text{const}. \tag{4.22}$$

Using these relationships, Volterra[2] showed in particular, that when $\mathbf{q} > \mathbf{0}$, all the species are bounded from above and below by positive constants and, if the initial state $\mathbf{N}(0)$ is distinct from $\mathbf{N}^*$, then at least one population exhibits undamped oscillations. In this case the *asymptotic averages* for variables $N_i(t)$, i.e., the expressions

$$\frac{1}{t - t_0} \int_{t_0}^{t} N_i(\tau)\, d\tau, \qquad i = 1, \ldots, n, \tag{4.23}$$

have limits as $t \to \infty$, which coincide with their steady-state values (the *asymptotic averages law*).

Expressions (4.21) and (4.22) are nothing else than *constants of motion* well-known from theoretical mechanics, the motion proceeding in the phase space of $\mathbb{R}^n$ along closed orbits, which are level sets of the functions on the left.

Example: The above statements can all be illustrated by the classical pair of prey-predator Lotka–Volterra equations, namely,

$$\begin{cases} \dot{N}_1 = N_1(a - bN_2), \\ \dot{N}_2 = N_2(-c + dN_1), \end{cases} \tag{4.24}$$

with coefficients a, b, c, and d being positive and having the biological sense as adopted in (4.4). Interaction matrix $(-\Gamma)$ has a sign skew-symmetric form

$$(-\Gamma) = \begin{bmatrix} 0 & -b \\ d & 0 \end{bmatrix},$$

so that quadratic form (4.19),

$$F(N_1, N_2) = \alpha_1 b N_1 N_2 - \alpha_2 d N_2 N_1,$$

can meet conditions (4.20), hence be identically zero, by a proper choice of α_i, e.g. $\alpha_1 = d$, $\alpha_2 = b$.

The unique feasible equilibrium is obvious:

$$N_1^* = c/d, \quad N_2^* = a/b, \quad \text{or } \mathbf{N}^* = \begin{bmatrix} c/d \\ a/b \end{bmatrix}, \tag{4.25}$$

and traditional manipulation consists of multiplying the first of equations

$$\begin{cases} \dot{N}_1/N_1 = a - bN_2, \\ \dot{N}_2/N_2 = -c + dN_1, \end{cases} \tag{4.26}$$

by $(dN_1 - c)$, the second by $(bN_2 - a)$, and adding them together, which results in

$$\frac{d}{dt} [dN_1 - c \ln N_1 + bN_2 - a \ln N_2] = 0. \tag{4.27}$$

If we denote now

$$G(\mathbf{N}) = d[N_1 - N_1^* \ln N_1] + b[N_2 - N_2^* \ln N_2], \tag{4.28}$$

then it follows from the previous equation that $dG(\mathbf{N}(t))/dt = 0$ everywhere along trajectories of (4.24). On the other hand, integrating in t both sides of (4.27), taking the inverse logarithms, and rearranging them results in

$$\frac{\left(e^{N_1}\right)^d \left(e^{N_2}\right)^b}{N_1^c \; N_2^a} = \text{const}, \tag{4.29}$$

which represents a particular case of (4.22) and gives a practical way to calculate the constant of motion in that case. Using these calculations Volterra plotted his famous ovals—closed orbits in plane (N_1, N_2) concentric to point $\mathbf{N}^*$ (4.25) and corresponding to different initial conditions, hence to different values of the constant (Figure 12).

The *isoclines* of system (4.24), i.e., the lines where the right-hand sides of equations become zero, are straight lines here, with equations

$$N_1 = c/d \quad \text{and} \quad N_2 = a/b,$$

which intersect at $\mathbf{N}^*$ and divide the positive quadrant into four regions where the derivatives have definite signs. Inspecting those signs in consecutive regions (marked 1 through 4 in Figure 12) shows the direction of motion along closed orbits to be counter-clockwise.

Since the orbits are closed, the population sizes oscillate periodically, with both the amplitudes and the period, T, of the oscillations depending on the initial conditions. However, the *cycle averages*, i.e., the quantities

$$\frac{1}{T} \int_0^T N_1(\tau) \, d\tau \quad \text{and} \quad \frac{1}{T} \int_0^T N_2(\tau) \, d\tau,$$

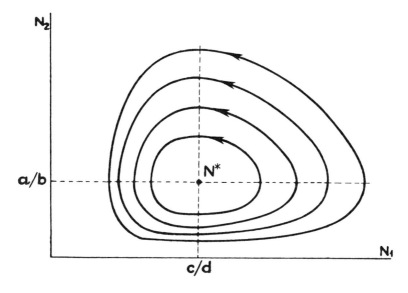

Figure 12.   Phase portrait of a conservative prey-predator model.

remain constant and equal to the corresponding equilibrium values. This can be seen from integrating equations (4.26) over the period; for instance,

$$\int_0^T \frac{d}{dt} N_1(\tau)\, d\tau = \int_0^T (a - b N_2(\tau))\, d\tau,$$

whereby

$$\ln N_1(T) - \ln N_1(0) = aT - b \int_0^T N_2(\tau))\, d\tau,$$

and, since $N_1(T) = N_1(0)$, we have

$$\frac{1}{T} \int_0^T N_2(\tau)\, d\tau = a/b = N_2^*.$$

Derivation for the N_1-average is similar.

In terms of stability, the dynamical pattern shown in Figure 12 is one of a *neutral* type: small variations of initial conditions generate small but nonvanishing changes in the trajectory, which simply shifts into another oval.   □

Technically, the trajectory behavior of a conservative system (4.4) can be also looked at from the viewpoint of the first Lyapunov method to analyze stability of equilibrium $\mathbf{N}^*$. Indeed, it can be shown that the following function:

$$G(\mathbf{N}) = \sum_{i=1}^{n} \alpha_i (N_i - N_i^* \ln N_i) \qquad (4.30)$$

is a Lyapunov function for the point $\mathbf{N}^*$ within the domain $\mathbb{R}^n_+$: the minimal value of $G(\mathbf{N})$ over int $\mathbb{R}^n_+$ is attained at $\mathbf{N}^*$ (see the Appendix, Statement 4.2) and the derivative along system trajectories, $dG(\mathbf{N}(t))/dt$, equals quadratic form F defined in (4.19) (see the Appendix, Statement 4.3), which is zero everywhere in $\mathbb{R}^n_+$. Hence, the equilibrium $\mathbf{N}^*$is Lyapunov stable (but not asymptotically).[13]

The second Lyapunov method, based on system linearization at equilibrium point and the spectral analysis of the ensuing matrix, gives insight into trajectory behavior in a small neighborhood of the equilibrium. Assuming the deviations from equilibrium,

$$x_i = N_i - N_i^*,$$

to be sufficiently small, we come to the linear system

$$\frac{d\mathbf{x}}{dt} = -\text{diag}\{N_1^*, \ldots, N_n^*\}\, \Gamma \mathbf{x}; \tag{4.31}$$

but if we consider deviations in the form of

$$y_i = N_i/N_i^* - 1,$$

then the linear system will take on the form

$$\frac{d\mathbf{y}}{dt} = -\Gamma\, \text{diag}\{N_1^*, \ldots, N_n^*\}\, \mathbf{y}. \tag{4.32}$$

Since a linear change of variables must not change stability, hence the matrix spectrum, in a linear system, the matrices in (4.31) and (4.32) should have equal spectra. This can be shown in formal terms by the observation that matrices $-D\Gamma$ and $-\Gamma D$ are similar, the matrix $D = \text{diag}\{\mathbf{N}^*\}$ being a matrix of their similarity transformation: $D^{-1}(-D\Gamma)D = -\Gamma D$.

The spectrum can be shown to consist of pure imaginary numbers (see Appendix, Statement 4.1), so that the well-known first approximation stability theorem does not apply. Note that the spectrum of matrix Γ itself also consists of pure imaginary numbers, which provides one more necessary condition for system (4.4) to be conservative.

Thus the linear approximation of trajectories in the vicinity of equilibrium $\mathbf{N}^*$ is a superposition of $n/2$ sine curves with generally different periods.

This is a typical example illustrating the potentialities of both Lyapunov's methods for a pure imaginary spectrum: direct linearization gives no definite answer to whether or not $\mathbf{N}^*$ is stable or unstable, which can only be established by examination of an appropriate Lyapunov function.

If there is one or more nonpositive components q_i in the solution to system (4.18), then after Volterra[2] we can state that one or more species either vanish or grow infinitely. Otherwise, i.e., if all the populations were bounded above and below by positive constants, there would exist the positive asymptotic means for all $N_i(t)$, which can be shown always to equal the corresponding roots q_i of system (4.18). This contradiction proves the statement.

Thus the existence of a positive equilibrium serves not only as a sufficient but also as a necessary condition for a conservative community to be stable.

Volterra used the pattern of phase trajectories in his predator-prey model to explain a particular phenomenon observed in fishery practice after the First World War.[2,11] But a general moral can also be gained from the model that an oscillation behavior of an ecological system can be caused not only by fluctuations in the environmental conditions, but also by dynamical laws intrinsic to the system in the constant environment.

Note, however, that conservative systems, treated as a "frictionless" idealization of population models, has only a limited ability to explain practically observed effects. A "dynamical" reason is that Volterra cycles lack *robustness*: any small variation in initial conditions (except those moving along the orbit) switches the system from one cycle to another, exhibiting the *neutral* type of stability. Clearly, models to explain reality should possess the property of being *robust*, retaining their dynamical behavior under, at least small, variations in initial states, parameters, etc., for otherwise they cannot give a definite answer to a practical question. But from the theoretical point of view, this drawback of conservative models should be considered as an advantage: for better understanding of the effect that a particular phenomenon exerts on stability or other features of the dynamical behavior, we should better begin the study by including the phenomenon into a neutral-case model. This can be seen in the next section.

D. Dynamical Properties of Dissipative Systems

A necessary condition for a system to be dissipative follows immediately from the definition and the fact that the diagonal entries of a positive definite matrix must be positive:

$$\gamma_{ii} > 0, \qquad i = 1, 2, \ldots, n, \tag{4.33}$$

i.e., all species are to be self-limited. If we recall that in conservative systems all $\gamma_{ii} = 0$, it becomes clear that the definitions for conservative and dissipative systems are totally incompatible. If, furthermore, we include any set of n positive diagonal entries (4.33) into matrix Γ of a conservative system, then it can be easily shown by definition that the result will be a dissipative system.

It is also clear that there may be systems which are neither conservative nor dissipative, say, systems in which only some, but not all, species feature self-limitation.

Another necessary condition of dissipativeness is that $\det[\gamma_{ij}] \neq 0$, thus implying uniqueness of the solution to system (4.18). Furthermore, all the principal minors (i.e., those symmetric about the main diagonal) of matrix $\Gamma = [\gamma_{ij}]$ should be positive.[2]

For a dissipative system (4.4), similarly to (4.22), Volterra[2] derived the relationship

$$\prod_{r=1}^{n} \left\{ \frac{1}{n_r} \exp\left(\frac{|q_r|}{q_r} n_r \right) \right\} \alpha_r q_r$$
$$= C \exp\left\{ -\int_{t_0}^{t} F\left(\ldots \left(1 - \frac{|q_r|}{q_r} n_r \right) q_r, \ldots \right) d\tau \right\}, \quad (4.34)$$

where $\mathbf{q} = [q_1, \ldots, q_n]^T$ is a solution to system (4.18) with no zero components, $n_r = N_r / |q_r|$, $C = \text{const} > 0$, and the quadratic form $F[\ldots]$ is positive definite. If $\mathbf{q} = \mathbf{N}^* > \mathbf{0}$, then expression (4.34) reduces to

$$\left(\frac{e^{n_1}}{n_1} \right)^{\alpha_1 N_1^*} \ldots \left(\frac{e^{n_n}}{n_n} \right)^{\alpha_n N_n^*} = C \exp\left\{ -\int_{t_0}^{t} F\left(\ldots, (1 - n_r) N_r^*, \ldots \right) d\tau \right\}. \quad (4.35)$$

Using this relationship, Volterra proved that as $t \to \infty$ the state $\mathbf{N}^* > \mathbf{0}$ is the limiting state for all trajectories of (4.4) initiating at positive values of $\mathbf{N}(0)$.

So any trajectory, whenever starting in int $\mathbb{R}^n_+$, converges eventually to the steady state. The result is quite interpretable, though almost trivial: self-limiting in each individual species brings about stabilization of the whole community.

As to the case where some q_r are negative, if there is a single negative component, then the corresponding species becomes extinct; if there are several negative components, then at least one of the species must vanish in the sense that

$$\underline{\lim}_{t\to\infty} N_r(t) = \lim_{t\to\infty} \left[\inf_{\tau \leq t} N_r(\tau) \right] = 0. \quad (4.36)$$

Example: The nearest example of a dissipative system is a Lotka–Volterra prey-predator model which now includes self-limitation effects in both prey and predator populations. These modify equations (4.24) by adding self-limitation terms eN_1 and fN_2 as follows:

$$\begin{cases} \dot{N}_1 = N_1(a - eN_1 - bN_2), \\ \dot{N}_2 = N_2(-c + dN_1 - fN_2), \end{cases} \quad e, f > 0. \quad (4.37)$$

Interaction matrix $(-\Gamma)$ takes now the form of

$$(-\Gamma) = \begin{bmatrix} -e & -b \\ d & -f \end{bmatrix},$$

but the same choice of α_i as before makes the quadratic form

$$F(N_1, N_2) = \alpha_1 e N_1^2 + \alpha_2 f N_2^2 + (\alpha_2 d - \alpha_1 b) N_1 N_2$$

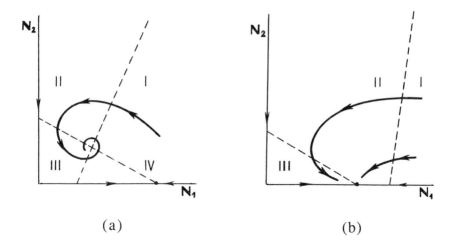

(a)   (b)

Figure 13.   Phase portraits of dissipative prey-predator models.

positive definite. A unique feasible equilibrium, if it exists, can easily be found and traditional manipulating would result in the proper particular case of relationship (4.35). Whether the equilibrium exists is equivalent to whether the isoclines, i.e., the straight lines of equations

$$a = eN_1 - bN_2 \quad \text{and} \quad c = dN_1 - fN_2,$$

intersect in int $\mathbb{R}_+^2$.

Depending on the parameters $a, \ldots, f$, this may or may not occur in particular cases of (4.37), corresponding to two major types of phase patterns (shown in Figure 13). Both of them illustrate general statements derived from (4.35), the one in Figure 13a illustrating also the general idea of global stability.

Note that in conservative system (4.24) a feasible equilibrium does always exist. Hence it also exists in the dissipative system (4.37) if the parameters e, f are small enough. But if they are even arbitrarily small, the phase pattern will neverthe-less change from the neutral ovals of Figure 12 to the globally stable spirals of Figure 13a. □

Similar to the conservative case, a study of trajectory behavior near $\mathbf{N}^*$ in the linear approximation relies upon evaluating the spectrum of matrix $-D\Gamma$ (or $-\Gamma D$, which is the same), where $D = \text{diag}\{N_1^*, \ldots, N_n^*\}$. All the eigenvalues can similarly be shown to have negative real parts. (The same is true for matrix Γ which gives another necessary condition for system (4.4) to be dissipative.) Hence, by the first approximation stability theorem, the equilibrium $\mathbf{N}^*$ is locally asymptotically stable, and the way the perturbed trajectories approach $\mathbf{N}^*$ depends on whether or not there are complex numbers in the spectrum of $(-D\Gamma)$.

For function $G(\mathbf{N})$ introduced in (4.30), by Statement 4.3 of the Appendix, we have, along trajectories of system (4.4),

$$\frac{d}{dt}G[\mathbf{N}(t)] = -F(\ldots,N_r(t) - N_r^*,\ldots) \leq 0,$$

since the quadratic form is now positive definite. More rigorously, $dG[\mathbf{N}(t)]/dt < 0$ everywhere within $\mathbb{R}_+^n$, and $dG[\mathbf{N}(t)]/dt = 0$ only when $\mathbf{N}(t) = \mathbf{N}^*$. This implies that $G(\mathbf{N})$ is a Lyapunov function for equilibrium $\mathbf{N}^*$ in the interior of $\mathbb{R}_+^n$. Therefore, $\mathbf{N}^*$ is globally asymptotically stable within the positive orthant, thus explaining the general pattern of trajectory behavior in the dissipative system, derived by Volterra from other considerations.

Volterra did not consider the case of there being zero components among those of the solution to (4.18). In that case, the linearization approach does not lead to a definite conclusion on stability of a rest point $\mathbf{N}^\nabla$ lying on the boundary of $\mathbb{R}_+^n$ (there are zero values in the spectrum of $-D\Gamma$). But we can consider for such points, e.g., for point $\mathbf{N}^\nabla = [0, N_2^\nabla, \ldots, N_n^\nabla]$, the following function:

$$G_1(\mathbf{N}) = \alpha_1 N_1 + \sum_{i=2}^{n} \alpha_i(N_i - N_i^\nabla \ln N_i) \tag{4.38}$$

as a Lyapunov function. Indeed, the minimum is attained at $\mathbf{N} = \mathbf{N}^\nabla$ and the derivative in virtue of the system can be proved, by analogy with Statement 4.3 of the Appendix, to be nonpositive as before:

$$\frac{d}{dt}G_1[\mathbf{N}(t)] = F(N_1^\nabla(t), N_2^\nabla(t) - N_2^\nabla, \ldots, N_n^\nabla(t) - N_n^\nabla) \leq 0. \tag{4.39}$$

Since the equality sign holds only at $\mathbf{N}(t) = \mathbf{N}^\nabla$, the rest point is asymptotically stable, globally in $\mathbb{R}_+^n$, i.e., species 1 becomes extinct from the community under any initial conditions.

Therefore, it follows from Volterra's findings, complemented with the interpretation in terms of Lyapunov methods, that for dissipative systems, as well as for conservative ones, the existence of stable equilibrium $\mathbf{N}^* > \mathbf{0}$ serves as not only a sufficient condition but also as a necessary one for stable dynamics of the model community.

The major point in which dissipative systems apparently differ from conservative systems is the negativity of all entries in the principal diagonal of the community matrix, interpreted as self-limitation effects in all populations. This, as we see, changes drastically the pattern of dynamic behavior: from neutral stability in the conservative case to equilibrium global stabilization in the dissipative system. The fact is not surprising and is a good reason to regard dissipative Lotka–Volterra models, particularly those used to describe species of a single trophic level, as multispecies analogs to the logistic growth of a single population. In this respect it is interesting to note that in further generalizations of Lotka–Volterra models,[14–15] the most meaningful results of stability studies were obtained for cases similar to dissipativeness in the Volterra sense.

IV. STABLE MATRICES AND GLOBALLY STABLE EQUILIBRIA: WHAT LIES IN BETWEEN?

When nothing is a priori known about the properties of the community matrix, at least local stability analysis can be implemented in system (4.4) (or any other one admitting linearization) to verify the stability of a feasible equilibrium. A *stable* community matrix A (in the sense that all its eigenvalues have negative real parts) brings about locally stable equilibrium, while an *unstable* matrix (which has an eigenvalue with a positive real part) indicates instability of the equilibrium. Further investigation of the community matrix may reveal a hierarchy in asymptotic properties of model trajectories, up to their global stability in case of dissipative system (4.4). What can the levels be in that hierarchy is considered in the remainder of this chapter.

A. *D*-stability of Matrices

As shown in the previous section, the linearization matrix of system (4.4) assumes the following form:

$$A = D(-\Gamma),\qquad(4.40)$$

where $D = \text{diag}\{N_1^*,\ldots,N_n^*\}$ and $\Gamma = [\gamma_{ij}]$ is the interaction matrix of a Lotka–Volterra model. Note that entries of the matrix depend both on the pattern of interactions among (and within) species and on equilibrium values of all population sizes, N_i^*. These, if matrix Γ is given and is nonsingular, are completely determined by parameters ε_i, i.e., those which are intrinsic to individual species outside of any interactions. According to (4.40) both the pattern of species interactions and a set of individual species characteristics contribute to the equilibrium stability. With a stable community matrix, we can state that once an equilibrium exists it is (at least locally) asymptotically stable.

But can we say anything of matrix Γ if stability is required to be an outcome of species relations alone, while being independent of the intrinsic rates of the species natural increase or decrease? In other words, what should matrix Γ be in order that stability of $\mathbf{N}^*$ might be determined by the matrix properties only and still be retained under any pertinent values of ε_i?

Since a proper choice of ε_i ($i = 1,\ldots,n$) can always (for a nonsingular Γ) guarantee any, a priori given, values of N_i^* (as system $\Gamma^{-1}\varepsilon = \mathbf{N}^*$ has a unique solution in ε), the above-stated requirement means that matrix Γ has to be stable with any values of $N_i^* > 0$. In other words, matrix Γ has to retain its stability when multiplied from the left by any diagonal matrix D of positive entries in its main diagonal. In mathematical economics such a matrix property is termed *D-stability*.[16–17]

Definition 4.2 *An $n \times n$ matrix A is called D-stable if matrix DA is stable in the sense that* Re $\lambda_i(DA) < 0$ $(i = 1, \ldots, n)$ *with any diagonal matrix D of set $\mathbb{D}_n^+$ (the set of all diagonal $n \times n$ matrices with positive entries in their main diagonals).*

The notion of *D*-stability makes sense, since there might be quite stable matrices lacking *D*-stability. For example, for a 2×2 matrix $A = [a_{ij}]$ to be stable, it is necessary that $a_{11} + a_{22} < 0$; clearly, when a_{11} and a_{22} have opposite signs, then there exist matrices $D = \text{diag}\{d_1, d_2\}$ such that $d_1 a_{11} + d_2 a_{22} > 0$, so that *DA* lacks stability.

An important property of *D*-stable matrices ensues immediately from the definition: if *A* is *D*-stable, then so also is *BA* for any positive diagonal matrix *B*, that is, the class of *D*-stable matrices of a given size is closed under multiplication by a positive diagonal matrix. This means that once a community matrix turns out *D*-stable, so too is the interaction matrix $(-\Gamma)$. Therefore, interpretation of *D*-stability in terms of Lotka–Volterra models is clear: *D*-stable matrices provide, at least local, stability of a feasible equilibrium irrespectively of what are the (positive) equilibrium sizes of the populations.

Any matrix $\mathcal{D} \in \mathbb{D}_n^+$ is nonsingular, so that matrix $A\mathcal{D} = \mathcal{D}^{-1}(\mathcal{D}A)\mathcal{D}$ is similar to, hence being stable simultaneously with, matrix $\mathcal{D}A$. An equivalent definition of *D*-stability may therefore require multiplication by the diagonal matrix on the right. Since, furthermore, the transpose of any matrix has the same spectrum as the matrix, while the inverse matrix has the reciprocal spectrum, all three matrices are well-known to be stable simultaneously, and can be easily shown to be *D*-stable simultaneously too (see, e.g., Svirezhev and Logofet[7]).

Important in applied fields, *D*-stable matrices were also intensively studied in matrix theory. For instance, several sufficient conditions for a matrix to be *D*-stable are known.[18–20] Most of those sufficient conditions turn out however to be some particular implications of a general hierarchy of matrix stability notions we will consider below. The following are two examples of known *D*-stability conditions.

Necessary Condition 4.1. *If matrix A is D-stable, then all its odd-order minors are nonpositive, while all even-order minors are nonnegative.*[21]

A necessary condition may help practically by making the investigator abandon any effort to prove a property once he (or she) has found the condition broken. The above 2×2 matrix is not *D*-stable just because one of its first-order minors (a_{11} or a_{22}) has the wrong sign.

Sufficient Condition 4.2. *If there exists a positive diagonal matrix B such that matrix $BA + A^T B$ is positive definite, then matrix A is D-stable.*[18]

D-stable is, for instance, a (symmetric) positive definite matrix *A*, since one can take matrix $I/2$ as matrix *B* of Condition 4.2. Furthermore, *D*-stable is also matrix $(-\Gamma)$ of any dissipative system (4.4). Indeed, by definition of dissipativeness, there exists matrix $\mathcal{A} = \text{diag}\{\alpha_1, \ldots, \alpha_n\}$ with all positive α_is such that quadratic form

$$F(\mathbf{x}) = \langle \mathcal{A}\mathbf{x}, \mathbf{x} \rangle \tag{4.41}$$

is positive definite. This, in turn, is obviously equivalent to (symmetric) matrix

$$\frac{1}{2}\left[\mathcal{A}\Gamma + (\mathcal{A}\Gamma)^T\right] = \frac{1}{2}\mathcal{A}\Gamma + \Gamma^T \frac{1}{2}\mathcal{A}$$

being positive definite, whereby it follows that matrix $\mathcal{A}/2$ can be taken as B for matrix $(-\Gamma)$, i.e., the sufficient condition of D-stability is satisfied for $(-\Gamma)$.

The fact that dissipative systems do not exhaust the class of D-stable matrices can be easily verified by an example of a second-order matrix $(-\Gamma)$ whose sign pattern is

$$S = \begin{bmatrix} - & - \\ + & 0 \end{bmatrix},$$

corresponding to a prey-predator pair with self-limitation among prey. It is D-stable by definition but cannot be dissipative due to a zero entry in the main diagonal (see condition (4.33)). Any feasible equilibrium will, therefore, be (asymptotically) stable at least locally in such models, while the domain of its global stability needs further investigation, which normally relies upon a particular form of equations.

Note that the set of D-stable matrices is not open in any natural topology of the space of real $n \times n$ matrices. To verify this, consider the following

Example: Let $n \times n$ matrix A have the form of

$$A = E \oplus (-I_{n-2}), \tag{4.42}$$

where

$$E = \begin{bmatrix} 0 & -1 \\ 1 & -2 \end{bmatrix},$$

$\oplus$ designates the *direct* (i.e., block-diagonal) sum of matrices, and I_{n-2} is the identity matrix of size $(n - 2) \times (n - 2)$. Matrix A is obviously D-stable by definition, but in any small vicinity within it there always exists a matrix of the form $A_\varepsilon = E_\varepsilon \oplus (-I_{n-2})$ where

$$E_\varepsilon = \begin{bmatrix} \varepsilon & -1 \\ 1 & -2 \end{bmatrix}$$

($\varepsilon > 0$ and is sufficiently small), which cannot be D-stable because of the positive diagonal entry.   □

B. Totally Stable Matrices

An important subclass of D-stable matrices, the so-called *totally stable matrices*, was also studied first in mathematical economics.[21]

Definition 4.3 *A matrix A is called* totally stable *if any of its principal submatrices (of any order) is D-stable too.*

As follows directly from the definition, any principal submatrix of a totally stable matrix is totally stable too. Therefore, total stability can be interpreted in the following terms: a community that remains after any group of species has been excluded from the initial composition, still retains the total stability. This apparently relates to the so-called "species-deletion" stability concept,[22] although the latter implies just stability of the remaining species composition. As a mathematical concept, however, the total stability is quite logical, requiring that any reduction in dimensionality by species deletion should not affect the stability property of the community matrix.

Dealing with any particular system (4.4), we still have to be concerned about whether "any remaining species composition" can emerge in the model as a feasible solution of the reduced system. First, it makes sense to speak of the "reduced system" properties as the restriction of a Lotka–Volterra system to any coordinate subspace is again of the Lotka–Volterra type. Then, the principal property of a totally stable matrix $(-\Gamma)$, as proved by Hofbauer and Zigmund,[23] Theorem 21.6, is that, for any set of intrinsic growth rates ε_i, system (4.4) has exactly one *saturated* rest point, either in the interior or on a boundary face of $\mathbb{R}_+^n$, and this rest point is asymptotically stable within its face. "Saturation" here means mathematically that zero components of the rest point have nonpositive growth rates $(dN_i/dt \le 0)$, and biologically, that a small number of migrants of any missing species can not grow at this composition.

Any totally stable matrix is *D*-stable by definition (the matrix is trivially a submatrix of its own). This subclass of *D*-stable matrices is also closed under transposition or multiplication by a positive diagonal matrix. Thus, total stability is also a property that holds simultaneously for both the community and interaction matrices.

Logically, a necessary condition of total stability should be more stringent than that of *D*-stability.

Necessary Condition 4.2. *For a matrix A to be totally stable it is necessary that all its principal minors of odd orders be negative and those of even orders be positive.*

The condition follows from the fact that all principal minors are themselves determinants of *D*-stable matrices. Formulated for matrix $(-A)$, this condition will require all the principal minors to be positive (the matrix is then said to be a *P-matrix*). Thus, for a totally stable community matrix it is necessary that its matrix Γ of interaction coefficients be a *P*-matrix.

Topological considerations have revealed an even more subtle implication. It turns out that the set of totally stable matrices is an open set in a topology of the $n \times n$ matrix space and, moreover, any *interior point* of the set of *D*-stable matrices (i.e., a point entering the set together with its nonempty vicinity) represents a totally stable matrix.[20] In other words, totally stable matrices represent the topological interior to the set of *D*-stable matrices. A practical conclusion is that, in many cases excepting exotic ones like example (4.42), a D-stable matrix we deal with will be also totally stable.

It is interesting to see the relation between total stability of a matrix $(-\Gamma)$ and dissipativeness of system (4.4) with the matrix Γ. Popular examples of typical Lotka–Volterra models reveal no certainty about the issue, which is considered in the next section.

As concerns the characterization of matrix stability properties in terms of matrix entries (hereafter just *characterization*), it is quite simple for $n = 2$, but already the case of $n = 3$ poses certain problems, and generally there exist quite rare examples where the characterization problem is solved at the general level n.

C. From Volterra Dissipative Systems to Dissipative Matrices

We notice that, in spite of the definitions having been given for systems of equations, the properties of being conservative or dissipative are in essence the properties of the corresponding matrices Γ. They are given below with the definitions referring directly to the matrices.

Definition 4.4 *A matrix A is called* dissipative *in the Volterra sense, or just* dissipative *if there exists a positive diagonal matrix $\mathcal{A}$ such that its quadratic form $\langle \mathcal{A}A\mathbf{x}, \mathbf{x} \rangle$ is negative definite.*

Matrix A being *conservative* is defined in a similar way.

Equivalent definitions were proposed elsewhere in the following, somewhat different, form.

Definition 4.4′. *A matrix A is called* diagonally stable,[24] Volterra–Lyapunov stable,[23,25] *or even* positively D-dissipative[15,26] *if there exists a positive diagonal matrix $\mathcal{A}$ $(\mathcal{A} \in \mathbb{D}_n^+)$ such that matrix $\mathcal{A}A + A^T\mathcal{A}$ is negative definite.*

In what follows we still use the Volterra original term *dissipative* matrix. As we have seen, it brings about global (in $\mathrm{int}\mathbb{R}_+^n$) asymptotic stability of a feasible equilibrium. Moreover, for any set of intrinsic growth rates ε_i system (4.4) with a dissipative matrix $(-\Gamma)$ is proved to have exactly one globally stable rest point, either in the interior or on a boundary face of $\mathbb{R}_+^n$ (see, e.g., Hofbauer and Zigmund,[23] Theorem 21.3). We can say that statements like "once an equilibrium exists, it is globally stable" are characteristic of dissipative systems.

It can be shown that dissipative matrices are again closed under multiplication by $\mathcal{D} \in \mathbb{D}_n^+$, transposition, or taking the inverse. Indeed, if there exists a matrix $\mathcal{A}$ which by definition provides for a given matrix A being dissipative, then matrix $\mathcal{A}\mathcal{D}^{-1} \in \mathbb{D}_n^+$ will do so for matrix $\mathcal{D}A$. By well-known properties of the inner product, the negative definite quadratic form can be written as

$$\langle A\mathcal{A}\mathbf{x}, \mathbf{x} \rangle = \langle \mathbf{x}, \mathcal{A}A^T\mathbf{x} \rangle = \langle \mathcal{A}A^T\mathbf{x}, \mathbf{x} \rangle,$$

which proves A^T to be dissipative. Then there exists matrix $\mathcal{B} \in \mathbb{D}_n^+$ such that

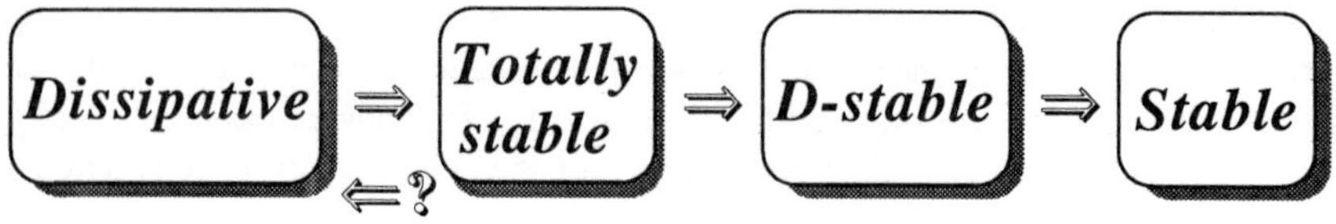

Figure 14. A simple hierarchy among several notions of matrix stability.

negative definite is the quadratic form $\langle A^T B \mathbf{y}, \mathbf{y} \rangle$, which is transformed, by letting $\mathbf{y} = A^{-1}\mathbf{x}$, into

$$\langle A^T B(A^{-1}\mathbf{x}), A^{-1}\mathbf{x} \rangle = \langle A^{-T}A^T B(A^{-1}\mathbf{x}), \mathbf{x} \rangle = \langle BA^{-1}\mathbf{x}, \mathbf{x} \rangle,$$

verifying the dissipativeness of A^{-1}.

If, in a sign definite quadratic form, some variables are always set to zero, the form will still be sign definite with respect to the rest of the variables. Therefore, any principal submatrix A' of a dissipative matrix A is dissipative too; hence, it is also D-stable, whereby we conclude that any dissipative matrix is totally stable.

All the above conclusions can now be summarized in the diagram shown in Figure 14. The above-mentioned examples assure that the two implications on the right are irreversible. Whether or not the first implication on the left is irreversible is quite far from being evident. For example, it can be easily shown that total stability of a general 2×2 matrix is equivalent to its being dissipative. But whether the sets of dissipative and totally stable matrices coincide in a general case of $n \geq 3$ was for some times unclear.[7,24] To clarify the point, if only for 3×3 matrices, the exact characterizations of dissipative and totally stable matrices are of crucial importance. These are given below.

D. Characterizations for 3 × 3 Matrices

To verify whether a particular matrix is D-stable, totally stable, or dissipative we need a constructive criterion, or *characterization*, for each of these properties expressed in terms of matrix entries. While these characterizations are not known in the general case, they are trivial for the case of $n = 2$ but quite substantive already for $n = 3$. Three theorems of this section give the characterizations respectively for D-stability, total stability, and dissipativeness of a general matrix of size 3×3.

What should be a "general" form of a 3×3 matrix is also nontrivial in this context. As noted above, matrix $(-A)$ being a *P-matrix* (i.e., having all principal minors positive) is a necessary condition for total stability or dissipativeness of matrix A, and matrix $(-A)$ being, say, a P_0-*matrix* (i.e., having all principal minors nonnegative with at least one minor in each order being positive) is a necessary condition for D-stability of matrix A. It makes sense therefore to consider only matrices with negative and nonpositive principal diagonals, respectively.

Since, furthermore, multiplication by a positive diagonal matrix cannot change any of these stability categories, we can always reduce, with no loss of generality, a *P-matrix* (respectively a P_0-*matrix*) to that of the unitary principal diagonal (respectively to units instead of nonzero diagonal entries). Also, no one simultaneous permutation of matrix lines and the same columns (i.e., *permutation of rows*) can change a stability category, so that any one particular entry, say, a_{12} can be fixed nonzero. If, indeed, $a_{12} = 0$, then the permutation of matrix rows $1 \Leftrightarrow i$, $2 \Leftrightarrow j$ can always transfer a nonzero off-diagonal element a_{ij} into the position of a_{12}.

So, a special form of the 3×3 matrix occurring in the formulations below is actually the general one for a problem under consideration. Notation $d(-A)$ stands for the determinant and $E_2(-A)$ for the sum of principal second-order minors of matrix $(-A)$.

Theorem 4.1 *A matrix*

$$A = - \begin{bmatrix} \delta_1 & a & b \\ c & \delta_2 & d \\ e & f & \delta_3 \end{bmatrix}, \qquad \delta_i = \begin{Bmatrix} 0 \\ 1 \end{Bmatrix} \quad (i = 1, 2, 3),$$

is D-stable if and only $(-A)$ is a P_0-matrix and the following condition holds true:

$$\left[\sqrt{\delta_1(\delta_2\delta_3 - df)} + \sqrt{\delta_2(\delta_1\delta_3 - be)} + \sqrt{\delta_3(\delta_1\delta_2 - ac)} \right]^2 \geq d(-A),$$

with equality implying that at least one of the products under radical signs has exactly one co-factor equal to zero.

The proof is given by Cain.[27]

Theorem 4.2 [28] *A matrix*

$$A = - \begin{bmatrix} 1 & a & b \\ c & 1 & d \\ e & f & 1 \end{bmatrix}, \qquad a \neq 0, \qquad (4.43)$$

is totally stable if and only if $(-A)$ is a P-matrix and the following condition holds:

$$d(-A) - E_2(-A) \equiv ade + bcf - 2 \qquad (4.44)$$

$$< 2 \left[\sqrt{(1 - ac)(1 - be)} + \sqrt{(1 - ac)(1 - df)} + \sqrt{(1 - be)(1 - df)} \right].$$

The proof is given in the Appendix.

Theorem 4.3 [28]   *A matrix A of the form (4.43) is dissipative if and only if* $(-A)$ *is a P-matrix and the following condition holds true:*

$$d(-A) + E_2(-A) > \inf_{\substack{x,y>0 \\ x \in I_2(a,c)}} \left\{ Q(x,y) = (a^2 - abf)x + (c^2 - cde)/x \right. \tag{4.45}$$

$$\left. + (d^2 - bed)y + (f^2 - aef)/y + (b^2 - abd)xy + (e^2 - cef)/xy \right\},$$

where $d(-A)$ *is the determinant and* $E_2(-A)$ *the sum of principal second-order minors of matrix* $(-A)$, *while*

$$I_2(a,c) = \{x > 0 : (1 - \sqrt{1 - ac})^2 < a^2 x < (1 + \sqrt{1 - ac})^2\}. \tag{4.46}$$

The proof is given in the Appendix.

Now, to find an example of a totally stable but nondissipative matrix means to find a *P*-matrix $(-A)$ which satisfies condition (4.44) but breaks condition (4.45). Below is one of such matrices

$$A = - \begin{bmatrix} 1 & 1 & 2 \\ 0.95 & 1 & 1.92 \\ 0.49 & 0.495 & 1 \end{bmatrix}, \tag{4.47}$$

that can be interpreted as a community matrix for 3 species competing for a resource in common under prescribed ratios among coefficients of intra- and interspecies competition.

Since the conditions of Theorems 4.2 and 4.3 represent strict inequalities with some algebraic expressions about matrix entries, they describe some open sets in the linear space of 3×3 matrices: small enough variations of any matrix element will neither disturb the conditions, nor will they annul the property of being a *P*-matrix. (The set of dissipative $n \times n$ matrices is also proved to be open in the topology of $n \times n$ matrix space.[20]) In particular, there exists a (open) vicinity of matrix A (4.47), say, $\mathbb{U}_1(A)$, that belongs entirely to the set of totally stable matrices. On the other hand, there also must exist a vicinity of A, say, $\mathbb{U}_2(A)$, consisting entirely of matrices that break, together with A, condition (4.45), therefore escaping the set of dissipative 3×3 matrices. Intersection of both vicinities, $\mathbb{U}_1(A) \cap \mathbb{U}_2(A)$, which is nonempty once containing at least matrix A, gives an open set of totally stable but nondissipative matrices.

Thus, the first implication in Figure 14 is also irreversible as well as the further ones, the set of totally stable matrices being topologically wider than the set of dissipative matrices. The diagram in Figure 14 shows a true hierarchy of stability properties in community matrices. In terms of Lotka–Volterra systems (4.4), it displays attenuation in the stability behavior of trajectories as described in the preceding sections.

E. Normal Matrices

There exists however the whole class of matrices within which all the above stability properties turn out equivalent. Those are the so-called *normal* matrices, for which $AA^* = A^*A$, with A^* denoting the *Hermitian adjoint* (i.e., transpose and complex conjugate) of A ($AA^T = A^TA$ in the case of real-valued matrices). Normal are, for example, symmetric and skew-symmetric matrices, interpreted in community terms, respectively, as the competition matrix for mutually equal effects of competitors and the prey-predator interaction matrix under the so-called "hypothesis of encounters and equivalents".[2,11]

The proof of the equivalence is based upon the fundamental fact of linear algebra that any normal matrix A can be reduced to the diagonal form by a unitary similarity transformation:

$$A = U^* \operatorname{diag}\{\lambda_1, \ldots, \lambda_n\} U, \qquad (4.48)$$

where $U^* = U^{-1}$, and $\lambda_1, \ldots, \lambda_n$ are the eigenvalues of A with their complex conjugates $\overline{\lambda_1}, \ldots, \overline{\lambda_n}$.[29-30] Then, for a real matrix A,

$$A^T = A^* = U^* \operatorname{diag}\{\overline{\lambda_1}, \ldots, \overline{\lambda_n}\} U,$$

whereby

$$(A + A^T)/2 = U^* \operatorname{diag}\{\operatorname{Re} \lambda_1, \ldots, \operatorname{Re} \lambda_n\} U. \qquad (4.49)$$

Equality (4.49) means that real parts of eigenvalues of a normal matrix A are equal to the eigenvalues of its "symmetric part" $(A + A^T)/2$. If a normal matrix A is stable, then matrix $A+A^T$ is negative definite. Therefore, A is dissipative, reversing all implications in Figure 14. ■

Thus, for normal community matrices, local stability of equilibrium $\mathbf{N}^* > \mathbf{0}$ already guarantees its global stability too. Whether we alter the equilibrium sizes of populations by 5% or remove 99% of all populations, the community will nevertheless restore itself eventually close to the equilibrium composition. The reader may decide whether this is "normal" for his (or her) own view of community stability.

But in the general case, between stable matrices and globally stable equilibria, there is a hierarchy of quite distinguishable subsets of stable matrices with their own types of asymptotic behavior.

V. SELF-LIMITATION EFFECTS VERSUS SPECIES INTERACTIONS IN COMMUNITY STABILITY

Stability studies in theoretical models of population dynamics, either in the linear approximation, or by means of Lyapunov functions similar to (4.30), are very often reduced to the issue of whether the community matrix is dissipative.[7,15,31-32] Due to the uniqueness and global stability of equilibrium, Lotka–Volterra dissipative systems can be regarded as multidimensional analogs to the logistic dynamics of a single population, where exponential growth is stabilized by intraspecies regulation, i.e., self-limiting by the population size or density. This may perhaps explain why sufficient conditions of dissipativeness have been mostly searched for and interpreted in terms of intraspecies regulation predominating over the interactions among species. These kinds of formulations are especially popular in models of competition communities,[9,33–35] in studying the so-called connective stability,[7,36–37] and in stochastic versions of population dynamics equations.[38] From the results of the present section, however, it follows that there may be quite another mechanism to provide the same kind of global stabilization.

A. Diagonal Dominance and *M*-Matrices

In matrix theory[20,24] the problem of constructive sufficient conditions for a matrix to be dissipative was reduced mainly to the notion of *diagonal quasi-dominance*[39] or *positive dominating diagonal.*[40]

Definition 4.5 *A real $n \times n$ matrix $A = [a_{ij}]$ is called* diagonally quasi-dominant *(formally, $A \in q\mathbb{D}$) if there exists a set of n positive numbers $\pi_1, \pi_2, \ldots, \pi_n$ such that the condition*

$$\pi_i a_{ii} > \sum_{j \neq i}^{n} \pi_j |a_{ij}| \qquad (4.50)$$

holds for every $i = 1, 2, \ldots, n$. In particular, when all $\pi_i = 1$, the matrix is just diagonally dominant.

Speaking of stable matrices, we should restrict the definition only to matrices without zero entries in the main diagonal; otherwise, the whole lines could consist of zeros, depriving the matrix of its stability.

Interpretation of diagonal dominance in terms of the community matrix $(-A)$ is obvious: the intraspecies pressure in each species exceeds the total impact from all other species on it. To verify this fact in practical cases is a matter of simple inspection of the community matrix. It is one of a few rare cases where the definition of matrix property also gives its characterization in terms of matrix elements.

It can be shown (see below) that if A is quasi-dominant ($A \in q\mathbb{D}$), then so too is its transpose, i.e., the self-regulation effect of each species exceeds the total impact the species exerts upon all others. This simplifies the interpretation of diagonal dominance by reducing it merely to the intraspecies regulation prevailing over the interactions among species.

It is known[41] that quasi-dominance of a matrix A implies dissipativeness of the matrix $(-A)$, so that quasi-dominant matrices (more precisely, their negations) constitute a subclass of the dissipative matrices. In that subclass, the global stability is provided by interspecies regulation dominating over interactions between species.

If we consider as an example the extreme case of controversy between the intra- and interspecies relations, then we come to the following sign structure of the community matrix A:

$$
S(A) = \begin{bmatrix}
- & \overset{+}{0} & \overset{+}{0} & \cdots & \overset{+}{0} \\
\overset{+}{0} & - & \overset{+}{0} & \cdots & \overset{+}{0} \\
\vdots & & & & \vdots \\
\overset{+}{0} & \overset{+}{0} & \overset{+}{0} & \cdots & -
\end{bmatrix}, \tag{4.51}
$$

where symbol $\overset{+}{0}$ stands for either + or 0. This represents a community of only mutualistic type of interspecies relations, with each species exhibiting density self-limitation. If, furthermore, matrix $(-A)$ turns out to be quasi-dominant, then $(-A)$ is what is called an *M-matrix*.

M-matrices are characterized by such a fundamental property as the equivalence among all the four conditions formulated below.[40]

Definition 4.6 *Matrix $M = [m_{ij}]$ with nonpositive off-diagonal entries $m_{ij} \leq 0$ $(i \neq j)$ is called a (nonsingular) M-matrix if any one of the following conditions holds:*

(M1) *there exists vector $\pi > 0$ such that $M\pi > 0$ (quasi-dominance of M);*

(M2) *the system of linear algebraic equations*

$$
\sum_{\substack{j \neq i}}^{n} m_{ij} x_j = c_i, \qquad i = 1, \ldots, n, \tag{4.52}
$$

is solvable in nonnegative numbers x_i for any set of nonnegative numbers $c_i \geq 0$ $(i = 1, \ldots, n)$;

(M3) *all nested (Sylvester) principal minors of M are positive;*

(M4) *all principal minors of M are positive (M is a P-matrix).*

From definitions 4.5 and 4.6 it can be seen that matrix A is quasi-dominant if and only if matrix

$$
\hat{A} = [\hat{a}_{ij}], \quad \hat{a}_{ij} = \begin{cases} a_{ii}, & i = j, \\ -|a_{ij}|, & i \neq j \end{cases}
$$

is an *M-matrix*. Hence, quasi-dominance of matrix A^T easily follows if A is quasi-dominant (by condition (M3) or (M4)).

B. Stability in Mutualistic Community Models

If the dynamics of a mutualistic community with sign structure (4.51) is described by a Lotka–Volterra system (4.4) or by its generalization in the form of

$$\frac{dN_i}{dt} = r_i(N_i) \left[\varepsilon_i - \sum_{j=1}^{n} \gamma_{ij} V_j(N_j) \right], \qquad j = 1, \ldots, n, \qquad (4.53)$$

with continuously differentiable functions

$$r_i(N) > 0, \quad N > 0, \quad r_i(0) = 0 \qquad (i = 1, \ldots, n), \qquad (4.54)$$

and

$$V_j(N) > 0, \quad dV_j/N > 0, \quad N \geq 0, \quad V_j(0) = 0 \qquad (j = 1, \ldots, n) \qquad (4.55)$$

(that was proposed earlier for competitive communities[14,42]), then a pertinent change of variables can, as before, reduce searching for a positive equilibrium in the model to solving the linear system (4.18) with matrix $\Gamma = [\gamma_{ij}]$;[26] stability of the equilibrium then follows from properties of a matrix similar to (4.31). In case all $\varepsilon_i > 0$, the existence of a positive solution to system (4.18) is equivalent to matrix Γ being an *M*-matrix. Multiplying by a positive diagonal matrix does not disturb this property (as seen from condition (M1)), so that the matrix of the linear approximation to system (4.53) at the positive equilibrium $\mathbf{N}^*$ turns out quasi-dominant, hence, dissipative. This brings about the global pattern of $\mathbf{N}^*$'s stability (which is also true for generalized Lotka–Volterra systems with dissipative matrices $(-\Gamma)$[15]).

    This general result can explain stability established in many particular models of mutualism, like, e.g., those of V. A. Kostitzin[43] for symbiosis between two species, and about half of the more than 20 cases of stability reported by Boucher[44] for two-species models of mutualism under density self-regulation.

C. Diagonally Recessive Matrices

Diagonal dominance has long been known as a sufficient condition to localize matrix eigenvalues to the right of the imaginary axis (or to the left when the main diagonal is negative), see, e.g., "Geršgorin discs,"[30] to verify at least mere stability in a matrix. As mentioned in the previous section, if for a community matrix *A*, matrix $(-A)$ is quasi-dominant, then *A* is also dissipative, with a certain consequence of globally stable behavior in the community model. Shown schematically in Figure 15, the question however remained open whether the set of dissipative matrices was actually exhausted by the subset of quasi-dominant ones, or, in other words, do there exist dissipative matrices whose principal diagonals are not dominating?

    To answer the question we introduce a class of matrices whose properties are opposite, in a certain sense, to diagonal dominance.

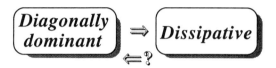

Figure 15.   Not an easy question.

Definition 4.7 *A matrix A will be called* diagonally recessive in columns *(in rows)* *if in any j-th column (in any i-th row) there is an entry a_{ij} such that $|a_{ij}| \geq |a_{ii}|$.*

Definition 4.8 *A matrix A will be called* diagonally recessive, *if A is diagonally recessive in columns or in rows.*

As a justification of the term, it can be proved that none of the diagonally recessive matrices can be quasi-dominant (Statement 4.4 of the Appendix). In terms of the community matrix, the diagonal recessiveness means that either each species experiences an influence from another species, stronger than its own intraspecies pressure, or, among the effects each species exerts upon the others, there always exists such an effect upon a species that prevails (in magnitude) over the self-regulation of the latter. In other words, this is the property which is diagonally recessive that corresponds to species interactions prevailing over the effects of species self-regulation.

In analogy to the notion of quasi-dominance, that of *quasi-recessiveness* can also be defined.[45]

Definition 4.9 *A matrix A will be called* diagonally quasi-recessive *(formally $A \in q\mathbb{R}$) if there exists a diagonal matrix $B = diag\{\beta_1, \ldots, \beta_n\}$ with all $\beta_i > 0$ such that the matrix AB is diagonally recessive.*

It follows immediately from the definitions given, that if $A \in q\mathbb{R}$, then also $A^T \in q\mathbb{R}$ and $A\mathcal{D} \in q\mathbb{R}$, where $\mathcal{D}$ is a diagonal matrix with any positive entries in its diagonal. Then, clearly, $\mathcal{D}A \in q\mathbb{R}$ also for all $\mathcal{D} \in \mathbb{D}_n^+$.

It can now be seen that the sets of diagonally quasi-dominant ($q\mathbb{D}$) and diagonally quasi-recessive ($q\mathbb{R}$) matrices do not intersect. If, indeed, a matrix $A \in q\mathbb{R}$, then, by definition, there exists a matrix $B \in \mathbb{D}_n^+$ such that $AB \in \mathbb{R}$ (i.e., AB is diagonally recessive). Consequently, AB cannot be quasi-dominant as mentioned above (Statement 4.4 in Appendix), i.e., $AB \in \overline{q\mathbb{D}}$, the set of matrices defined as the negation of diagonal quasi-dominance. It can be shown that set $\overline{q\mathbb{D}}$ is also closed with respect to transposition or multiplication by a positive diagonal matrix, as is the class $q\mathbb{D}$. This shows us immediately that $A = ABB^{-1} \in \overline{q\mathbb{D}}$, or, due to A being an arbitrary quasi-recessive matrix, $q\mathbb{R} \subset \overline{q\mathbb{D}}$, which proves the following

Theorem 4.4 $q\mathbb{D} \cap q\mathbb{R} = \varnothing$, *i.e., the set of quasi-dominant matrices and the set of quasi-recessive matrices do not intersect.*[45]

Hence, the concept of quasi-recessiveness is logically correct. Class $q\mathbb{R}$ is fairly broad: in particular, some singular or even unstable matrices (e.g., those with positive traces) may enter the class. Reverting to the question, whether there exist dissipative matrices without dominating diagonals, we determine whether the class $q\mathbb{R}$ contains a dissipative matrix.

Example: The matrix

$$A = - \begin{bmatrix} 1 & 7/6 & 1/2 \\ 3/4 & 1 & 3/2 \\ 3/2 & 1/2 & 1 \end{bmatrix} \tag{4.56}$$

belongs to the class $q\mathbb{R}$ (even to $\mathbb{R}$), for it is recessive both in rows and in columns. Since, furthermore, the inequalities that provide matrix (4.56) as recessive hold strictly, there is a (yet small, but open) vicinity of matrix A, say $\mathbb{U}_R(A)$, that consists entirely of recessive matrices: $\mathbb{U}_R(A) \subset q\mathbb{R}$. On the other hand, all principal minors of $(-A)$ are positive and all the remaining conditions of Theorem 4.3 about dissipativeness of 3×3 matrices hold also true for matrix (4.56). Together with this matrix, some (open) vicinity $\mathbb{U}_D(A)$ within it is contained in the open set of dissipative 3×3 matrices. The intersection, $\mathbb{U}_D(A) \cap \mathbb{U}_R(A)$, of these two vicinities gives a set of dissipative, diagonally recessive matrices, that is open in the natural topology of the space of 3×3 matrices.   □

The example shows that the set of dissipative matrices is far from being exhausted by the diagonal quasi-dominant property alone and allows us to give a positive answer to the question posed in the beginning of this section. The true hierarchy of matrix stability notions shown in Figure 14 can now be supplemented with the irreversible implication in Figure 15.

The fact there are diagonally recessive matrices among dissipative ones (the above example is not extraordinary), shows that the pattern of global stability typical of dissipative community matrices can be guaranteed not only by strong intraspecies regulation (quasi-dominance) but also by a pattern of stronger species interactions under weak self-regulation (quasi-recessiveness).

VI. "MATRIX FLOWER" GROWING

Considered in this chapter, several notions of matrix stability have both rigorous mathematical definitions and clear "biological" interpretations, i.e., interpretations in terms of corresponding models for the dynamics of biological communities. Modeling practice however requires something more, namely, the characterization of individual matrix properties in terms of signs and magnitudes of matrix elements. In other words, the practice must have a constructive criterion, rather than a definition alone, according to which one can test any given matrix as either having or not having the matrix property in question.

A classical example of such a criterion is the well-known Routh–Hurwitz criterion for Lyapunov stability of a given matrix.[46] When the size of a matrix is not too large, the criterion can really allow us to verify the stability by calculating and examining the signs of the so-called Hurwitz determinants, which have finite expressions in terms of the characteristic polynomial coefficients, hence, eventually in terms of matrix elements. Unless pure computational problems makes the task impracticable, one can be sure in the result. One can be much less sure when the result is obtained by direct computer calculation and by testing the sign of each real part of the eigenvalue, i.e., when the definition is being directly applied.

Unfortunately, the definitions, except for some rare examples like those of diagonal dominance or recessiveness, are of little help in the task of practical verification, especially when $n > 3$. What can really help in tackling the characterization problem when the characterization is unknown, is considered in the remainder of the chapter.

A. Diagram of Logical Relations: Conventions and Examples

At least some help in the characterization problem may be offered by a diagram of logical relations among the various subsets of stable matrices if it is arranged by means of simple conventions related somehow to matrix elements. Figure 16

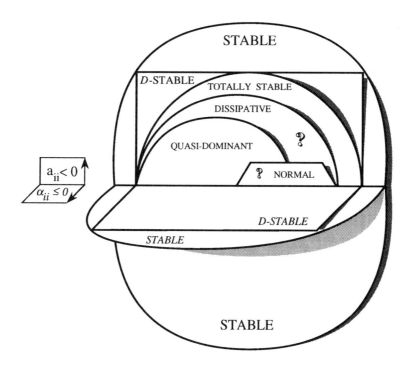

Figure 16. Several more conventions in the same diagram.

shows such a diagram covering the stability properties considered before. It has been designed under the following three basic conventions.

First of all, each subset is pictured as one or more contours, among which a visible inclusion means logical inclusion, either proved, or hypothesized. Moving by arrows of logical implication, like those in Figures 14 and 15, corresponds to moving outwards in Figure 16. A visible overlap should naturally mean a logical intersection.

Second, the upper vertical half-plane "contains" matrices with negative principal diagonals, the lower one contains those having some positive entries on the diagonal, while the horizontal half-plane corresponds to matrices with nonpositive principal diagonals (and with negative traces to avoid instability).

Third, the sets which are open in n^2-dimensional space are conventionally pictured as curvilinear contours, while the sets which are not open are pictured by polygonal contours.

With all these conventions in mind, a careful reader will agree that Figure 16 summarizes all that which was stated above about the matrix properties selected. (The set of normal matrices is not open, as the definition requires some equality-type relations to hold among matrix entries.)

But still there are two question marks in Figure 16. The upper one has been inherited from Figure 15; it signifies the question that has been answered comprehensively in the previous section. Speaking imaginatively, it stands for the question whether in the diagram there is enough room to put a question mark. If the matter is about open sets, then "enough room" should mean a nonempty open subset of matrices possessing certain properties (dissipative but not quasi-dominant in Figure 16). But if the question mark refers to a nonopen set, it should mean just a nonempty subset of pertinent matrices.

For example, the lower question mark in Figure 16 has arisen from a "practical" need to picture somehow the set of normal stable matrices. While any such normal matrix is already dissipative, it is still a question whether there are normal matrices among the quasi-dominant ones. The answer is as simple as the task to find a diagonally dominant matrix which is, for instance, symmetrical or skew-symmetric. Thus, the two subsets do intersect, while, on the other hand, the fact there should be normal dissipative matrices outside $q\mathbb{D}$ is confirmed by the following

Example: The symmetric matrix

$$A = - \begin{bmatrix} 1 & 0.66 & 0.5 \\ 0.66 & 1 & 0.66 \\ 0.5 & 0.66 & 1 \end{bmatrix} \tag{4.57}$$

of a competition community is stable (see, e.g., sufficient conditions in Section 7.II.B) and is apparently not diagonally dominant. That it can yet be quasi-diagonally dominant is not so evident but is verifiable by means of the M-matrix characterization. If, indeed, matrix A were in $q\mathbb{D}$, then, by definition, there would exist a vector $\pi > \mathbf{0}$ such that $\pi_i|a_{ii}| > \sum_{i \neq j}^{3} \pi_j|a_{ij}|$ ($i = 1, 2, 3$). By condition (M1) this would signify that matrix

$$\hat{A} = \begin{bmatrix} 1 & -0.66 & -0.5 \\ -0.66 & 1 & -0.66 \\ -0.5 & -0.66 & 1 \end{bmatrix}$$

be an M-matrix. Fortunately, among equivalent definitions of M-matrices there are those, like (M3) and (M4), which are typical characterizations; they show readily by calculation of the determinant that $\hat{A}$ is not an M-matrix. Hence, matrix A (4.57) is actually not in $q\mathbb{D}$.   □

Consequently, there is really "enough room" for the question marks posed in Figure 16 and both can therefore be safely deleted.

B. Further Relations, or How to Cut the Matrix Flower

Using an imaginative terminology, we could say that the diagram in Figure 16 represents a "flower" of matrix stability subsets, or a *matrix flower*. However, as we have seen above and will see below, to determine a correct form and to find a right place for a particular contour, or *petal*, within the flower is not a matter of imagination alone but also requires certain mathematical problems to be solved. If, for example, we would like to see how the flower will look like in relation to the set of quasi-recessive matrices ($q\mathbb{R}$), then the findings of Section V.C can be used to identify that:

(i) set $q\mathbb{R}$ is not open (hence polygonal contour(s));

(ii) $q\mathbb{R}$ contains unstable matrices (hence the matter is about its cutting across, or *cross-section* of, the flower);

(iii) being in $q\mathbb{R}$ imposes no constraints on the main diagonal signs (hence the cross-sections must be in all three "half-planes");

(iv) $q\mathbb{R}$ does not intersect with $q\mathbb{D}$ (nor do the cross-sections), but

(v) $q\mathbb{R}$ does intersect with the dissipative set (so the cross-section does).

This already is almost sufficient to draw a flower like that in Figure 17, where an additional, pure "artistic," convention supposes the transecting $q\mathbb{R}$-"planes" to be "transparent," with the "transparency" still having no special mathematical sense. What has not yet been clear is the intersection with the set of normal matrices. To clarify this, consider the following

Example: The so-called *circulant* matrix[29]

$$- A = \begin{bmatrix} 1 & a & b \\ b & 1 & a \\ a & b & 1 \end{bmatrix} \tag{4.58}$$

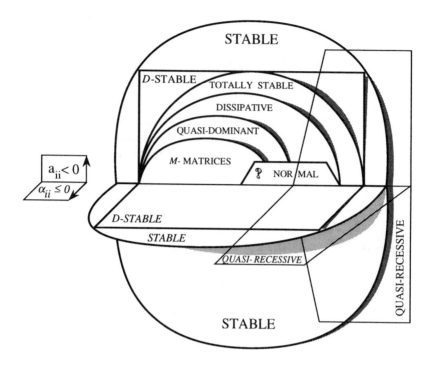

Figure 17. Cross-section by the "planes" of quasi-recessive matrices (the "planes" are drawn as if they were "transparent," with no special mathematical meaning).

is normal and has the eigenvalues explicitly calculable:

$$\lambda_1(-A) = 1 + a + b, \quad \text{Re } \lambda_2(-A) = \text{Re } \lambda_3(-A) = 1 - (a + b)/2.$$

It follows that, in the plane (a, b), the region where matrix A is stable (hence dissipative due to its normality) is the (infinite) strip defined by $-1 < a + b < 2$ (see Figure 18). Then, within the strip, for any point with either $|a| \geq 1$ or $|b| \geq 1$ (the shaded region), matrix A is normal, dissipative, and diagonally recessive. □

Since the region specified in the example contains an open domain within which matrix (4.48) retains its properties, it might seem at the first glance that the intersection of stable normal and $q\mathbb{R}$-matrices should contain an open subset of matrices. (Intersection of nonopen sets may generally result either in an open set, or in a nonopen, or even in a closed set.) But the above-mentioned domain is actually open only in the plane (a, b), i.e., in a two-dimensional subspace, rather than in the whole nine-dimensional space of 3×3 matrices. The example is thus unable to prove the intersection open, but it is able, in fact, to demonstrate at least its being nonopen: any matrix corresponding to a point on the boundary within the strip in

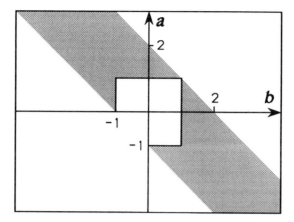

Figure 18.   The region where 3×3 circulant matrices are dissipative and diagonally recessive.

Figure 18, e.g., the point $(a = 1, b = -1)$, has an arbitrarily close neighbor already outside $q\mathbb{R}$.

That is why the intersection is pictured as a polygonal (trapezoidal) petal in Figure 18.

A pedantic reader could then ask whether there really exist any matrices which should lie in a small petal (implicit in Figure 18) between the $q\mathbb{D}$-set and the dissipative set within the trapezoid of normal matrices? The positive answer is given by matrix (4.57) which is normal and dissipative but lies neither in $q\mathbb{D}$, nor in $q\mathbb{R}$.

Finally, the question mark in the "normal" petal means whether a stable normal matrix may be an M-matrix, and the positive answer is general enough. Consider a matrix

$$A = S - I, \tag{4.59}$$

where S is a nonnegative symmetrical matrix whose main diagonal is zero and each row sum totals less than one in modulus (by its sign structure, A corresponds to a mutualistic n-species community). Constructed in this way, matrix $(-A)$ is a diagonally dominant M-matrix (or A is a $(-M)$-matrix). Mere multiplication shows that the equality $AA^T = A^TA$ is valid whenever $SS^T = S^TS$, which holds true by definition.

Thus, how the quasi-recessive "planes" cut the matrix flower in Figure 18 is quite consistent with all the conventions adopted. The flower itself summarizes the community matrix properties which are crucial for stability behavior in multispecies Lotka–Volterra models and in some of their modern extensions.

Appendix

Statement 4.1. *All the eigenvalues of the linear system matrix (4.31) are pure imaginary numbers.*

Proof: If $\lambda = a + ib$ is an eigenvalue of matrix $\text{diag}\{\ldots\}\Gamma$ with an eigenvector $\mathbf{z} = \mathbf{u} + i\mathbf{w}$, then $\overline{\lambda} = a - ib$ is also an eigenvalue with the eigenvector $\overline{\mathbf{z}} = \mathbf{u} - i\mathbf{w}$. By definition of an eigenvector we have

$$(a + ib)z_k = \sum_{l=1}^{n} N_k^* \gamma_{kl} z_l,$$

$$(a - ib)\overline{z_k} = \sum_{l=1}^{n} N_k^* \gamma_{kl} \overline{z_l}, \qquad k = 1, 2, \ldots, n,$$

hence

$$(a + ib) \sum_k \alpha_k N_k^* z_k \overline{z_k} = \sum_k \sum_l \alpha_k \gamma_{kl} N_k^* N_l^* \overline{z_k} z_l,$$

$$(a - ib) \sum_k \alpha_k N_k^* \overline{z_k} z_k = \sum_k \sum_l \alpha_k \gamma_{kl} N_k^* N_l^* z_k \overline{z_l}.$$

Adding these equations term by term, rearranging the right-hand side, and referring to (4.20), we come to the equality

$$2a \sum_k \alpha_k N_k^* |z_k|^2 = \sum_k \sum_l (\alpha_k \gamma_{kl} + \alpha_l \gamma_{lk}) N_k^* N_l^* \overline{z_k} z_l = 0.$$

Hence, since $\alpha_k > 0$, $N_k^* > 0$, and not all of z_k are zero, it immediately follows that $a = 0$.   ∎

Statement 4.2. *Function*

$$G(\mathbf{N}) = \sum_{i=1}^{n} \alpha_i (N_i - N_i^* \ln N_i), \qquad \alpha_i > 0, \quad i = 1, \ldots, n,$$

as defined in (4.30) over int $\mathbb{R}_+^n$, *attains its global minimum at* $\mathbf{N}^*$.

Proof: If we make the change of variables

$$y_i = N_i / N_i^* - 1, \qquad i = 1, \ldots, n,$$

then expression

$$G(\mathbf{N}) - G(\mathbf{N}^*) = \sum_{i=1}^{n} \alpha_i \left[(N_i - N_i^*) - (N_i^* \ln N_i - N_i^* \ln N_i^*) \right]$$

reduces to

$$\sum_{i=1}^{n} \alpha_i N_i^* \left[y_i - \ln(1 + y_i) \right],$$

which is always nonnegative since inequality $y \geq \ln(1+y)$ holds true for any $y > 1$, turning into the equality only at $y = 0$. Therefore, $G(\mathbf{N}) > G(\mathbf{N}^*)$ everywhere in int $\mathbb{R}_+^n$ except $\mathbf{N}^*$, which gives the global minimum of $G(\mathbf{N})$. ∎

Statement 4.3. *If system (4.4) has an equilibrium $\mathbf{N}^*$ with no zero components, then the time derivative of function*

$$G(\mathbf{N}) = \sum_{i=1}^{n} \alpha_i (N_i - N_i^* \ln N_i), \qquad \alpha_i > 0, \quad i = 1, \ldots, n,$$

along system trajectories equals

$$\frac{d}{dt} G(\mathbf{N}(t)) = -F(\mathbf{N}(t) - \mathbf{N}^*),$$

where F is a quadratic form defined in (4.19) as

$$F(N_1, \ldots, N_n) = \sum_{i,j=1}^{n} \alpha_i \gamma_{ij} N_i N_j.$$

Proof: Since $\mathbf{N}^*$ is a solution to system (4.18) of linear algebraic equations, equations (4.4) can be rewritten in the following form:

$$\frac{dN_i}{dt} = -N_i \sum_{j} \gamma_{ij}(N_j - N_j^*), \tag{A4.1}$$

whereby

$$\frac{d}{dt}(\ln N_i) = -\sum_{j} \gamma_{ij}(N_j - N_j^*), \qquad i = 1, \ldots, n. \tag{A4.2}$$

Forward differentiation then yields

$$\frac{d}{dt} G(\mathbf{N}) = \sum_{i=1}^{n} \alpha_i (\dot{N}_i - N_i^* \dot{N}_i / N_i),$$

which transforms by (A4.1) and (A4.2) into

$$-\sum_{i=1}^{n}\left[\alpha_i N_i \sum_j \gamma_{ij}(N_j^* - N_j) - \alpha_i N_i^* \sum_j \gamma_{ij}(N_j^* - N_j)\right]$$

$$= \sum_{i,j=1}^{n} \alpha_i \gamma_{ij}(N_i - N_i^*)(N_j^* - N_j) = -F(\mathbf{N} - \mathbf{N}^*)$$

by the definition of quadratic form F. ∎

Theorem 4.2 _A matrix_

$$A = -\begin{bmatrix} 1 & a & b \\ c & 1 & d \\ e & f & 1 \end{bmatrix}, \quad a \neq 0, \tag{4.43}$$

is totally stable if and only if $(-A)$ _is a P-matrix and the condition holds_

$$d(-A) - E_2(-A) \equiv ade + bcf - 2$$
$$< 2\left[\sqrt{(1 - ac)(1 - be)} + \sqrt{(1 - ac)(1 - df)} + \sqrt{(1 - be)(1 - df)}\right], \tag{4.44}$$

where $d(-A)$ _is the determinant and_ $E_2(-A)$ _the sum of second-order principal minors of matrix_ $(-A)$.

Proof: Let

$$D = \text{diag}\{d_1, d_2, d_3\} \in \mathbb{D}_3^+$$

be a matrix with an arbitrary positive diagonal. The Hurwitz matrix for the characteristic polynomial of matrix DA takes on the form

$$H = \begin{bmatrix} -E_1(DA) & d(DA) & 0 \\ 1 & E_2(DA) & 0 \\ 0 & E_1(DA) & -d(DA) \end{bmatrix}. \tag{A4.3}$$

By the Routh–Hurwitz criterion, for matrix DA to be stable it is necessary and sufficient that all nested principal minors of H be positive:

$$\Delta_1 = E_1(DA) > 0;$$
$$\Delta_2 = d(DA) - E_1(DA)E_2(DA) > 0;$$
$$\Delta_3 = -\Delta_2 d(DA) > 0.$$

From the theorem condition and matrix D properties it follows that both Δ_1 and $-d(DA)$ are always positive, so that stability of DA is equivalent to the condition that $\Delta_2 > 0$. The latter can be equivalently reduced to the inequality

$$
\begin{aligned}
d(-A) - E_2(-A) \quad < \quad & (1 - be)d_1/d_2 + (1 - df)d_2/d_1 \\
& + (1 - ac)d_2/d_3 + (1 - be)d_3/d_2 \\
& + (1 - ac)d_1/d_3 + (1 - df)d_3/d_1, \quad\quad (A4.4)
\end{aligned}
$$

which must hold for any values of $d_i > 0$ and, in particular, for those which give the minimum value to the right-hand side of (A4.4). If we consider the first and the second pair of terms independently, then the minimum of the first pair is attained at $d_1/d_2 = \sqrt{(1 - df)/(1 - be)}$ and equal to $2\sqrt{(1 - be)(1 - df)}$, while the minimum of the second is attained at $d_2/d_3 = \sqrt{(1 - be)(1 - ac)}$ and equal to $2\sqrt{(1 - ac)(1 - be)}$. For these particular values of variables we have

$$
d_1/d_3 = (d_1/d_2)(d_2/d_3) = \sqrt{(1 - df)/(1 - ac)},
$$

which automatically gives the independent minimum of the third pair of terms in (A4.4) equal to $2\sqrt{(1 - ac)(1 - df)}$. Hence, the minimum of the total sum equals the sum of the above three pair-wise minima, whereby the inequality (A4.4) turns into the statement of (4.44).   ∎

Theorem 4.3 *A matrix A of the form (4.43) is dissipative if and only if $(-A)$ is a P-matrix and the following condition holds true:*

$$
d(-A) + E_2(-A) > \inf_{\substack{x,y>0 \\ x \in I_2(a,c)}} \Big\{ Q(x,y) = (a^2 - abf)x + (c^2 - cde)/x \quad\quad (4.45)
$$

$$
+ (d^2 - bed)y + (f^2 - aef)/y + (b^2 - abd)xy + (e^2 - cef)/xy \Big\},
$$

where

$$
I_2(a,c) = \{ x > 0 : (1 - \sqrt{1 - ac})^2 < a^2 x < (1 + \sqrt{1 - ac})^2 \}. \quad\quad (4.46)
$$

Necessity. It is obvious that all principal minors of $(-A)$ must be positive, as they are the determinants of dissipative, hence totally stable, submatrices. Let $\mathcal{A} = \text{diag}\{\alpha_1, \alpha_2, \alpha_3\}$ be a diagonal matrix of Definition 4.4. Then the matrix $\mathcal{A}(-A) + (-A)^T \mathcal{A}$ is positive definite and the positiveness of its Sylvester minors is equivalent to the following two inequalities:

$$
4 - (a^2 \alpha_1/\alpha_2 + c^2 \alpha_2/\alpha_1 + 2ac) > 0, \quad\quad (A4.5)
$$

$$
\begin{aligned}
d(-A) + E_2(-A) \quad > \quad & (a^2 - abf)\alpha_1/\alpha_2 + (c^2 - cde)\alpha_2/\alpha_1 \\
& + (d^2 - bed)\alpha_2/\alpha_3 + (f^2 - aef)\alpha_3/\alpha_2 \\
& + (b^2 - abd)\alpha_1/\alpha_3 + (e^2 - cef)\alpha_3/\alpha_1. \quad\quad (A4.6)
\end{aligned}
$$

If we denote

$$\alpha_1/\alpha_2 = x > 0, \quad \alpha_2/\alpha_3 = y > 0, \tag{A4.7}$$

then inequality (A4.5) becomes a quadratic trinomial which retains its sign everywhere in the interval (4.46). The required inequality (4.45) then follows from (A4.6) and the necessity is proved.

Sufficiency. If a P-matrix $(-A)$ satisfies condition (4.45), then there exists a pair of particular values x°, y° such that

$$d(-A) + E_2(-A) > Q(x^\circ, y^\circ).$$

Then any of the admissible changes of variables inverse to (A4.7) will generate the desired diagonal matrix $\mathcal{A}^\circ = \text{diag}\{\alpha_1^\circ, \alpha_2^\circ, \alpha_3^\circ\}$ which meets both condition (A4.5) and (A4.6). These two confirm matrix $\mathcal{A}^\circ(-A) + (-A)^T \mathcal{A}^\circ$ to be positive definite.
∎

Statement 4.4. *If an $n \times n$ matrix A is diagonally recessive (Definition 4.8), then it cannot be diagonally quasi-dominant (in the sense of Definition 4.5).*

Proof: Let $\pi = [\pi_1, \pi_2, \ldots, \pi_n] > \mathbf{0}$ be an arbitrary vector of $\mathbb{R}^n$ with

$$\pi_m = \min_i \{\pi_i\}, \quad \pi_M = \max_i \{\pi_i\}, \quad i = 1, \ldots, n.$$

If A is recessive in columns, then in its Mth column there is an entry a_{iM} $(M \neq i)$ such that $|a_{ii}| \leq |a_{iM}|$. Then $\pi_i |a_{ii}| \leq \pi_M |a_{iM}|$ and a fortiori

$$\pi_i |a_{ii}| \leq \sum_{i \neq j} \pi_j |a_{ij}|,$$

which means the absence of quasi-dominance. If matrix A is recessive in lines, then in its mth line there is an entry a_{mj} $(m \neq j)$ such that $|a_{mm}| \leq |a_{mj}|$. Then $\pi_m |a_{mm}| \leq \pi_j |a_{mj}|$ and a fortiori

$$\pi_m |a_{mm}| \leq \sum_{i \neq m} \pi_j |a_{mj}|,$$

that is, no quasi-dominance again.   ∎

Additional Notes

To 4.I. Possible ways used to represent graphically the structure of a biological community or an ecosystem are numerous, perhaps even more numerous than opinions about what should be understood as the structure itself. No attempt to review all of them is known to the author. However, the conventions adopted in constructing a signed directed graph (SDG) are useful as they not only lead to a visible perception of species interactions, but also conform to the correspondence between a directed graph and its adjacency matrix or a matrix and its associated digraph, standard respectively for the graph and matrix theories (see, e.g., Harary et al.[47] or other texts[30,48−49] and a more thorough review article[50]). The conventions are basically the same as those used in constructing the graph of age transitions in Leslie-type models of age-structured populations (Chapter 2), since, in terms of model equations, the transitions can be also viewed as "effects" on the given age group from the preceding one(s).

Some authors, e.g. Hofbauer and Sigmund[23] with their treatment of the dynamical systems theory applied to models of community dynamics, use the term *time-independent* instead of *autonomous* systems in the sense that the right-hand side of equations does not explicitly depend on the time variable.

The idea to classify the types of species interactions by the effects the interaction has on both interacting populations, rather than by interaction mechanisms at the level of individuals, dates back to the middle of the century, the works by Haskell,[51] Burkholder,[52] and especially E. P. Odum[53], who had introduced the idea into textbooks. He did not however stay within the six possible combinations of 2 signs out of the 3 (+, −, or 0 with the order being irrelevant), but expanded the basis of classification with knowledge about interaction biology and came up with 9 major types.[54−55] Also 9 types have recently been obtained by Patten[56] in his formalizing the idea of indirect species interactions.

To 4.II. Despite historical evidence[11] that both A. Lotka and V. Volterra independently had made basic contributions to the theory of models (4.4), it was V. Volterra who developed it as a complete mathematical theory. This is probably the reason why some authors[7,10,26,57−58] still refer to this kind of systems as just Volterra models, particularly in multispecies contexts.

While definition (4.2) is general enough and, at least mathematically, unambiguous, the term *community matrix* in Lotka–Volterra models has its own history, where there are both definitions contradicting the term *interaction matrix* and attempts to clear up the point (see Hallam and Clark[59] and references therein).

The idea of deriving the Lotka–Volterra description from the conservation principle was presented in Svirezhev[58] and developed further in Svirezhev and Logofet[7]. Special form of Lotka–Volterra equations not only challenged matrix algebra to be applied in analysis, but also induced attempts to apply standard methods of theoretical mechanics. For example, Antonelli[60] (see also the references therein) obtained a standard Hamiltonian form of equations by combining original Lotka–Volterra equations for N_i with those for $x_i = \int_0^t N_i \, dt$ by a group of the so-called holonomic bonds. While the theory guaranteed the existence and uniqueness of solutions, practice may be satisfied with the interpretation of x_i as an index of species i's in the accumulation of a toxic substance.

To 4.III. The term *interior* equilibrium is also used in parallel with *positive* or *feasible* ones.[23] As stated by an elegant Theorem 1 of Hofbauer and Sigmund[23], existence of such an equilibrium means also the existence of interior limit points for model trajectories. But to verify the existence in practical cases will still be a matter of algebraic routine to find the solution of a system of linear equations.

Although Volterra did not use the Lyapunov terminology, he did analyze the spectrum of conservative and dissipative community matrices, the behavior of small deviations from the equilibrium, and his form (4.21) for the integral of motion had predetermined the subsequent forms of Lyapunov functions.

Graphical methods to analyse phase behavior in two- and three-dimensional systems, even after Volterra[2] and Kolmogoroff,[61–62] have not been less popular in population models than in other areas (see e.g. Svirezhev and Logofet,[7] Farkas and Budt[63], and references therein).

To 4.IV. To avoid awkwardness in formulations due to alternating signs of principal minors when all the eigenvalues are to the left of the imaginary axis, some authors define matrix stability and further properties on the right of the imaginary axis. This is not the case in the formulations presented, where we have had recourse to matrix $(-A)$ whenever needed.

The distinction betweeen *D*-stable and merely stable matrices should be borne in mind in any study of structure and stability in multispecies systems. Bodini and Giavelli[64] investigated the response of equilibrium levels to changes in the growth rates from the standpoint of stablity-instability issues. The authors dealt with "a system of moving equlibrium" (p. 290) in the sense that the parameters assumed as constants in the system equations, change at a rate (sufficiently) slower than that of the state variables. As follows from Section III, at least within the class of Lotka–Volterra systems with *D*-stable interaction matrices, those changes are unable to change stability of a feasible equilibrium into its instability, but they do can make the equilibrium nonfeasible. So "moving equilibria" may well move out of the positive orthant of the phase space, or in more general terms, may appear or disappear in the phase space—a phenomenon which is well-known in bifurcation theory.[65]

More on "species deletion" stability can be found in Pimm & Gilpin;[66] see also the review article by Pimm, Lawton, and Cohen.[67]

In somewhat different form, a theorem on characterization of 3×3 dissipative matrices was proved by Cross.[25] A dimensionality-reducing algorithm to solve the problem in a general case is proposed by Redheffer.[68] A number of general criteria of diagonal stability have been proved by Kraaijevanger[69] using Hadamar (element-wise) matrix products. Though being in essence some equivalents to the definition rather than element-based characterizations, those theorems have nevertheless resulted in one more of such characterizations for 3×3 matrices.[69] Example (4.47) is taken from Logofet.[70]

To 4.V. *M*-matrices were first introduced by A. Ostrowski,[71] and were intensively studied in mathematical economics as a basis of the Leontief model for multisector economy.[72–73] For applications in ecological modeling see DeAngelis, Post and Travis.[74]

The class of quasi-recessive matrices and the matrix example (4.56) were proposed in Logofet.[45]

To 4.VI. The idea of a "flower" diagram was proposed in Logofet;[28,45] example (4.58) was considered in Logofet.[70]

Although the hierarchy of matrix stability subsets has been developed from stability considerations within Lotka–Volterra systems, the matrix properties and their relations do not depend on the origin, which rather determines only the interpretation. So, the hierarchy itself makes sense generally.

REFERENCES

1. Lotka, A. J. *Elements of Physical Biology*, Williams & Wilkins, Baltimore, MD, 1925 (reissued as *Elements of Mathematical Biology*, Dover, New York, 1956, 465 pp.)

2. Volterra, V. *Lecons sur la Théorie Mathématique de la Lutte pour la Vie*, Gauthier-Villars, Paris, 1931. 214 pp.

3. Paine, R. T. Food web complexity and species diversity, *The American Naturalist*, 100, 65–75, 1966.

4. Jeffries, C. Qualitative stability and digraphs in model ecosystems, *Ecology*, 55, 1415–1419, 1974.

5. Jeffries, C. *Mathematical Modeling in Ecology: A Workbook for Students*, Birkhäuser, Boston, 1988, Chap. 5.

6. Logofet, D. O. To the issue of ecosystem qualitative stability, *Zhurnal Obshchei Biologii* (Journal of General Biology), 39, 817–822, 1978 (in Russian).

7. Svirezhev, Yu. M. and Logofet, D. O. *Stability of Biological Communities* (revised from the 1978 Russian edition), Mir Publishers, Moscow, 1983, Chaps. 3 and 4.

8. May, R. M. Stability in model ecosystems, *Quantifying Ecology*, 6, 18–56, 1971.

9. May, R. M. *Stability and Complexity in Model Ecosystems*, Princeton University Press, Princeton, NJ, 1973, Chap. 2.

10. Goel, N. S., Maitra, S. C. and Montroll, E. W. On the Volterra and other nonlinear models of interacting populations, *Reviews of Modern Physics*, 43, 231–276, 1971.

11. Scudo, F. M. and Ziegler, J. R. *The Golden Age of Theoretical Ecology: 1923–1940. A Collection of Works by V. Volterra, V. A. Kostitzin, A. J. Lotka and A. N. Kolmogoroff* (*Lecture Notes in Biomathematics*, Vol. 22), Springer, Berlin, 1978, 490 pp.

12. Roberts, A. The stability of a feasible random ecosystem, *Nature* (London), 251, 607–608, 1974.

13. Barbashin, E. A. *Introduction to Stability Theory*, Nauka, Moscow, 1967, Chap. 1 (in Russian).

14. Grossberg, S. Pattern formation by the global limits of a nonlinear competitive interaction in n dimensions, *J. Math. Biol.*, 4, 237–256, 1977.

15. Pykh, Yu. A. *Equilibrium and Stability in Models of Mathematical Ecology and Genetics*. Doctoral dissertation. Agrophysical Research Institute, Leningrad, 1981 (in Russian).

16. Metzler, L. Stability of multiple markets: the Hicks conditions, *Econometrica*, 13, 277–292, 1945.

17. Arrow, K. J. and McManus, M. A note on dynamical stability, *Econometrica*, 26, 448–454, 1958.

18. Johnson, C. R. Sufficient conditions for *D*-stability, *J. Econom. Theory*, 9, 53–62, 1974.

19. Datta, B. N. Stability and D-stability, *Linear Algebra and Its Applications*, 21, 135–141, 1978.

20. Hartfiel, D. J. Concerning the interior of D-stable matrices, *Linear Algebra and Its Applications*, 30, 201–207, 1980.

21. Quirk, J. P. and Ruppert, R. Qualitative economics and the stability of equilibrium, *Rev. Econom. Studies*, 32, 311–326, 1965.

22. Pimm, S. L. *Food Webs*, Chapman & Hall, London, 1982, 219 pp.

23. Hofbauer, J. and Sigmund, K. *The Theory of Evolution and Dynamical Systems*, Cambridge University Press, Cambridge, 1988, 341 pp., chap. 9.2.

24. Barker, G. P, Berman, A. and Plemmons, R. J. Positive diagonal solutions to the Lyapunov equations, *Linear and Multilinear Algebra*, 4, 249–256, 1978.

25. Cross, G. W. Three types of matrix stability, *Linear Algebra and Its Applications*, 20, 253–263, 1978.

26. Pykh, Yu. A. *Equilibrium and Stability in Models of Population Dynamics*, Nauka, Moscow, 1983, 182 pp. (in Russian).

27. Cain, B. E. Real, 3×3, D-stable matrices, *J. Res. Nat. Bur. Standards U.S.A.*, 80B, 75–77, 1976.

28. Logofet, D. O. On the hierarchy of subsets of stable matrices, *Soviet Math. Dokl.*, 34, 247–250, 1987.

29. Marcus, M. and Minc, A. *A Survey of Matrix Theory and Matrix Inequalities*, Allyn & Bacon, Boston, 1964, Chap. 4.

30. Horn, R. A. and Johnson C. R. *Matrix Analysis*, Cambridge University Press, Cambridge, 1990, Chap. 6.

31. Goh, B. S. Stability in many-species systems, *Amer. Natur.*, 111, 135–143, 1977.

32. Takeuchi, Y., Adachi, N. and Tokumara, H. Global stability of ecosystems of the generalized Volterra type, *Math. Biosci.*, 42, 119–136, 1978.

33. May, R. M. On the theory of niche overlap. *Theor. Pop. Biol.*, 5, 297–332, 1974.

34. Yodzis, P. *Competition for Space and the Structure of Ecological Communities*, Springer, Berlin (*Lecture Notes in Biomathematics*, Vol. 25), 1978, 191 pp.

35. Abrosov, N. S. *Analysis of the Species Structure in a Single Trophic Level (Applied to a Continuous Polyculture of Microorganisms)*. Doctoral dissertation. Institute of Physics, Krasnoyarsk, 1979 (in Russian).

36. Šiljak, D. D. Connective stability of complex ecosystems, *Nature* (London), 249, 280, 1974.

37. Šiljak, D. D. When is a complex system stable? *Math. Biosci.*, 25, 25–50, 1975.

38. Ladde, G. S. and Šiljak, D. D. Stochastic stability and instability of model ecosystems, *Proc. IFAC 6th World Congr. Boston-Cambridge*, Part 3, Pittsburgh, 1975, pp. 55 4/1– 55 4/7.

39. Quirk, J. and Saposnick, R. *Introduction to General Equilibrium and Welfare Economics*, McGraw-Hill, New York, 1968, 221 pp.

40. Nikaido, H. *Convex Structures and Economic Theory*, Academic Press, New York, 1968, Chap. 7.

41. Moylan, P. J. Matrices with positive principal minors, *Linear Algebra and Its Applications*, 17, 53–58, 1977.

42. Rescigno, A. and Richardson, I. W. On the competitive exclusion principle, *Bull. Math. Biophys.*, 27, 85–89, 1965.

43. Kostitzin, V. A. *Simbiose, Parasitisme et Evolution (Etude Mathematique)*, Hermann, Paris, 1934, 47 pp.

44. Boucher, D. H. Lotka–Volterra models of mutualism and positive density-dependence, *Ecological Modelling*, 27, 251–270, 1985.

45. Logofet, D. O. Do there exist diagonally stable matrices without dominating diagonal? *Soviet Math. Dokl.*, 38, 113–115, 1989.

46. Gantmacher, F. R. *The Theory of Matrices*, Vol. 2. Chelsea, New York, 1960, Chap. 16.

47. Harary, F., Norman, R. Z. and Cartwright, D. *Structural Models: An Introduction to the Theory of Directed Graphs*, John Wiley, New York, 1965, Chap. 7.

48. Voyevodin, V. V. and Kuznetsov, Yu. A. *Matrices and Calculations*, Nauka, Moscow, 1984, Chap. 18 (in Russian).

49. Lancaster, P. *Theory of Matrices*, Academic Press, New York–London, 1969, Chap. 9.

50. Maybee, J. S., Olesky, D. D., Van den Driessche, P. and Wiener, G. Matrices, digraphs, and determinants, *SIAM J. Matrix Anal. Appl.*, 10, 500–519, 1989.

51. Haskell, E. F. A clarification of social science, *Main Currents in Modern Thought*, 7, 45–51, 1949.

52. Burkholder, P. R. Cooperation and conflict among primitive organisms, *Am. Sci.*, 40, 601–631, 1952.

53. Odum, E. P. *Fundamentals of Ecology*, Saunders, Philadelphia, 1953, Chap. 7.

54. Odum, E. P. *Fundamentals of Ecology*, 3rd ed., Saunders, Philadelphia, 1971, Chap. 7.16.

55. Odum, E. P. *Basic Ecology*, Saunders College Publishing, Philadelphia, 1983, Chap. 7.

56. Patten, B. C. Concluding remarks: Network ecology: Indirect determination of the life-environment relationships in ecosystems. In: *Theoretical Ecosystem Ecology: The Network Perspective*, Higashi, M., and Burns, T. P., eds., Cambridge, 1990, 117–154.

57. Poluektov, R. A., ed. *Dynamical Theory of Biological Populations*, Nauka, Moscow, 1974, Chap. 2 (in Russian).

58. Svirezhev Yu. M. Vito Volterra and the modern mathematical ecology. The postscript to the book: Volterra, V., *Mathematical Theory of Struggle for Existence*, Nauka, Moscow, 1976, 245–286 (in Russian).

59. Clark, C. E. and Hallam, T. G. The community matrix in three species community models, *J. Math. Biology*, 16, 25–31, 1982.

60. Antonelli, P. L. A brief introduction to Volterra-Hamilton theory of ecological modelling, *Math. Comput. Modelling*, 13, 19–23, 1990.

61. Kolmogoroff, A. N. Sulla teoria di Volterra de la lotta per l'esistenza, *Giorn. Instituto Ital. Attuari*, 7, 74–80, 1936.

62. Kolmogoroff, A. N. Qualitative study of mathematical models of populations, *Problemy Kibernetiki* (Problems of Cybernetics), 25, 100–106, 1972 (in Russian).

63. Farkas, M. and Budt, H. On the stability of one-predator two-prey systems, *Journal of Mathematics*, 20, 909–916, 1990.

64. Bodini, A. and Giavelli, G. The qualitative approach in investigating the role of species interactions in stability of natural communities, *BioSystems*, 22, 289–299, 1989.

65. Marsden, J. E. and McCracken, M. *The Hopf Bifurcation and Its Applications*, Springer, New York, Chap. 1.

66. Pimm, S. L. and Gilpin, M. E. Theoretical issues in conservation biology. In: *Perspectives in Ecological Theory*, Roughgarden, J., May, R. M., and Levin, S. A., eds., Princeton University Press, Princeton, NJ, 1989, Chap. 20.

67. Pimm, S. L., Lawton, J. H. and Cohen, J. E. Food web patterns and their consequences, *Nature*, 350, 669–674, 1991.

68. Redheffer, R. Volterra multipliers II, *SIAM J. Alg. Disc. Meth.*, 6, 612–623.

69. Kraaijevanger, J. F. B. M. A characterization of Lyapunov diagonal stability using Hadamard products, *Linear Algebra and Its Applications*, 151, 245–254, 1991.

70. Logofet, D. O. *Matrices and Graphs: The Stability Problem of Mathematical Ecology.* Doctoral dissertation. Computer Center, USSR Acad. Sci., Moscow, 1985 (in Russian).

71. Ostrowski, A. Uber die Determinanten mit uberwiegender Hauptdiagonale, *Comment. Math. Helv.*, 10, 1937–1938, 1937.

72. Leontief, W. W. *The Structure of American Economy*, Academic Press, London, 1949, 181 pp.

73. Leontief, W. W. *Input–Output Economics*, Oxford University Press, New York, 1966, 561 pp.

74. DeAngelis, D. L., Post, W. and Travis, C. C. *Positive Feedbacks in Natural Systems* (*Biomathematics*, Vol. 15), Springer-Verlag, Berlin, 1986, 305 pp.

CHAPTER 5

Stability Analysis on Community Graphs

Knowledge about which particular stability subset a community matrix belongs to may give us valuable information on stability in the model. But to obtain this knowledge we must often apply an entry-based criterion to characterize the matrix. In other words, the quantitative magnitude of the interaction coefficients may be crucial for the community matrix to have a particular stability property. The question of whether there are stability properties which do not depend on quantitative magnitudes leads us naturally to formulations which are based exclusively on graph representations of the community models. If such a representation happens to be capable of yielding some knowledge about the stability properties, then the intrinsic relation between community structure and stability becomes more explicit.

We consider in this chapter methods and results of stability analysis using the "graph" approach. The most striking findings of these methods are in so-called "qualitative stability" analysis, which provides a convenient tool for studying beforehand the structure of intra- and interspecific relations in a multispecies community or a multi-component ecosystem model.

I. STABILITY IN CONSERVATIVE SYSTEMS

As noted before (Section 4.III.B), one can consider only those communities to be conservative which comprise exclusively prey-predator species with no self-limitation. It follows from conditions (4.20) that symmetrical entries of the interaction matrix Γ must have opposite signs, while Γ becomes skew-symmetric in its product with a positive diagonal matrix (Volterra's "hypothesis of equivalents"). Stability in a Lotka–Volterra conservative system was shown to be equivalent to the existence of a positive solution to the steady-state equations (4.18) with nonsingular matrix Γ. Clearly, this can always be achieved by a pertinent choice of ε_i values, where it would be sufficient to have, for instance, $\varepsilon = \Gamma[1, \ldots, 1]^T$. Stability of a conservative community is thus reduced to the nonsingularity of the interaction matrix Γ if the formulation of the problem allows ε_i to vary. Then it turns out that stability of the community can be ascertained within its trophic (nondirected) graph alone.

A. Nondirected Graph and Nonsingularity of Its Matrix

Although the graph associated with a community matrix is in general a directed graph, the nondirected graph is quite sufficient to display the structure of a conservative community. Indeed, by (4.31) and (4.20) the conservative community matrix takes on the form of a skew-symmetric matrix

$$A = \mathcal{A}\Gamma, \quad \mathcal{A} = -\text{diag}\{\mathbf{N}^*\}. \tag{5.1}$$

Thus, if there is an arc $j \to i$ in the (directed) associated graph, then there is also an arc in the opposite direction and both can be displayed by the nondirected arc, or *link* (examples are shown in Figure 19). Constructed in this way, the nondirected graph will be denoted by $G(\Gamma)$. Obviously, a particular graph is to be associated with the whole set of matrices Γ which differ only with respect to the magnitudes of their symmetrical entries.

Definition 5.1 *A subset P of the set of all links in* $G(\Gamma)$ *is said to give a* predator-prey pairing *if neither of the vertices belonging to the links of P is a vertex of more than one link.*[1] *The pairing is called* complete *if it comprises all vertices of the graph.*

The graphs in Figures 19a and 19b have no complete pairings; pairing arises in the latter when the link $2 \to 3$ is added (Figure 19c); for the graph of Figure 19d a complete pairing can be carried out in three different ways.

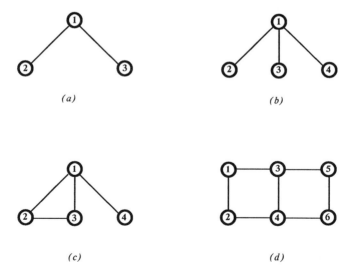

(a)

(b)

(c)

(d)

Figure 19. Examples of nondirected graphs for prey-predator communities (from Yorke and Anderson,[1] with permission).

The following theorem of Yorke and Anderson[1] enables one to check whether a matrix associated with a particular graph G can be *nonsingular* in the sense that $\det A \neq 0$.

Theorem 5.1 *Let A be a skew-symmetric matrix and $G(A)$ its associated (nondirected) graph. If A is nonsingular, then $G(A)$ has a complete predator-prey pairing. If, conversely, $G(A)$ has a complete predator-prey pairing, then there exists a nonsingular skew-symmetric matrix B such that $G(B) = G(A)$ and the entries in B can be chosen arbitrarily close to those of matrix A.*

Theorem 5.1 indicates that the existence of a complete pairing is necessary for a skew-symmetric matrix $A = \mathcal{A}\Gamma$, hence matrix Γ itself, to be nonsingular, that is, for a corresponding conservative community to be stable. The condition is also sufficient in the sense that if it is true but matrix $A = \mathcal{A}\Gamma$ (hence matrix Γ) is singular, then a small variation of its skew-symmetric entries can result in a nonsingular matrix B; consequently, also nonsingular will be the matrix $\tilde{\Gamma} = \mathcal{A}^{-1}B$, arbitrarily close to Γ, thus corresponding to a stable conservative community.

By Theorem 5.1 we can state immediately that any conservative community with a graph like that of Figure 19a is unstable (as was noted already by Volterra, stability of conservative systems is possible only when the number of species is even); any conservative community with the graph of Figure 19b is unstable as well. For the other two graphs in Figure 19, one can always find a stable conservative system whose matrix is either coincident with or close to a corresponding skew-symmetric matrix.

When the existence of a complete pairing is not so evident, the formal notion of the *trophic graph* discussed below may be of some use.

B. Complete Pairing in Trophic Graphs

To define a *trophic graph* we need to know "who eats whom" in the community structure, so that a nondirected representation should become insufficient. Let the arrow of each prey-predator link now indicate the predator, i.e., the direction of biomass transfer.

Definition 5.2 *A graph $G(\Gamma)$ is said to be* trophic[1] *if there exists a subset T of its vertices such that none of the vertices of T is linked (by arrow) to another vertex of T and all of the vertices outside T are linked (by arrows) to vertices of T.*

In terms of real trophics the subset of species T is such that no member of the subset feeds on a species of T and all of the species outside T feed only upon species of T. In the trophic graph of Figure 20 the vertices of T are depicted by squares. This graph also illustrates a more specific case where the trophic structure is a hierarchy of *trophic levels*, i.e., aggregates of species that feed only upon species of the next level, while being the food for species of the previous level.

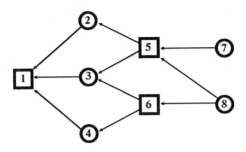

Figure 20.   A graph with trophic levels. Arrows go from prey to the predators (from Yorke and Anderson,[1] with permission).

A hierarchy of trophic levels is obviously a trophic graph, since the subset T can be composed of all those levels which are at odd "distances" from prey at the initial level.

Theorem 5.2 *If A is a nonsingular matrix and graph $G(A)$ is trophic, then it has a complete predator-prey pairing.*

The formal proof of Theorems 1 and 2 can be found in Yorke and Anderson[1] or Svirezhev and Logofet.[2] It ensues from the Euler theorem on the structure of a (directed) graph (see, e.g., Berge[3]) applied to the graph that corresponds to a nonzero term in the standard expansion of $\det A$, or equivalently, from the fundamental property of n-element substitutions, namely, their representability as a finite number of non-intersecting cycles (see, e.g., Mal'tsev[4]).

Theorem 2 provides for some necessary conditions that facilitate the search for a complete predator-prey pairing in graphs with several trophic levels. Let a nonsingular matrix have an associated graph with several trophic levels, labeled by $i = 1, 2, \ldots, l$, and with n_i species in the i-th level. It follows that:

if $l = 2$, then $n_1 = n_2$ as a complete pairing is otherwise not possible;

if $l = 3$, then $n_1 + n_3 = n_2$ (clearly, $n_1 < n_2$ and $(n_2 - n_1)$ species of the 2nd level must be paired to n_3 species of the 3rd level;

if $l > 3$, then $n_i = \alpha_i + \beta_i$ $(i = 1, 2, \ldots, l)$, where $\alpha_i \geq 0$, $\beta_i \geq 0$, $\alpha_1 = \beta_l = 0$, $\beta_i = \alpha_{i+1}$.

The final condition does apparently not hold for the graph in Figure 20, so that it cannot have a complete pairing, hence the pertinent skew-symmetric matrix is singular and the conservative community cannot be stable in this case.

A large number of species in the predator-prey community can make the search for a complete pairing rather difficult. Special algorithms however exist that can solve the problem by searching for a pairing that covers the maximum number of vertices, hence the complete pairing, if it ever exists.

II. QUALITATIVE STABILITY IN MODEL ECOSYSTEMS AND SIGN STABILITY OF THE COMMUNITY MATRIX

Conservative systems of population dynamics represent a high degree of idealization, so that the results and methods of the previous section have mostly only theoretical value. On the other hand, the theory which we explain further on does not require any assumptions about the interaction matrix or even about the model equations themselves, except that a feasible equilibrium does exist and the system can be linearized at the equilibrium point.

A. Qualitative Stability in Model Systems

The notion of *qualitative stability* is defined in intuitive rather than formal terms for a system of interacting components of any kind. In a general context, this means that a system holds its stability under any of those quantitative variations in the strength of linkage among the system components which however keep unchanged the qualitative types of all interactions in the system.

In other words, the qualitative stability is a property to be determined by the qualitative pattern of systems interactions alone, but not to depend on their quantitative expressions. In ecological terms, qualitative stability means that an ecosystem or community holds its stability under any quantitative variations in intensity of interactions among components (species) which however keep unchanged the type of intra- and interspecies relations for all components (all species).

The appeal this concept has to an ecologist is apparent: while it is always a hard problem to estimate quantitatively the "strength of linkage," there is generally more certainty in what concerns the types of interactions within a community. And already from this qualitative knowledge, i.e., from the signs of interactions between each pair of species in the community, it may be possible to speculate on stability in a whole class of models describing the dynamics in the community of a given structure.

It is clear that "any variations in intensity of interactions" carries too much uncertainty, and even to assess whether there exists any qualitatively stable system requires the notion to be defined in more formal terms. The natural way to do this for models of community dynamics is to use the fundamental relation between the structure and stability as expressed in the community matrix. The formalization that is considered below relies upon Lyapunov stability of a feasible equilibrium in the system of model equations, and the qualitative stability then reduces to the so-called *sign stability* of the community matrix, which means that the eigenvalues must stay on the left of the imaginary axis as long as no changes occur in the allocation of pluses, minuses, and zeros within the matrix.

Clearly, the results obtained by this approach are applicable not only to community models but to any case where the model admits linearization. Actually, the first conditions relevant to qualitative stability were formulated in mathematical economics[5-6] and then reformulated by R. May[7] in ecological terms. A kind of characterization was also proposed but no conditions were obtained at that moment which would be both necessary and sufficient for an $n \times n$ matrix to be sign-stable. Rigorous characterizations were obtained later[8-12] in terms of *signed directed graphs* and types of species pair-wise interactions as defined in Chapter 4.I. All these conditions, which have brought about a complete description of the class of sign-stable systems, are considered in further sections of this chapter.

B. Equivalence Classes and Sign Stability

Let the dynamics of population sizes (or densities) in an n-species community be governed by system (4.1) with a community matrix A (4.2). Then the well-known fact of the Lyapunov stability theory states that equilibrium $\mathbf{N}^* > \mathbf{0}$ is asymptotically stable if it is true for the zero solution of the linearized system

$$\dot{\mathbf{x}} = A\mathbf{x}, \tag{5.1}$$

where $\mathbf{x}(t) = \mathbf{N}(t) - \mathbf{N}^*$. The latter is equivalent to matrix A being *stable*, i.e., Re $\lambda_i < 0$, $i = 1, \ldots, n$. As noted in Section 4.I, the qualitative structure of species interactions can be represented by the sign structure (4.3) of matrix A or the corresponding *signed directed graph* (SDG).

Quantitative changes in intensity of relations among and within species of the community result naturally in variations of the entries in the community matrix. As long as we argue in terms of linear approximation, these two fields of variations are equivalent. If the community stability is identified with the Lyapunov asymptotic stability of equilibrium $\mathbf{N}^*$, then the qualitative stability reduces to the stability of the community matrix that holds for any variations in absolute values of matrix entries a_{ij} which do not disturb the sign pattern $S(A)$ of the matrix. (Bearing in mind a nonlinear prototype of system (5.1), we must certainly restrict "any variations" to those which do not deprive the system of its equilibrium $\mathbf{N}^* > \mathbf{0}$.) This matrix property was termed *sign stability*.[5]

A mathematically trivial, although ecologically absurd, example of qualitative stability is a "community" of n self-regulating species with no interactions among species at all. The community matrix is then nothing but a diagonal matrix with negative entries in the diagonal, showing that the set of sign stable matrices is at least nonempty. Less trivial is an example of the prey-predator pair with self-regulation in prey as modeled by the Lotka–Volterra system (4.37) with $e > 0$, $f = 0$; its community matrix of sign pattern

$$\begin{bmatrix} - & - \\ + & 0 \end{bmatrix}$$

is always stable too.

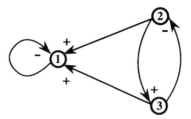

Figure 21.   A signed directed graph: self-limiting species 1 is commensalistic to prey-predator pair 2-3.

In more formal terms, the notion of sign stability is defined as follows.[11]

Definition 5.3 *Matrix B* = [b_{ij}] *is called* sign-equivalent *(or just* equivalent*) to a matrix A (notation A ∼ B) if* Sign a_{ij} = Sign b_{ij} *for any i, j* = 1, ..., *n (which means S(A) = S(B)).*

   The relation of sign equivalence is clearly *reflexive* (A ∼ A), *symmetrical* (if B ∼ A, then A ∼ B), and *transitive* (if A ∼ B and B ∼ C, then A ∼ C), hence this is an *equivalence relation* in the general sense of modern algebra (see, e.g., Van der Waerden[13] or Krantz[14]). As an equivalence relation, the sign equivalence determines a partition of the set of all real $n \times n$ matrices into equivalence *classes*, each class consisting of all equivalent matrices. Since the sum of two sign-equivalent matrices and the product of a matrix with any positive number remain obviously in the same equivalence class, the classes represent convex (unpointed) cones in the linear space of real $n \times n$ matrices. More precisely, they represent coordinate "semi-planes" in which certain of the coordinates (i.e., matrix entries) are zero and the rest have fixed signs.

   Each equivalence class can be uniquely presented by its sign structure of the form (4.3) or by the corresponding SDG. For example, any matrix of the following form:

$$A_1 = \begin{bmatrix} -a & b & c \\ 0 & 0 & -d \\ 0 & e & 0 \end{bmatrix}, \quad a, b, c, d, e > 0 \tag{5.2}$$

corresponds to the SDG shown in Figure 21; any matrix equivalent to

$$A_2 = \begin{bmatrix} 0 & -1 & 0 & 0 & 0 \\ 1 & 0 & -1 & 0 & 0 \\ 0 & 1 & -1 & -1 & 0 \\ 0 & 0 & 1 & 0 & -1 \\ 0 & 0 & 0 & 1 & 0 \end{bmatrix} \tag{5.3}$$

Figure 22.   SDG of a 5-species community: species 1 is a food to species 2, species 2
to 3, and so on; species 3 is self-limited.

corresponds to the SDG shown in Figure 22. Now *sign stability* can be given the
following

Definition 5.4 *Matrix A is called* sign-stable *if stable is any matrix B equivalent
to A.*

If a matrix $A = [a_{ij}]$ is not sign-stable, then, under its fixed sign pattern, there
exist values of a_{ij}, say a_{ij}°, such that the spectrum of matrix $A^{\circ} = [a_{ij}^{\circ}]$ contains
eigenvalues with Re $\lambda_i \geq 0$; it is also possible that A is stable for other particular
sets of a_{ij}.

Sign stability is clearly a property of the whole class of equivalent matrices, or
the corresponding SDG, which represents essentially the structure of trophic and
other intra- and interspecies relations in the community. Any formal test for sign
stability should therefore allow formulation either in terms of matrix sign structure
or in terms of SDG. In what follows we shall use both of these possibilities.

C. Necessary Conditions of Sign Stability

The following conditions[6] are necessary for a real $n \times n$ matrix $A = [a_{ij}]$ to be
sign-stable:

(1) $a_{ii} \leq 0$ for all i and $a_{i_0 i_0} < 0$ for some i_0;

(2) $a_{ij} a_{ji} \leq 0$ for any $i \neq j$;

(3) for any sequence of 3 or more indices $i_1 \neq i_2 \neq \cdots \neq i_m$ the inequalities
$a_{i_1 i_2} \neq 0,\ a_{i_2 i_3} \neq 0, \ldots, a_{i_{m-1} i_m} \neq 0$ imply $a_{i_m i_1} = 0$;

(4) there exists a nonzero term in the standard expansion of the determinant

$$\det A = \sum_{\sigma} \operatorname{sgn} \sigma \prod_{i=1}^{n} a_{\sigma(i)i} \tag{5.4}$$

where σ is a permutation of numbers $\{1, 2, \ldots, n\}$, sgn$\sigma = \pm 1$ is the sign
of σ.

The following argument is typical for proving necessity of these conditions. As follows from the Routh–Hurwitz criterion, for all roots of the characteristic polynomial $p(\lambda) = \lambda^n + k_1\lambda^{n-1} + \cdots + k_n$ to have negative real parts, all its coefficients k_i must be positive. In particular, the inequalities

$$k_1 = -\sum_{i=1}^{n} a_{ii} > 0, \tag{5.5}$$

$$k_2 = \sum_{i \neq j}(a_{ii}a_{jj} - a_{ij}a_{ji}) > 0 \tag{5.6}$$

must hold for any magnitudes of entries a_{ij} within a fixed sign pattern. If Condition 1 were violated in a matrix A, i.e., there existed $a_{pp} > 0$, then one could always find a matrix $\tilde{A}$ equivalent to A and having $\tilde{a}_{pp} > 0$ great enough for k_1 becoming negative, hence matrix $\tilde{A}$ being unstable. The necessity of Condition 2 ensues similarly from inequality (5.6).

The proof of Condition 3 is given in the Appendix, Statement 5.1.

The necessity of Condition 4 is obvious: if it is violated, then $\det A = 0$, hence $\lambda = 0$ is an eigenvalue of matrix A, which therefore cannot be stable.

Conditions 1 through 4 can be easily verified for particular community matrices as they have clear interpretations in terms of matrix sign patterns or/and corresponding SDGs. The first three of them forbid some visible features in the sign pattern and SDG, while the forth can also be checked in the graph.

Condition 1 states that there cannot be any self-stimulating species in a qualitatively stable community, and at least one species must be self-limited. In SDG terms: no 1-loops of the plus sign and at least one 1-loop of the minus sign. All examples cited up to now (Figures 10, 21, and 22) apparently meet this condition.

Condition 2 means that no relations of competition $(--)$, nor mutualism $(++)$ can be in a qualitatively stable community. This is obviously violated in Figure 10 but holds in Figures 21 and 22.

Condition 3 forbids any directed loop (or *cycle*) of length three or more in the community structure. This is probably the most severe restriction for ecological systems. In particular, it excludes all "omnivory" cases, where a predator feeds on two prey species one of which also serves as a food to another, or in more general terms, when a predator feeds on more than one trophic level.[15–17] The SDG in Figure 10 contains a lot of such cycles (e.g., $2 \to 4 \to 5 \to 2$ or $2 \to 5 \to 4 \to 2$) but both Figure 21 and Figure 22 contain only 1- and 2-cycles.

Finally, Condition 4 is more technical but still interpretable in the graph terms. Recall that a *skeleton* subgraph of a graph is its subgraph which involves all vertices of the graph. Condition 4 is then equivalent (see the Appendix, Statement 5.2, for the proof) that the SDG contains a skeleton subgraph which splits into a number of disjoint directed cycles (whose total length also equals n). With regard to Condition 3, forbidding dicycles longer than 3, this means that a sign-stable pattern must contain some m prey-predator pairs (i.e., 2-cycles) such that the rest $(n - 2m)$ species possess self-regulation (i.e., 1-cycles). Such "skeletons" for the SDGs of Figures 21 and 22 are shown respectively in Figures 23 and 24.

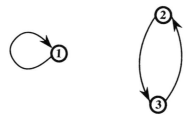

Figure 23.   A skeleton subgraph: splitting up the SDG in Figure 21 into 1- and 2-cycles.

Figure 24.   A skeleton subgraph: splitting up the SDG in Figure 22 into 1- and 2-cycles.

In some publications Condition 4 was cited as a simpler one, namely, that $\det A \neq 0$. This must certainly be true for any stable matrix but contradicts the very idea of sign stability since, to be verified, it may require the knowledge of particular numeric values of the matrix entries.

On the other hand, Conditions 1 through 4 already imply that $\det A \neq 0$. If, indeed, the permutation σ generating a nonzero term in (5.4) consists of m 2-cycles and $(n - 2m)$ 1-cycles, then, due to Conditions 1 and 2, we have the sign of the term equal to

$$\mathrm{sgn} \left[\mathrm{sgn}\ \sigma \prod_{i=1}^{n} a_{\sigma(i)i} \right] = (-1)^m \mathrm{sgn} \left[\prod_{i=1}^{n} a_{\sigma(i)i} \right] = (-1)^m(-1)^m(-1)^{n-2m} = (-1)^n.$$

So, the terms of the sum in (5.4) are either all positive or all negative, with no chance for $\det A$ to be zero.

Thus, all of the Conditions 1 through 4 of a matrix sign stability can be checked in the corresponding SDG, the violation of any one indicating that the matrix is not sign-stable. As seen above, for example, the SDG in Figure 10 violates Conditions 2 and 3, so that the equivalence class of matrices is not sign-stable.

On the contrary, matrices (5.2) and (5.3) are seen to meet the Conditions but whether they are sign-stable is still a question. As a matter of fact, A_1 has a pair of pure imaginary numbers $\lambda_{1,2} = \pm i\sqrt{de}$ ($\lambda_3 = -a$) in its spectrum, thus it is not even stable. The spectrum of A_2 also contains a pair of $\lambda = \pm i$ depriving of sign stability the equivalence class of A_2.

Thus, Conditions 1 to 4 are insufficient for a matrix to be sign-stable, and what are the conditions that guarantee sign stability is considered in the next section.

III. "COLOR TEST" AND THE CRITERION OF SIGN STABILITY

As the notions of local and global Lyapunov asymptotic stability are equivalent within linear systems, the question of sign stability is actually the question of when, given only signs of matrix entries, can one be sure that all the trajectories of system (5.1) converge to the origin? This conjecture will be used below in proving a graph theoretic criterion of sign stability; also, the criterion appears to be closely related to the property of the matrix being indecomposable.

A. Indecomposability and Matrices of 0-Symmetry

Indecomposability of a matrix, as defined in Chapter 1.III, is an important property in formulations of sign stability conditions. Recall that the matrix being inde-composable is equivalent to its associated graph being strongly connected, thus signifying a kind of integrity in the community structure. Originating from the associated graph by the addition of signs, the SDG must obviously show the same property of being or being not strongly connected as its origin does. By this reason, for example, it can be seen immediately that matrix (5.2) is decomposable as its digraph (Figure 21) is not strongly connected, while matrix (5.3) is indecomposable as its digraph (Figure 22) is strongly connected.

Verifiable by means of digraphs, indecomposability of a matrix is also a property of the whole class of sign-equivalent matrices, thus making sense of using it in formulations of sign stability conditions. Moreover, when combined with a certain restriction on the matrix sign pattern, indecomposability gives rise to a nontrivial result stated in the definition and lemma below.

Definition 5.5 *A matrix $A = [a_{ij}]$ is said to be* 0-symmetric *(or possess* 0-symmetry*) if for any pair of subscripts $i \neq j$ the condition $a_{ij} = 0$ implies $a_{ji} = 0$; otherwise the matrix is* 0-asymmetric.

Matrix (5.3), for instance, is 0-symmetric, whereas matrix (5.2) is 0-asymmetric.

Lemma 5.1 *Let a matrix $A = [a_{ij}]$ satisfy Condition 3 (i.e., no cycles in its SDG longer than 2). If, in addition, A is 0-asymmetric, then A is decomposable.*

The proof, whose main idea was set forth by Quirk and Ruppert,[6] is given in the Appendix.

If an indecomposable matrix (or strongly connected SDG) satisfies Condition 3, then the matrix must be 0-symmetric, for it would otherwise contradict Lemma 5.1. Thus, in a strongly connected SDG with no cycles longer than 2, once any pair of vertices is linked by an arc, the pair is also linked by the arc in the opposite direction (see Figure 22). In other words, any arc of such an SDG is either a self-influence loop (i.e., 1-cycle), or is a link of mutualism, competition or predation between a pair of species (i.e., half of a 2-cycle). In view of Condition 2, only those of predation can enter a sign-stable SDG. As exemplified by matrix (5.3), Conditions 1 to 4 turn out to be insufficient even for an indecomposable matrix to be sign-stable. However, indecomposable prey-predator matrices and their SDGs prove to be fundamental to the further theory of sign stability.

If a decomposable matrix A satisfies Condition 3, then each of its indecomposable diagonal blocks also does. By Lemma 5.1 each of them represents a square matrix of symmetric 0-pattern. Since the spectrum of a decomposable matrix is composed of the spectra of its diagonal blocks, matrix A is stable if and only if all of the blocks are stable. Therefore, to investigate sign stability in an arbitrary square matrix, one must do so in each of its indecomposable diagonal blocks, or equivalently, in each strong component of the associated SDG. Therefore, we concentrate further on sign stability conditions for the indecomposable community matrix and its strongly connected SDG.

B. Predation Graphs and the "Color Test" Failure: Sufficiency

It makes sense to search for sufficient conditions of sign stability among those matrices or SDGs which meet the necessary conditions. In this section it will be shown how Conditions 1 to 4 can be extended to provide for sign stability in an indecomposable community matrix and its strongly connected SDG.

As we noted above, Conditions 2 and 3 together imply that only prey-predator links are allowed in a strong, sign-stable SDG. Therefore, it is easy to find groups of species that form indecomposable diagonal blocks in the community matrix (and strong components in its SDG): one should just delete all arrows of amensalism (0−) or commensalism (0+) in the SDG. Then any strong subgraph of the SDG will consist either of an isolated vertex, or a number of prey-predator pairs linked by +− arrows (mutualism and competition are forbidden by Condition 2).

In the following passages, these strongly connected subgraphs are called *predation subgraphs* (or *predation substructures*). For the sake of uniform terminology, isolated vertices are also called *trivial predation subgraphs* (or *trivial substructures*). For example, the SDG in Figure 21 contains two predation subgraphs: $\{1\}$ and $\{2 \leftrightarrow 3\}$, while the SDG in Figure 22 consists entirely of a single predation subgraph.

From the comments made above it follows that sign stability of a matrix is equivalent to sign stability in all its predation substructures if any. The criterion given below for a predation graph to be sign-stable, is associated with the so-called "color test."[8,10]

Definition 5.6 *A predation graph is said to* pass the color test *if each of its vertices can be colored* black *or* white *in such a way that:*

(a) *all self-regulated vertices are black;*

(b) *there exist white vertices, any one of which is linked to at least another one white vertex;*

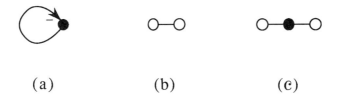

(a)   (b)   (c)

Figure 25.   Rules (a), (b), (c) of the color test.

(c) if a black vertex is linked to a white one, then it is also linked to at least another one white vertex.

If the coloring is impossible, then the graph does not pass, or *fails* the test.

Theorem 5.1 *If an indecomposable matrix meets all Conditions 1 to 4 and the following Condition 5:*

(5) The predation graph fails the color test,
   then the matrix is sign-stable.

In particular, a trivial structure (an isolated species) is sign-stable if and only if it is self-regulated: in this case rule (a) of the color test always contradicts rule (b). But if an isolated species has no self-regulation, then the necessary Conditions 1 and 4 are violated.

The proof of Theorem 5.1 is given in the Appendix. Its main idea (proposed by Jeffries[8]) relies upon constructing a Lyapunov function for the zero solution to system (5.1), whereby its stability ensues, either asymptotic or neutral. The color test then proves to be nothing else than a convenient tool to examine the conditions necessary for oscillating components to exist in a perturbed solution, i.e., necessary for nonasymptotic stability. Failure of the test then makes possible the case of asymptotic stability alone.

Note that to check and implement the color test rules in a predation (sub)graph it is sufficient that the graph be nondirected; it can be abstracted from the SDG by substituting a nondirected link for each pair of directed prey-predator arcs and deleting the signs. Also the test rules themselves can be conventionally shown as in Figure 25.

In view of Condition 5 it is easy to explain the lack of sign stability in the SDGs of Figures 21 and 22. Predation subgraph $\{1\}$ in Figure 21 violates rule (b), whereas predation subgraph $\{2 \leftrightarrow 3\}$ passes the color test. Predation graph in Figure 22 actually passes the test, but should self-regulation be transferred to any other vertex, the SDG would fail the test, and hence acquire sign stability.

When all $n \geq 2$ vertices of a predation graph are colored white, then the test is obviously passed. Hence, the presence of self-limited species, i.e., Condition 1,

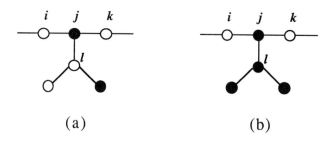

(a)   (b)

Figure 26.   Additional condition (c′) to the color test.

is, in fact, a consequence of Condition 5. Thus, not only the presence of self-limited species is crucial for sign stability, but also their special allocation within the community structure as specified by Condition 5.

The above-mentioned examples, where the conditions of Theorem 5.1 were violated, as well as many other examples, show these sufficient conditions to be sufficiently "delicate": there is no case where a matrix were sign-stable when passing the color test. The actual reason is quite simple: together with the first four, Condition 5 is in fact the necessary one for sign stability of an indecomposable matrix. This is proved below.

C. Predation Graphs and the "Color Test" Failure: Necessity

A logic to prove that Condition 5 is also necessary for a predation matrix to be sign-stable may be quite simple: if, on the contrary, the SDG of A passes the color test, then we have to find an equivalent matrix $\bar{A}$ which is not stable. But to implement the logic will require some additional constructions below.

First, note that if a predation graph passes the color test, then it also passes the test with a more stringent rule (c′):

(c′) if a black vertex is linked to a white one, then it is linked to exactly two white vertices.

If, indeed, a black vertex j (Figure 26a) is linked to more than two white ones (for instance, to vertices i, k, and l in Figure 26), then "excessive" white vertices (vertex l) and the whole proper branches of the graph (i.e., those which are connected with vertex j via l) can be recolored black (see Figure 26b). Rules (a)–(b) still hold as there are no cycles in the graph.

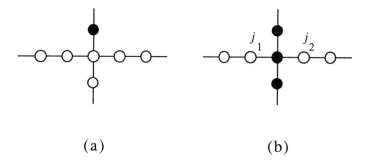

(a)   (b)

Figure 27.   Additional condition (d) to the color test.

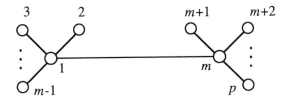

Figure 28.   Canonical structure of a maximal connected white subgraph.

Second, if a graph passes the test with conditions (a)-(c′), then the coloring can be changed to meet the following additional rule:

(d) there are no *white branches* (i.e., those consisting of white vertices) that contain more than 4 vertices.

If, indeed, there is a branch of 5 or more vertices in a graph (see Figure 27a), then the third of those vertices can be recolored black (as shown in Figure 27b) together with all the other branches (if any) passing through the third vertex. Rules (a) to (c) are evidently not affected.

If, at last, the graph is colored by the rules (a)–(c′), (d), then any of its maximal connected *white subgraphs* (i.e., subgraphs consisting of white vertices) looks like that shown in Figure 28, where $m \geq 2$, $p \geq m$. Two maximal connected white subgraphs can be linked (by a branch of length 2) through only one black vertex and exactly two white ones belonging to the different subgraphs.

If, indeed, a white subgraph spans 2, 3, or 4 vertices, then it represents a special case of Figure 28. If there are more vertices, then by rule (d) no more than four of them can compose a branch, which gives rise to the pattern shown in Figure 28. If a connection of two maximal white subgraphs were realized in more than one branch, it would mean the existence of long cycles in the associated graph. If the connecting branch spanned more than one black vertex, the predation graph would break rules (c) or (c′) of the color test.

Let now the submatrix B of A, that corresponds to a maximal strong white subgraph, be formed by p rows and columns labeled as shown in Figure 28 (any other labeling can be changed by a pertinent permutation of rows and columns). Then it takes on the following form:

$$
B = \left[
\begin{array}{ccccc|cccc}
0 & b_2 & \cdots & b_{m-1} & b_m & 0 & \cdots & 0 \\
c_2 & & & & 0 & & & \\
\vdots & & \mathbf{0} & & \vdots & & \mathbf{0} & \\
c_{m-1} & & & & 0 & & & \\
\hline
c_m & 0 & \cdots & 0 & 0 & b_{m+1} & \cdots & b_p \\
0 & & & & c_{m+1} & & & \\
\vdots & & \mathbf{0} & & \vdots & & \mathbf{0} & \\
0 & & & & c_p & & &
\end{array}
\right], \tag{5.7}
$$

where $b_j c_j < 0$, $p \geq m \geq 2$.

An idea underlying the proof that Condition 5 is necessary, is to show that these are the maximal strong white subgraphs which are responsible for pure imaginary eigenvalues occurring in the spectrum of a "color-valid" matrix. The idea is grounded on the following two lemmas proved in the Appendix.

Lemma 5.2 *For any positive number ω there exists a matrix $\tilde{B}$ equivalent to B of the form (5.7) and such that $\lambda = \pm i\omega$ are eigenvalues of $\tilde{B}$.*

Lemma 5.3 *For any matrix B of the form (5.7) there exists an equivalent matrix $\tilde{B}$ such that the system of equations*

$$
\dot{\mathbf{x}} = \tilde{B}\mathbf{x}, \quad \mathbf{x} \in \mathbb{R}^p, \tag{5.8}
$$

has a particular solution

$$
x_j(t) = D_j \cos \omega t + E_j \sin \omega t, \quad j = 1, \ldots, p, \tag{5.9}
$$

with constants D_{j_0}, E_{j_0} taking on any prescribed values for a fixed subscript j_0.

By means of these lemmas the following theorem can now be proved.

Theorem 5.2 *If an indecomposable matrix is sign-stable, then it meets all the Conditions 1 to 5.*

Conditions 1 to 4 were proved to be necessary before and, before strengthened with Condition 5, they were also shown to restrict the set of matrices under consideration to predation matrices alone. If a predation matrix passes the color test, then Lemmas 5.3 and 5.4 will allow us to construct a nonvanishing partial solution to system (5.1) with an equivalent matrix, signifying the lack of sign stability. The proof is given in the Appendix in technical detail.

D. Characterization of Sign-Stable Communities

We can now attenuate the indecomposability restriction on the community matrix, bearing in mind that a decomposable matrix is sign-stable if and only if all its inde-composable blocks are sign stable; it is with these blocks that strongly connected components of the original SDG, or predation subgraphs, are associated. Thus, we come to the general criterion of sign stability that is given in the following theorem.

Theorem 5.3 *A real matrix $n \times n$ is sign-stable if and only if it meets Conditions 1 to 4 and each of its predation subgraphs fails the color test. (*A shorter formulation is just *Conditions 1 to 5.)*

As strong components of the SDG, the predation subgraphs can be linked to one another by arcs of either amensalism or commensalism, because any other kind would combine them into a single strong component. Besides, those arcs must not form any directed cycles in order that the SDG might have a chance of being sign-stable. Interpreting now Theorem 5.3 in ecological terms, we obtain a complete description, or characterization, for the class of sign-stable community structures: these are

- *structures of only prey-predator or similar types of interspecies relations (resource-consumer, host-parasite) with special allocation of self-limited species to fail the color test*, and also

- *any collections of these substructures linked by amensalistic and/or commensalistic relations forming no directed cycles in the whole community graph.*

Described so, the class of qualitatively stable communities appears to be fairly narrow. By no means, for example, can it contain models of nutrient turnovers so typical in modeling real ecosystems, nor omnivorous cases, where a predator feeds on more than one trophic level—all of them induce cycles in the community graphs. But still the criterion of sign stability may be of some use in both the theory and practice of ecological modeling, and this is considered in the next section.

IV. APPLICATIONS: MATRIX THEORY AND MODELING PRACTICE

Established in Theorem 5.3 the sign stability criterion gives a convenient means to prove some theorems relevant to matrix stability and to develop further the logical hierarchy of stability subsets. In the practice of ecological modeling, the

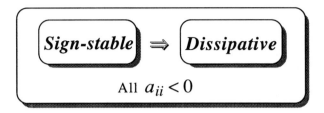

Figure 29.   Theorem 5.5.

criterion brings about deeper insight into cause-effect relations holding between the structures of interactions and stability in dynamic behavior.

A. New Petals in the Matrix Flower

An immediate consequence of Theorem 5.3 is a theorem on the conditions of sign stability for matrices with all negative entries in the main diagonal, which formerly required special constructions.[6]

Theorem 5.4 *A real $n \times n$ matrix $A = [a_{ij}]$ with all $a_{ii} < 0$ is sign-stable if and only if it meets Conditions 2 and 3. In other words, there must be neither mutualism, nor competition in the community, and no cycles longer than 2 in its SDG.*[2]

The proof now is as easy as to see that the rest of the Conditions already hold true: the product of all the diagonal entries produces a nonzero term in the determinant and the color test stops at rule (b).

Theorem 5.4 can, in turn, identify logical relations among sign stability and the other matrix properties we considered before. First, as multiplying matrix A by a positive diagonal matrix does not change the sign pattern of A, the sign stability clearly implies D-stability. Furthermore, if A is sign-stable, then it is at least totally stable.[2] Indeed, by Theorem 5.4, matrix A satisfies Conditions 2 and 3, hence so does any of its principal submatrices, because deleting any subset of vertices and the arcs connected with them can neither change the signs of interactions, nor result in any new cycles.

Actually, the above statement is a consequence of an even stronger theorem given below.

Theorem 5.5 *A sign-stable matrix $A = [a_{ij}]$ with all $a_{ii} < 0$ is dissipative.*[18]

Proof is given in the Appendix, and Figure 29 shows the theorem in the fashion of Figures 14 and 15.

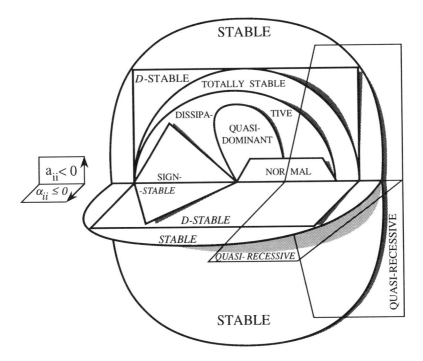

Figure 30.   Sign stability "petals" in the matrix flower.

In the more developed fashion, the sign-stability criterion allows the logical "flower" in Figures 16 and 17 to now be grown still further, developing into the form in Figure 30, while following the same conventions as before.

The set of sign-stable matrices, $s\mathbb{S}$, represents all those of "coordinate" subspaces in the space of real $n \times n$ matrices which bear sign-stable matrices. Despite the maximum freedom of variation within the subspaces, the smallest change in a zero entry may bring the matrix out of the set $s\mathbb{S}$, which, therefore, is not open in the whole space. Set $s\mathbb{S}$ and the set of normal stable matrices intersect by the (zero measure) subset of negative diagonal matrices, since normality (i.e., the property $A^T A = A A^T$) requires certain equality-type relationships to hold among matrix elements, which may be trivially disturbed in any nontrivial class of sign equivalence. For a similar reason the intersection of diagonally quasi-dominant matrices also contains only matrices with the negative main diagonals.

B. Sign Stability and Multispecies Structures

The criterion of sign stability can be also useful as a convenient tool in studying theoretical models of multispecies communities. The next Chapter will demonstrate its application to a mathematical theory of trophic chains. Another area of application lies within a traditional approach to investigating the traditional "stability vs. complexity" problem[19] in model ecosystems.

The approach is based on constructing a statistical ensemble of community matrices pertaining to certain properties of their real prototypes (see Pimm,[17] Svirezhev and Logofet,[20] and references therein). Calculating the percentage of stable matrices in the ensemble gives rise to speculations on the role of matrix stability in community organization. It is clear that, given the rules to construct an ensemble, one can determine the measure of sign-stable matrices within the ensemble, that is, to estimate the minimal share of stable matrices. This is important, since the computer routine to analyze matrix stability may be incomplete, falsely identifying the ill-conditionality of a matrix with its being unstable, thus resulting in erroneous conclusions. To examine the routine, the sign stability criterion can easily generate any number of nontrivial examples of stable matrices of an arbitrary size $n \times n$.

As already noted, the class of qualitatively stable systems described by Theorem 5.3 appears to be fairly narrow. But the lack of sign stability does not yet mean that the system cannot be stable at all: rather the matter is about a maximum domain of stability in the parameter space of the model. The lack of sign stability just signifies a greater vulnerability of stable dynamic behavior under variations of intra- and/or interspecies relations in the community.

Therefore, the criterion of sign stability provides a convenient tool, preliminary and quick enough, to analyze trophic and other structures from the viewpoint of stability in a proper dynamic model. To apply the tool, nothing else is needed than displaying the structure by means of the corresponding SDG. In particular, the criterion indicates that the presence of density self-regulated species and their special allocation within the structure is of critical importance. Besides, the criterion may indicate those links whose presence or absence is crucial for the whole structure to be stable.

The case study presented below is a sign stability investigation[21] of various graphs aggregated from the total scheme of biological interactions in the community of a cotton plant. It proves that a population-theory explanation can be proposed for a number of such observed phenomena as, for instance, the stabilizing effect of competition among insect pests, and the destabilizing effects of wilt disease pathogens and the competition between parasites and predators of pests.

C. Case Study: Biocenosis of a Cotton Plant

The structure of biological interactions in the community of a cotton plant[22] illustrates the general scheme of the turnover of matter in terrestrial ecosystems shown in Figure 11. Cotton culture as a primary producer is the basic component. Diverse phytophagans, cotton pests, are alive due to the culture. They include various genera of aphids (cotton, melon, alfalfa black), spider mites, a variety of Heliothis earworms (corn, winter, earthy), some hemiptera and locust species. Pests find abundant feeding resources and reproduce intensively.

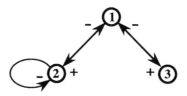

Figure 31.   Maximum-aggregated SDG of the cotton biocenosis.

Also, the cotton field is permanently inhabited by enormously great number of insect and arachnid species feeding on debris, plant remnants, weeds, nectar, and the sweet excretion of aphids (honeydew). These organisms (ants, bees, grasshoppers, spiders, ground beetles, etc.) are not directly linked to the cotton plant, but find favorable habitats in the field. The presence and diversity of this fauna is strongly believed to produce stability in the biocenosis by balancing its various components and often protecting the monoculture from sudden pest outbreaks[22] (see also Pimm et al.[23] and references therein).

Following the pests in the cotton there appear parasites which attack the pests and predators which feed on the pests. The predators include the thrips (Aeolothrips, Scolothrips, and other spp.), bugs (Nabis, Orius, and other spp.), ladybugs (variable, seven-spotted, mite-eating, and other spp.), golden-eyed flies, syrphus flies, ground beetles, ants, and spiders. Parasites of cotton pests are ichneumons, tachina flies, Braconidae, Trichogrammae, and other spp.

Under resource shortages, competition may occur among predators and parasites for the same prey-host. For example, ichneumon and braconid flies parasitizing older instars of corn earworm larvae were observed to be less abundant when predatory bugs preying on eggs and younger larvae of earworms were more abundant.[22]

In addition, there is a number of secondary predators and parasites that inhibit the growth of many entomophagous species. For example, some ichneumon fly species are known to parasitize larvae of syrphus flies; mite-eating thrips may often fall a prey to predatory Orius and Campylomma bugs, or predatory Aeolothrips; predatory Orius bugs, in turn, attack larvae of Aeolothrips.[22]

Besides the fauna mentioned, an important role in biocenosis formation is played by microorganisms and pathogens. Particularly important are Verticillium and Fusarium fungi, the wilt disease pathogens, which penetrate the plant through the root system, then propagate up the stem, reaching the leaves. First, light yellow, then dark brown spots emerge in the affected leaves, the majority of the wilt-affected plants lose their foliage, cease growth, and dry out. Developing inside the host (cotton plant), the pathogens intervene in the metabolism of the basic component of the cotton biocenosis and should therefore be considered among its main components as well.

So, the highest aggregated model of the cotton biocenosis should consist of at least three components as shown in Figure 31: (1) cotton, (2) pests, including

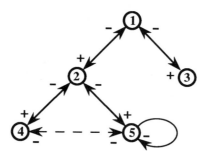

Figure 32.   Parasitism and predation on pests and among pest enemies.

their enemies, and (3) wilt pathogens. It is easy to see that the SDG meets all the sign stability conditions: interaction signs are $+-$; cycles longer than 2 are absent; splitting up into cycles that indicates a nonzero term in the determinant is given by vertices $\{1 \leftrightarrow 3\}$ and $\{2\}$; the SDG represents a single predation subgraph failing the color test (rule (c)). Hence, the highest aggregated model of the cotton biocenosis is sign-stable and its sign stability depends principally on competition and other regulatory phenomena taking place within the pest component ($a_{22} < 0$). Stabilizing effects of this competition were observed in field studies as well.[22]

To specify the effects which are depicted by the self-regulation loop in Figure 31, we consider a more complex, but still highly aggregated system of five components shown in Figure 32: parasites (4) and predators (5) of pests may sometimes compete for resources being temporally in shortage ("temporal" interaction is depicted by the dashed arrow $4 \leftarrow - \rightarrow 5$), while interference among the predators (self-regulation loop at vertex 5) is fairly typical. The community matrix takes on the following form:

$$
A = \begin{bmatrix}
0 & -a_{12} & -a_{23} & 0 & 0 \\
a_{21} & 0 & 0 & -a_{24} & -a_{25} \\
a_{31} & 0 & 0 & 0 & 0 \\
0 & a_{42} & 0 & 0 & -z_{45} \\
0 & a_{52} & 0 & -z_{54} & -a_{55}
\end{bmatrix},
\tag{5.10}
$$

where entries $z_{45} \geq 0$ and $z_{54} \geq 0$ correspond to the temporal competition between parasites and predators.

When $z_{45}z_{54} = 0$, matrix (5.10) is sign-stable by Conditions 1 to 5 : a single predation graph violating rule (c) of the color test; splitting up into cycles of a nonzero determinant term is given by vertices $\{1 \leftrightarrow 3\}$, $\{2 \leftrightarrow 4\}$, and $\{5\}$. But when $z_{45}z_{54} > 0$, that is, when the effects of competition between parasites and predators of pests are strong enough, rather than negligible, the matrix loses its sign stability. This means that some additional restrictions on positive parameters

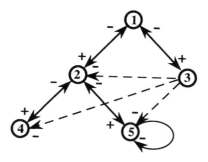

Figure 33.   Amensalistic effects from pathogenic fungi.

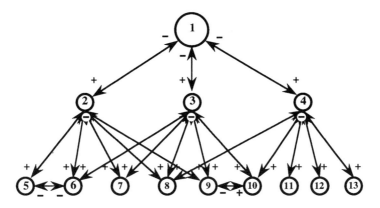

Figure 34.   Interactions among less aggregated components without pathogenic fungi.

a_{ij} and z_{kl} will be required to guarantee an equilibrium stability in a dynamic model with matrix (5.10).

Note that we have not yet considered any effects of pathogenic fungi on insects in the cotton field. Some examples are however known where the cotton disease pathogens exerted negative effects on insect population in the cotton biocenosis.[22] The opposite influence have not been reported, so that the relations are amensalistic. These are also shown by dashed arrows in Figure 33. The SDG is already not sign-stable as it now contains longer cycles, e.g., $1 \rightarrow 3 \rightarrow 2 \rightarrow 1$ and others.

Therefore, from the viewpoint of qualitative stability in the whole community dynamics, pathogenic fungi exert destabilizing effects, which are combined with the direct damage due to diseases.

If disaggregating further the chains $1 \leftrightarrow 2 \leftrightarrow 4$ and $1 \leftrightarrow 2 \leftrightarrow 5$ and neglecting the pathogenic fungi, we come to the structure shown in Figure 34. It includes 13, still aggregated, components at the following three major levels: cotton (1); cotton pests (2 - spider mites, 3 - aphids, 4 - cotton earworms); predators (5 - thrips, 6 - bugs, 7 - ladybugs, 8 - golden-eyed flies, 9 - syrphus flies) and parasites of pests, 10 - ichneumons, 11 - Braconidae, 12 - Trichogrammae, 13 - tachina flies).

It is clear that, in the presence of temporal links of competition or parasitism mentioned above, the structure does not possess sign stability, either because of forbidden sign pairs, or due to the links closing longer cycles. But even if there was no such links in the graph, self-regulation is still needed for sign stability and the self-regulated vertices must be allocated so as to fail the color test. The question which species in the biocenosis can be considered self-limiting and which cannot might be a subject of special study, but we may ascertain a priori that self-regulation at an arbitrary vertex in the graph would be insufficient. For example, self-regulation at vertex 5 (thrips) will result in failing the test, hence stabilization, while self-regularization at vertices 6 to 10, i.e., at predators (or parasites) with a wide enough spectrum of their prey (hosts), will not produce the stabilization effect.

From these theoretical exercises some practical conclusions can yet be drawn. Most of all they might be relevant to biological pest control projects (introduction of new parasites of pests, release of sterilized individuals of a pest species, etc.) as such projects disturb essentially the qualitative structure of biological interactions in the community. On the contrary, biocide application normally does not affect the structure, it just suppresses the population densities of both pests and their natural enemies. Since stability domains are always finite in reality, such intervention into regulatory mechanisms may result in the opposite effects, namely, pest outbreaks.[22] This general point can be illustrated by empirical studies in a variety of fields including cotton,[22] while mathematical illustrations require more specific models (see, e.g., the survey by Rosenzweig[24]).

Biological pest control projects clearly feature extreme specificity in their biological agents, but some general recommendations may still be derived from the sign stability analysis of the community matrix. A program should be designed in such a way that the resulting changes in community structure improve its properties as formulated in the sign stability conditions. For instance, introduction of sterilized individuals to increase competition for the breeding mate should be considered as self-regulation emerging at a given vertex. Those choices which result in color test failure should be advantageous. Similar arguments may be relevant to the introduction of new parasitic or predatory species, as well as to integrated methods of biological pest control involving various measures.

The recommendations thus derived may seem too trivial to experts engaged in the practical aspects of biological pest control. But they are scarcely more trivial than the qualitative-type information needed to formulate them.

D. Case Study: Koshka-Kum Desert Community

Even the first attempts to model the dynamics of a desert animal community relied on even earlier belief in the intrinsic relations between the structure of biological

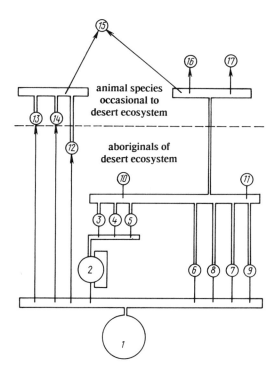

Figure 35.   Diagram for interactions among the dominant components of the Koshka-Kum Desert community. Legend: (1) herbaceous vegetation; (2) insects; (3) linear lizard *Eremias lineolata* (Nik.); (4) black-eyed lizard *Eremias nigrocollata* (Nik.); (5) steppe agama *Agama sanguinolenta* (Pallas); (6) red-tailed gerbal (jird) *Meriones* (Pallasimus) *erythrourus* (Gray); (7) thin-toed suslik *Spermophilopsis leptodactylus* (Licht.); (8) small five-toed jerboa *Allactaga elater* (Licht.); (9) tolai hare *Lepus tolai* (Pallas); (10) Fergana badger *Meles meles severzovi* (Heptner); (11) Turkmen fox *Vulpes vulpes flavescens* (Gray); (12) Persian gazelle *Gazella subgutturosa* (Gueld.); (13) Central-Asian wild boar *Sus scrofa nigripes* (Blanf.); (14) Bukhara deer *Cervus elaphus bactrianus* (Lyd.); (15) desert wolf *Canis lupus desermorum* (Bogdanov); (16) Turkestan jackal *Canis aureus aureus* (L.); (17) Turkestan wild cat *Felis silvestris caudata* (Gray). From Usmanov et al.,[25] with permission.

interactions and stability of system dynamic behavior (see Usmanov et al.[25] and references therein). Figure 35 shows the structure of interactions among dominant species as revealed by monitoring an "ecomodel" area of Koshka-Kum sandy desert, Tajikistan, in the former USSR.

Several trophic chains can be traced from the primary production level up to the top predators, and the SDG elicited from the structure is shown in Figure 36. With regard to the links depicted as dashed arrows, field experts were not certain

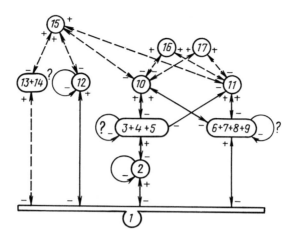

Figure 36.   Trophic chains to be elicited from the desert community structure. Species 3–5, 6–9, and 13–14 are aggregated into trophic levels; dashed lines show trophic links of species occasional to the desert community.

about whether or not those links are essential to the real community dynamics.[26−27] Also, as we have already seen from the previous exercises, there is no sense in speculating in terms of sign stability with respect to a disaggregated graph: too many cycles will make any conclusion quite uncertain.

To have trophic chains aggregated, it is logical to combine species 3 to 5 and 6 to 9, which have apparently similar positions in the trophic structure and probably compete for some common resources. If the corresponding competition communities reveal stable dynamics (all other levels being constant), then the aggregates of competing species can logically be assumed to evince negative feedbacks. Naturally, the existence or lack of density regulation is a subject of special field studies; however, from the standpoint of sign stability, there is a nontrivial interplay between the presence of "ephemeral" links and allocating self-limitation among vertices 3-5, 6-9, or 12. For example, in the chain $1 \leftrightarrow 2 \leftrightarrow (3\text{-}5) \leftrightarrow 10 \leftrightarrow 16$, the self-limitation at 2 is crucial even when it also acts on (3-5) (see the example in Figure 22); but without vertex 16 the chain is sign-stable irrespectively of where a (yet sole) "black" vertex is.

When aggregated as shown in Figure 36, the SDG contains dicycles longer than 2, e.g., $1 \rightarrow 2 \rightarrow (3\text{-}5) \rightarrow 11 \rightarrow (6\text{-}9) \rightarrow 1$, which appear due to broad food specialization of species 10 and 11, the Fergana badger and Turkmen fox. "Ephemeral" links, like $10 \leftrightarrow 17$, $11 \leftrightarrow 16$ and others, also result in cycle formation, thus violating the sign stability criterion.

However, the lack of sign stability means no more than that the model community will not maintain stability under large-scale variations in intensities of its intra- and interspecies relations. Rather, there might be (and normally are) finite regions in model parameter spaces where the corresponding dynamical model is stable. What the lack of sign stability does signify is a greater vulnerability of stable community functioning under variations in species interactions.

What kind of knowledge can be gained from the matrix and graph theory analysis of non-sign-stable structures is considered in later chapters of this book.

Appendix

Statement 5.1. *The Condition:*

> (3) for any sequence of 3 or more indices $i_1 \neq i_2 \neq \ldots \neq i_m$ the inequalities $a_{i_1 i_2} \neq 0$, $a_{i_2 i_3} \neq 0, \ldots, a_{i_{m-1} i_m} \neq 0$ imply that $a_{i_m i_1} = 0$,

is necessary for matrix A to be sign-stable.

Proof: Let Condition 3 be broken, i.e., let there exist a set of m different integers $i_1, i_2, \ldots, i_m$ such that the product

$$T = a_{i_1 i_2} a_{i_2 i_3} \ldots a_{i_{m-1} i_m} a_{i_m i_1}$$

is nonzero. Consider matrix $B = [b_{pq}]$ of the same size as A in which

$$b_{pq} = \begin{cases} a_{pq} & \text{if } p \neq q \text{ and } p, q \in \{i_1, \ldots, i_m\}, \\ 0 & \text{otherwise.} \end{cases}$$

The characteristic equation for matrix B takes on the form

$$\lambda^n \pm \lambda^m T = \lambda^{n-m}(\lambda^m \pm T) = 0, \quad T \neq 0.$$

Since $m \geq 3$, among the roots of the equation there is at least one root, λ_0, with a positive real part, Re $\lambda_0 > 0$. The roots of a polynomial depend continuously on its coefficients, so that the eigenvalues of a matrix depend continuously on its entries. Therefore, in the space of all real matrices $n \times n$ there exists a vicinity of matrix B, $\mathbb{U}_B \subset \mathbb{R}^{n^2}$, such that any matrix $B^\circ \in \mathbb{U}_B$ also has an eigenvalue with a positive real part. Among $n \times n$ matrices of $\mathbb{U}_B$ one can always find a matrix of any fixed sign pattern, in particular, a matrix $\bar{B}^\circ$ that is sign-equivalent to A. Thus, A cannot be sign-stable.   ∎

Statement 5.2. *Existence of a nonzero term in* det A *is equivalent to the associated digraph D(A) containing a skeleton subgraph that splits into a number of disjoint directed cycles.*

Proof: Corresponding to a particular permutation σ, a nonzero term of det $A = [a_{ij}]$ has the form of

$$\pm a_{\sigma(1)1} a_{\sigma(2)2} \ldots a_{\sigma(n)n},$$

which means that in the digraph there exists a set of arcs $1 \to \sigma(1), 2 \to \sigma(2), \ldots,$ $n \to \sigma(n)$ which includes each of n vertices exactly twice: as the beginning of an arc and as the end of an arc. The fundamental property of permutations is the following: any permutation of n integers $\{1, 2, \ldots, n\}$ is representable as the product of a finite number of *disjoint cycles*, i.e., cyclic permutations defined on non-intersecting subsets of $\{1, 2, \ldots, n\}$.[4] A cyclic permutation of a subset generates the directed cycle as a proper subgraph of the associated digraph, the number of arcs in the cycle coinciding with the number of its vertices. Any integer i which remains under the permutation (i.e., $\sigma(i) = i$) generates a 1-cycle. Thus, σ generates a subgraph which includes all n vertices, i.e., a *skeleton* subgraph, the total number of arcs in (the total length of) all 1- and longer disjoint cycles coinciding with the number of vertices.

Since Condition 3 forbids cycles longer than 2, a sign-stable SDG must therefore contain a skeleton subgraph that splits into a number of disjoint 1- or 2-cycles whose total length equals n, the number of all vertices.   ∎

Lemma 5.1 *Let a matrix* $A = [a_{ij}]$ *satisfy Condition 3 (i.e., no cycles in its SDG longer than 2). If, in addition, A is 0-asymmetric, then A is decomposable.*

Proof: In a 0-asymmetric matrix A there exists an entry $a_{ij} \neq 0$ $(i \neq j)$ such that $a_{ji} = 0$. By permutation of A's rows, or renumbering of SDG's vertices,

$$\begin{pmatrix} 1 & 2 & \cdots & i & j & \cdots & n \\ i & j & & 1 & 2 & & n \end{pmatrix}$$

we get $a_{12} \neq 0$, $a_{21} = 0$. If now A has no nonzero entries to the right of a_{22}, then it is decomposable. Otherwise, if there are any nonzero entries to the right of a_{22}, let $a_{2j} \neq 0$ be the first of them. Renumbering $3 \leftrightarrow j$ then yields $a_{23} \neq 0$, $a_{31} = 0$ by Condition 3. Then A takes on the form

$$A = \begin{bmatrix} * & \phi & * & * & \cdots \\ 0 & * & \phi & * & \cdots \\ 0 & * & \cdots & & \\ & \cdots & & & \\ \cdots & & & & \end{bmatrix},$$

where character $*$ denote an arbitrary and ϕ a nonzero entry.

If the procedure can be repeated for the 3rd, 4th, $\ldots$, nth row, i.e., if each of them has nonzero entries to the right of the diagonal, then the first column of A, with all the elements below a_{11}, consists solely of zeros, thus showing that A is decomposable. Otherwise, assume that the first row that has no nonzero entries to the right of the diagonal is the mth one, i.e., $a_{m,m+1} = a_{m,m+2} = \ldots = a_{mn} = 0$ and

$$A = \begin{bmatrix} * & \phi & * & \cdots & & * & \cdots \\ 0 & * & \phi & * & \cdots & * & \cdots \\ \vdots & & \ddots & & & & A^\circ \\ 0 & * & \cdots & * & \phi & * & \cdots \\ 0 & * & \cdots & & * & 0 & 0 & \cdots & 0 \\ * & * & \cdots & & & & \end{bmatrix} \begin{matrix} 1 \\ 2 \\ \\ m-1 \\ m \end{matrix}$$

If now the submatrix

$$A^\circ = A \begin{pmatrix} 2, & 3, & \ldots, & m-1 \\ m & +1, & \ldots, & n \end{pmatrix}$$

consists entirely of zeros, then renumbering $1 \leftrightarrow m$ results in the $(m-1)\times(n-m+1)$ block of zeros in the upper right corner of A, so that A is again decomposable. But if there is an $a_{ij} \neq 0$ in submatrix A°, then renumbering $j \leftrightarrow m+1$ will not affect the first m rows and columns and yield $a_{i,m+1} \neq 0$ ($2 \leq i \leq m-1$). Since $a_{12} \neq 0, a_{23} \neq 0, \ldots, a_{i-1,i} \neq 0$, it follows from Condition 3 that $a_{m+1,i} = 0$.

Consider further the $(m+1)$th row: either there are nonzero entries to the right of the diagonal and then $a_{m+2,1} = 0$, or all of those entries are zeros and then we turn to submatrix A°. If it still has any nonzero entries, we permute, as before, one of them to the $(m+2)$th column and, by Condition 3, get $a_{m+2,1} = 0$. Eventually, we either exhaust all nonzero entries in A° and obtain a block of zeros in the upper right corner, or find $a_{n1} = 0$ in the first column, the both resulting in decomposability of A.   ∎

Theorem 5.1 *If an indecomposable matrix A meets all Conditions 1 to 4 and the following Condition 5:*

    (5) The predation graph fails the color test,
        then the matrix is sign-stable.

Proof: As shown in Section III.A, Conditions 2 and to 3 imply that the indecomposable matrix A and its (strong) SDG are of the predation type. Let the matrix have size $m \times m$. We define m positive numbers $\alpha_1, \alpha_2, \ldots, \alpha_m$ as follows. Put $\alpha_1 = 1$. For each vertex i linked with vertex 1, let

$$\alpha_i = -a_{1i}/a_{i1},$$

which is positive due to the predation type of the pattern. In the same way, we define numbers α_j for all those vertices j which are linked to i but have not yet had α_j defined, from the relationship

$$\alpha_j a_{ji} = -\alpha_i a_{ij}, \qquad i \neq j, \tag{A5.1}$$

and so on. Due to the absence of cycles longer than 2 in the digraph, we shall thus have all m numbers defined. Then define the quadratic function

$$\varphi(x_1, \ldots, x_m) = \sum_{i=1}^{m} \alpha_i x_i^2, \tag{A5.2}$$

which is obviously positive definite in $\mathbb{R}^m$. Its derivative along trajectories of the linear system of differential equations

$$dx/dt = Ax \tag{A5.3}$$

equals, by differentiation rules and condition (A5.1),

$$d\varphi(\mathbf{x})/dt \;=\; 2\sum_{i=1}^{m}\sum_{j=1}^{m}\alpha_i x_i a_{ij} x_j$$

$$=\; 2\sum_{i=1}^{m}\alpha_i a_{ii} x_i^2. \tag{A5.4}$$

Since all $a_{ii} \leq 0$, it follows that $d\varphi/dt \leq 0$, and $\varphi(\mathbf{x})$ is therefore a Lyapunov function for the zero solution to (A5.3), ascertaining at least its local nonasymptotic stability. Two alternatives are now possible by the well-known theory of solutions to a constant matrix linear system (A5.3): either all Re $\lambda(A) < 0$ and the zero solution is in fact asymptotically stable, or some eigenvalues (of multiplicity 1) have zero real parts, the case sometimes called *neutrally stable*. The main motif of what follows is to determine the conditions separating these two cases.

Since $\det A \neq 0$ (guaranteed by Conditions 1 to 4), the alternative to asymptotic stability is the presence of pure imaginary numbers in the spectrum of A. Then each pair of them induces a pair of sinusoidal terms (of a constant amplitude) in the general solution, $\mathbf{x}(t)$, to system (A5.3). Those components $x_i(t)$ of the solution vector which really contain such terms will be called *oscillating*; oscillating and nonoscillating species must have special allocation in the structure of a predation community.

First, any component x_k with $a_{kk} < 0$ (a self-regulated species) cannot be oscillating. Indeed, any periodic solution of (A5.3) represents a finite closed orbit in the phase space; everywhere along the orbit we must have $d\varphi(\mathbf{x})/dt \equiv 0$, for otherwise, since $d\varphi/dt$ does not change the sign along any trajectory, the integral of $d\varphi(\mathbf{x})/dt$ along the closed contour would be nonzero. It now follows from (A5.4) that $x_k(t) \equiv 0$ everywhere along the periodic solution.

Second, any oscillating species i must be linked with at least one other oscillating species; otherwise, the oscillating components would add nothing to the sum of zero components in the (A5.3) equation for $x_i(t)$, so that its time derivative would be identically zero.

Third and final, any nonoscillating species x_k that is linked to an oscillating one must be also linked to another oscillating species; otherwise, dx_k/dt would have no chance to be identically zero.

If oscillating and nonoscillating components are colored respectively white and black in the predation graph, then the above three requirements coincide respectively with the rules (a), (b), and (c) of the color test. They are necessary for the case of neutral stability to be realized in matrix A. Should any one of the rules fail (Condition 5), the matrix can only remain with the alternative of asymptotic stability.   ∎

Lemma 5.2 *For any positive number ω there exists a matrix $\bar{B}$ equivalent to B of the form (5.7) and such that $\lambda = \pm i\omega$ are eigenvalues of $\bar{B}$.*

Proof: For any matrix $\bar{B}$ equivalent to B the coefficients of its characteristic equation

$$\det(\lambda I - \bar{B}) = \lambda^p + d_1 \lambda^{p-1} + \cdots d_p = 0 \qquad (A5.5)$$

can be determined by the following argument. Since there are no self-limited species (i.e., 1-cycles) among white vertices of the predation graph, whose links represent 2-cycles in the associated SDG, it can be seen from Figure 28 when there are any nonzero terms in $\det \bar{B}$. Indeed, a skeleton subgraph which splits into a number of disjoint 2-cycles exists only when $p = 2$ or $p = 4$ (respectively, $m = 2$ or $m = 3$). In matrix $\bar{B}$, therefore, among all principal minors of any order only those of the 2nd and 4th orders are distinct from zero. Thus, equation (A5.5) actually takes the form

$$\lambda^p + d_2 \lambda^{p-2} + d_4 \lambda^{p-4} = 0. \qquad (A5.6)$$

Direct calculation of d_2 and d_4 yields

$$d_2 = \alpha + \beta + \gamma, \quad d_4 = \alpha\gamma \qquad (A5.7)$$

where

$$\begin{cases} \alpha = -\bar{b}_2\bar{c}_2 - \cdots - \bar{b}_{m-1}\bar{c}_{m-1}, \\ \beta = -\bar{b}_m\bar{c}_m, \\ \gamma = -\bar{b}_{m+1}\bar{c}_{m+1} - \cdots - \bar{b}_p\bar{c}_{pp}. \end{cases} \qquad (A5.8)$$

Note that $\alpha, \beta, \gamma > 0$ by construction of matrix $\bar{B}$. Next, a direct verification shows that if $\omega > 0$ is an arbitrary number and if

$$d_2 = \omega^2 + \varepsilon > 0, \quad d_4 = \varepsilon\omega > 0, \qquad (A5.9)$$

then $\lambda = \pm i\omega$ are roots of equation (A5.6). Under conditions (A5.7-8) the roots are eigenvalues of $\bar{B}$. For any d_2 and d_4 determined in (A5.9), one can easily find $\alpha, \beta, \gamma > 0$ such that (A5.7) holds and then chose the entries of $\bar{B}$ of the proper signs so as to satisfy the equalities (A5.8). Thus, $\lambda = \pm i\omega$ will be eigenvalues of $\bar{B}$. ∎

Lemma 5.3 *For any matrix B of the form (5.7) there exists an equivalent matrix $\bar{B}$ such that the system of equations*

$$\dot{\mathbf{x}} = \bar{B}\mathbf{x}, \quad \mathbf{x} \in \mathbb{R}^p, \qquad (5.8)$$

has a particular solution

$$x_j(t) = D_j \cos \omega t + E_j \sin \omega t, \quad j = 1, \ldots, p, \qquad (5.9)$$

with constants D_{j_0}, E_{j_0} taking on any prescribed values for a fixed subscript j_0.

Proof: By Lemma 5.2 there exists a matrix $\tilde{B}$ equivalent to B and such that $\lambda = \pm i\omega$ are its eigenvalues. Let us denote the corresponding eigenvectors by $z = \eta + i\zeta$ and $\bar{z} = \eta - i\zeta$. Then there exists a set of particular solutions to system $dx/dt = \tilde{B}x$ of the following generic form:

$$x(t) = Cz \exp\{i\omega t\} + \overline{Cz} \exp\{-i\omega t\}, \qquad (A5.10)$$

where $C = C_1 + iC_2$ is an arbitrary constant (see, e.g., Arnold[28]). Solution (A5.10) can be shown to coincide with that of the form

$$x(t) = D \cos \omega t + E \sin \omega t, \qquad (A5.11)$$

where

$$\begin{cases} D = 2(C_1\eta - C_2\zeta), \\ E = -2(C_1\zeta + C_2\eta). \end{cases} \qquad (A5.12)$$

For any fixed subscript j_0, relationships (A5.12) generate a system of two linear equations,

$$\begin{cases} 2(C_1\eta_{j_0} - C_2\zeta_{j_0}) = D_{j_0}, \\ -2(C_1\zeta_{j_0} + C_2\eta_{j_0}) = E_{j_0}, \end{cases} \qquad (A5.13)$$

whereby, given D_{j_0} and E_{j_0}, the value of C_1 and C_2, hence a particular solution in the form (A5.11), can be determined uniquely if the determinant of (A5.13) is not zero.

Let it, on the contrary, be zero, i.e.,

$$-4 \left(\eta_{j_0}^2 + \zeta_{j_0}^2 \right) = 0.$$

Then $\eta_{j_0} = \zeta_{j_0} = 0$ and $x_{j_0}(t) \equiv 0$. Recall now that matrix $\tilde{B}$ is associated with a maximal strong white subgraph, which obeys, in particular, rule (d) of the color test (Section III.C). In order for matrix $\tilde{B}$ to have an oscillating solution (A5.11) with nonoscillating component j_0, it is necessary that coloring the oscillating components white and nonoscillating components black meet the color test. But it is easy to see that coloring black any white vertex of the graph in Figure 28 will immediately result in failure of the test. Hence, there cannot be identically zero components in the periodic solution (A5.11), i.e., $x_{j_0}(t) \neq 0$. This contradiction completes the proof. ∎

Theorem 5.4 *If an indecomposable matrix A is sign-stable, then it meets all the Conditions 1 to 5.*

Proof: Conditions 1 to 4 have already been proved necessary (Section II.C). Before strengthened with Condition 5, they restrict the set of indecomposable matrices under consideration to predation matrices alone.

For a trivial predation community, i.e., an isolated, self-regulated species, the theorem is trivial too: the graph of one vertex, which is black by rule (a), breaks rule (b) of the color test.

In the nontrivial case, suppose the contrary, namely, that a sign-stable predation graph passes the color test. Then we will find an equivalent matrix $\tilde{A}$ and construct an oscillating solution (A5.11) to system

$$dx/dt = \tilde{A}x, \quad x \in \mathbb{R}^n. \tag{A5.14}$$

This will show a lack of asymptotic stability in $\tilde{A}$, hence the lack of sign stability in A.

First, color the predation graph of A black and white, following the rules (a), (b), (c'), and (d), which is shown possible in Section III.C. The graph then represents a collection of its maximal white subgraphs, each one being linked to another through exactly one black vertex. In a particular case the predation graph may entirely consist of only one maximal white subgraph; then, by Lemma 5.2 it has a matrix $\tilde{A}$ with $\lambda = \pm i\omega$, and the contradiction is achieved.

In the general case, since the maximal white subgraphs do not intersect, we can, by Lemma 5.3, change proper entries in the corresponding submatrices $\tilde{B}$ (while keeping the sign pattern unchanged) in such a way that each system (5.8) has a solution (5.9) with $\omega > 0$ being the same for all submatrices $\tilde{B}$. Created by this procedure, matrix $\tilde{A}$ is equivalent to A. Let us now show that there exists a periodic solution to the total system (A5.14).

We can construct this solution by using solutions (5.9) for all white vertices combined into proper submatrices $\tilde{B}$ and by putting $x_i(t) \equiv 0$ for all black vertices i. Equation (A5.14) can, indeed, be rewritten in the component-wise form as

$$\dot{x}_k = \sum_{j=1}^{n} \tilde{a}_{kj}x_j(t), \quad k = 1, \ldots, n, \tag{A5.15}$$

the coefficients $\tilde{a}_{kj}$ being nonzero if and only if the vertices k and j are linked together. Since all black components are identically zero, only those white ones actually contribute to the sum (A5.15) which are linked to the vertex k. If, therefore, k is a white component, then its equation (A5.15) reduces to the corresponding equation of system (5.8) and, by Lemma 5.3, has a particular solution in the form (5.9). But if k is a black component, then either the sum in (A5.15) has no nonzero terms at all (the black vertex is linked to no one white), or, by rule (c'), there is exactly two such terms, namely, white components j_1 and j_2 linked to the black vertex k.

In the former case equation (A5.15) holds trivially, while in the latter it reduces to

$$\begin{aligned} 0 &\equiv \dot{x}_k = \tilde{a}_{kj_1}x_{j_1}(t) + \tilde{a}_{kj_2}x_{j_2}(t) \\ &= \tilde{a}_{kj_1}(D_{j_1}\cos\omega t + E_{j_1}\sin\omega t) + \tilde{a}_{kj_2}(D_{j_2}\cos\omega t + E_{j_2}\sin\omega t). \end{aligned} \tag{A5.16}$$

Since there are no cycles in the graph, vertices j_1 and j_2 belong to different maximal white subgraphs. By Lemma 5.3 we can chose numbers D_{j_2} and E_{j_2} (with no more changes in entries $\tilde{a}_{kj}$) equal to

$$D_{j_2} = -\tilde{a}_{kj_1} D_{j_1} / \tilde{a}_{kj_2}, \quad E_{j_2} = -\tilde{a}_{kj_1} E_{j_1} / \tilde{a}_{kj_2}, \tag{A5.17}$$

so that the right-hand side of (A5.16) turns identically into zero.

Since a pair of maximal white subgraphs can be connected through only one black vertex, we can consecutively combine solutions (5.9) of subsystems (5.8), "sewing" them by formulae (A5.17), into the periodic solution to the total system (A5.14).   ∎

Theorem 5.5 *A sign-stable matrix $A = [a_{ij}]$ with all $a_{ii} < 0$ is dissipative.*

Proof: First, consider the case where A is indecomposable and construct a collection of numbers $\alpha_i > 0$ $(i = 1, \ldots, n)$ such that A will satisfy the definition of dissipativeness. Let $\alpha_1 = 1$. Since the associate graph $D(A)$ is strongly connected, there must be a vertex, say j, with the outgoing arc $j \to 1$. For all such vertices j, we define $\alpha_j > 0$ from the equation

$$\alpha_1 a_{1j} + \alpha_j a_{j1} = 0, \tag{A5.18}$$

which is solvable due to Condition 2 of sign stability and Lemma 5.1.

If less than n numbers α_i have been defined up to the moment, then, among the vertices $j_1, j_2, \ldots$ already involved, there exist a vertex with the outgoing arc $j \to k$ to a vertex k not yet involved in the procedure (otherwise, A would be decomposable). Of all such vertices k, chose one with the minimal number k and define $\alpha_l > 0$ from the equation

$$\alpha_k a_{kl} + \alpha_l a_{lk} = 0, \tag{A5.19}$$

for all those vertices l linked to k which have not yet been involved, and so on. Due to the absence of dicycles longer than 2, the procedure can never result in a definition contradictory to any α_i, nor can any α_i escape it.

Let

$$\mathcal{A} = \operatorname{diag}\{\alpha_1, \alpha_2, \ldots, \alpha_n\} \in \mathbb{D}_n^+.$$

Then by conditions (A5.18)–(A5.19) we have

$$\mathcal{A}A + A^T \mathcal{A} = 2 \operatorname{diag}\{\alpha_1 a_{11}, \alpha_2 a_{22}, \ldots, \alpha_n a_{nn}\}$$

with all diagonal entries being negative. In other words, the quadratic form $\langle \mathcal{A}\, A\mathbf{x}, \mathbf{x} \rangle$ is negative definite, Q.E.D.

If now a decomposable sign-stable matrix takes on the form (after a proper renumbering of vertices in its associated graph)

$$A = \left[\begin{array}{c|c} B & C \\ \hline \mathbf{0} & D \end{array}\right],$$

where $B = [b_{ij}]$ and $D = [d_{kl}]$ are indecomposable matrices of sizes $p \times p$ and $q \times q$ respectively $(p+q) = n$, then B and D are sign-stable and, by what has already been proved, there exist $\mathcal{B} = \text{diag}\{\beta_1, \ldots, \beta_p\} \in \mathbb{D}_p^+$ and $\mathcal{D} = \text{diag}\{\delta_1, \ldots, \delta_q\} \in \mathbb{D}_q^+$ such that

$$
\begin{aligned}
(\mathcal{B}B + B^T\mathcal{B})/2 &= \text{diag}\{\beta_1 b_{11}, \beta_2 b_{22}, \ldots, \beta_p b_{pp}\}, \\
(\mathcal{D}D + D^T\mathcal{D})/2 &= \text{diag}\{\delta_1 d_{11}, \delta_2 d_{22}, \ldots, \delta_q d_{qq}\}.
\end{aligned}
$$

Consider a diagonal matrix

$$\mathcal{A} = \mathcal{B} \oplus \mu \mathcal{D},$$

where $\oplus$ denotes the direct sum of matrices, and μ is a large enough positive number. If an n-vector $\mathbf{x} = [y_1, \ldots, y_p, z_1, \ldots, z_q]^T \neq \mathbf{0}$, then the Cauchy–Schwarz inequality (see, e.g., Horn and Johnson[29]) yields the following estimation for the quadratic form $\langle \mathcal{A}A\mathbf{x}, \mathbf{x} \rangle$:

$$
\begin{aligned}
\langle \mathcal{A}A\mathbf{x}, \mathbf{x} \rangle &= \sum_{i=1}^{p} \beta_i b_{ii} y_i^2 + \sum_{i=1}^{p} b_{ii} y_i \sum_{j=1}^{q} \delta_j c_{ij} z_j + \mu \sum_{j=1}^{q} \delta_j d_{jj} z_i^2 \\
&\leq -m_\beta \|\mathbf{y}\|^2 + M_\beta \|C\| \, \|\mathbf{y}\| \, \|\mathbf{z}\| - \mu m_\delta \|\mathbf{z}\|^2, \quad\quad \text{(A5.20)}
\end{aligned}
$$

where

$$m_\beta = \min_i\{\beta_i \mid b_{ii}\}, \quad M_\beta = \max_i\{\beta_i\}, \quad m_\delta = \min_j\{\delta_j \mid d_{jj}\},$$

and $\|\cdot\|$ denotes a vector or matrix norm in a corresponding space.

It easy to see that if $\mu > M_\beta^2 \|C\|^2/(4m_\beta m_\delta)$, then the right-hand side of inequality (A5.20) is negative for any $\mathbf{x} \neq \mathbf{0}$, i.e., matrix A is dissipative.

In the general case the proof is completed by induction on the number of indecomposable diagonal blocks in matrix A.   ∎

Additional Notes

To 5.I. In the literature on graph theory, a predator-prey pairing is termed a *matching*, and a trophic graph is called *bipartite* or *simple*.[3] An algorithm to find a pairing of maximum cardinality was proposed by Edmonds.[30]

To 5.II. As noted in the extensive article by Jeffries, Klee, and Van den Driesche,[31] the history of sign stability ideas can be traced back to the 1947 book of Samuelson on mathematical economics.[5] In 1965 Conditions 1 to 4 were proposed by Quirk and Ruppert[6] in the same area and reformulated later by May[7] in an ecological context. In fact, they gave a characterization of sign semi-stable community matrices.

To 5.III. Mathematically rigorous formulations, with a due concern for the indecomposability restriction, appeared in the 1974 paper by Jeffries,[8] which is also the source of examples (5.2) and (5.3). Conditions 1 to 5 were proved there to be sufficient and elsewhere to be necessary by Jeffries, Klee, and Van den Driessche[9] (in a somewhat different but equivalent form) and Logofet and Ulianov.[11−12]

Probably the most elegant and impressive of all mathematical contributions to theoretical population biology, the theory of sign stability is quite definitive only in the linear autonomous case presented. (For possible non-autonomous extensions see Jeffries.[32]) As long as we assume the equilibrium formalization of community stability, the characterization of qualitatively stable systems as the class of sign-stable community matrices is quite adequate. Whether and to what an extent can this class be extended with a weaker formalization, like, e.g., *permanence*,[33] is an interesting question. Close linear analogs are the notions of *sign semi-* and *quasi-stability*.[31] If, in the linear case, the characterization of sign semi-stability can again be given in terms of graphs and colorings,[31] "the recognition of sign quasi-stability is a very delicate [in fact, algorithmic—D. L.] matter" (p.1 in ref.31). Unfortunately, the modern formulations of permanence theorems in (nonlinear) population models[33−36] are even farther from being of a "sign nature," even less capable of sign characterizations.

For the definition and some properties of matrix sign stability using Hadamard products see Johnson and Van den Driessche.[37]

REFERENCES

1. Yorke, J. A. and Anderson, W. N. Predator-prey patterns (Volterra–Lotka equations), *Proc. Natl. Acad. Sci. USA*, 70, 2069–2071, 1973.

2. Svirezhev, Yu. M. and Logofet, D. O. *Stability of Biological Communities* (revised from the 1978 Russian edition), Mir Publishers, Moscow, 1983, 319 pp., Chap. 4.4.

3. Berge, C. *Théorie des Graphes et ses Applications*, Dunod, Paris, 1958, Chap. 10.

4. Mal'tsev, A. I. *Foundations of Linear Algebra*, 2nd ed., Gostekhizdat, Moscow, 1956, 340 pp., Chaps. 1, 2 (in Russian).

5. Samuelson, P. A. *Foundations of Economic Analysis*, Harvard University Press, 1947; Antheneum, New York, 1971.

6. Quirk, J. P. and Ruppert, R. Qualitative economics and the stability of equilibrium, *Rev. Econom. Studies*, 32, 311–326, 1965.

7. May, R. M. Qualitative stability in model ecosystems, *Ecology*, 54, 638–641, 1973.

8. Jeffries, C. Qualitative stability and digraphs in model ecosystems, *Ecology*, 55, 1415–1419, 1974.

9. Jeffries, C., Klee, V. and Van den Driesche, P. When is a matrix sign stable? *Can. J. Math.*, 29, 315–326, 1977.

10. Logofet, D. O. To the issue of ecosystem qualitative stability, *Zhurnal Obshchei Biologii* (Journal of General Biology), 39, 817–822, 1978 (in Russian).

11. Logofet, D. O. and Ulianov, N. B. Sign stability in model ecosystems: A complete class of sign-stable patterns, *Ecol. Modelling*, 16, 173–189, 1982.

12. Logofet, D. O. and Ul'janov, N. B. Necessary and sufficient conditions for sign stability of matrices, *Soviet Math. Dokl.*, 25, 676–680, 1982.

13. Van der Waerden, B. L. *Moderne Algebra*, Vol. 1, Springer, Berlin, 1937, 272 pp.

14. Krantz, S. G. *Real Analysis and Foundations*, CRC Press, Boca Raton, FL, 1991, 295 pp., Chap. 22.

15. Pimm, S. L. and Lawton, J. H. The number of trophic levels in ecological communities, *Nature*, 268, 329–331, 1977.

16. Pimm, S. L. and Lawton, J. H. On feeding more than one trophic level, *Nature*, 275, 542–544, 1978.

17. Pimm, S. L. *Food Webs*, Chapman and Hall, London, 1982, 219 pp.

18. Logofet, D. O. On the hierarchy of subsets of stable matrices, *Soviet Math. Dokl.*, 34, 247–250, 1987.

19. May, R. M. *Stability and Complexity in Model Ecosystems*, Princeton University Press, Princeton, NJ, 1973, 265 pp.

20. Svirezhav, Yu. M. and Logofet, D. O. *Stability of Biological Communities* (revised from the 1978 Russian edition), Mir Publishers, Moscow, 1983, Chap. 9.

21. Logofet, D. O. and Junusov, M. K. The issues of qualitative stability and regularization in dynamic models of the cotton agrobiocenosis. In: *Control and Optimization in Ecological Systems* (*Issues of Cybernetics*, Vol. 52), Svirezhev, Yu. M., ed., USSR Acad. Sci., Moscow, 62–74, 1979 (in Russian).

22. Narzikulov, M. N. and Umarov, S. A. Theoretical grounds and practical prerequisites for integrated control of cotton pests. In: *Foundations of Integrated Methods for Pest and Disease Control of Cotton in the Middle Asia*, Donish, Dushambe, 8–46, 1977 (in Russian).

23. Pimm, S. L., Lawton, J. H. and Cohen, J. E. Food web patterns and their consequences, *Nature*, 350, 669–674, 1991.

24. Rosenzweig, M. L. Aspects of biological exploitation, *Quart. Rev. Biol.*, 52, 371–380.

25. Usmanov, Z. D., Sapozhnikov, G. N., Ismailov, M. A., Cherenkov, S. N., Blagoveshchenskaya, S. T., and Jakovlev, E. P. Modelling the dynamics of desert communities in "Tigrovaya Balka" reserve, *Dokl. Acad. Nauk. Tadzhik. SSR*, 32, 629–632, 1981 (in Russian).

26. Logofet, D. O. and Svirezhev, Y. M. Modeling of population and ecosystem dynamics under reserve conditions. In: *Conservation, Science and Society. Contributions to the First International Biosphere Reserve Congress, Minsk, Byelorussia/USSR, 26 Sep.– 2 Oct. 1983*, Vol. 2. UNESCO-UNEP, 331–339, 1984.

27. Logofet, D. O. and Svirezhev, Yu. M. Modeling the dynamics of biological populations and communities under reserve conditions. In: *Mathematical Modeling of Biogeocenotic Processes*, Svirezhev, Yu. M., ed., Nauka, Moscow, 25–37, 1985 (in Russian).

28. Arnold, V. I. *Ordinary Differential Equations*, Nauka, Moscow, 240 pp., Chap. 3 (in Russian).

29. Horn, R. A. and Johnson, C. R. *Matrix Analysis*, Cambridge University Press, Cambridge, 1990, Chap. 0.

30. Edmonds, J. Path, trees and flowers, *Can. J. Math.*, 17, 449–457, 1965.

31. Jeffries, C., Klee, V. and Van den Driessche, P. Qualitative stability of linear systems, *Linear Algebra and its Applications*, 87, 1–48, 1987.

32. Jeffries, C. Stability of ecosystems with complex food webs, *Theor. Pop. Biol.*, 7, 149–155, 1975.

33. Hofbauer, J. and Sigmund, K. *The Theory of Evolution and Dynamical Systems*, Cambridge University Press, Cambridge, 1988, 341 pp., Chap. 19.

34. Jansen, W. A permanence theorem for replicator and Lotka–Volterra systems, *J. Math. Biol.*, 25, 411–422, 1986.

35. Hofbauer, J., Hutson, V. and Jansen, W. Coexistence for systems governed by difference equations of Lotka–Volterra type, *J. Math. Biol.*, 25, 553–570, 1987.

36. Hofbauer, J. and Sigmund, K. On the stabilizing effect of predators and competitors on ecological communities, *J. Math. Biol.*, 27, 537–548, 1989.

37. Johnson, C. R. and Van den Driessche, P. Interpolation of *D*-stability and sign stability, *Linear and Multilinear Algebra*, 23, 363–368, 1988.

CHAPTER 6

Trophic Chains and Stability in Vertical-Structured Communities

In addition to special mathematical results, the previous chapter illustrates such a general idea that both a hierarchy of trophic levels and the interactions within a single level are of crucial importance to the stability of the whole community structure. The present chapter is devoted to what can be obtained from matrix analysis of stability in the former, i.e., the hierarchical, or "vertical," type of structure, while Chapter 7 treats the latter, or "horizontal," structure.

I. VERTICAL-STRUCTURED COMMUNITIES AND TROPHIC CHAINS

A. Trophic Chains and the Resource Flow

In the foregoing chapters there were no constraints imposed a priori on the structure of the systems under consideration—the only constraints were incidental to the relevant analyses. But when the matter relates to biological communities or ecosystems, one can distinguish at least two major types of characteristic patterns, namely, systems of *vertical* and *horizontal structures*.[1] By the *horizontal structure* we mean interactions among species within a single trophic level, while the *vertical structure* implies interactions only among the levels.

A *trophic chain* is a structure that displays typical paths for the resource flow through the ecosystem. This flow is realized in food energy transfers from one species (or a group of species) to another one both of which are linked by a prey-predator (or resource-consumer) relationship. We could see such chains in the case studies of the previous chapter. Each transfer is associated with losing a major part (up to 90%) of the energy, which is spent for respiration, being converted into heat. This restricts the number of "links" in the chain normally to four or five,[2-4] although in an ecosystem there often exist (albeit at insignificant densities), or there may be brought from outside, individuals of those species that could form the next trophic level. When the energy input highly increases or when, as a result of an impact like fertilization, the primary level production gains sufficient impetus, the potential to form a new level is realized and the level becomes fixed in the system.

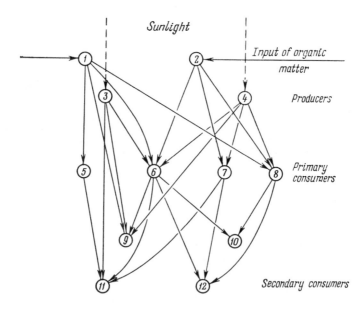

Figure 37. A part of the trophic web observed in a stream ecosystem in South Wales (after Jones[5] and E. P. Odum[6]).

Particular paths of the resource flow through an ecosystem may be numerous. Trophic chains are often not isolated from each other, rather, they interweave, forming a *trophic web*. An example of such a trophic web is given in Figure 37, showing a part of the ecosystem of a small stream[5] in the form of a trophic graph. It is an open system with a part of the basic resource inflow coming with leaf fragments (1) and other organic residues (2) carried by the flow. According to E. P. Odum,[6] the web comprises three basic trophic levels. Green algae (3) and diatoms (4) make up the producer level. The primary consumer level consists of stoneflies (Protonemura), 5; blackflies (Simulium), mayflies (Ephemeridae and Baetis), and midges (Chinoromidae) combined into one component, 6; caddis (Philopotamidae), 7, and mayflies (Ecdyonuridae), 8. The secondary consumer level comprises stoneflies (Perla), 11, and Dinocras spp., 12, while caddis (Rhyacophilidae), 10, and net-spinning caddis (Hydropsyche), 9, occupy an intermediate position. Trophic chains are formed apparently by species 3 → 6 → 12, or 2 → 7 → 11, or other paths.

Even this case study, though fairly deep in detail, had to combine several species into one component due to their similar roles in the trophic relations. This feature is principally used in subsequent mathematical constructions.

Since any natural community presents a fairly complicated trophic web, its model becomes complicated too. There are fundamentally two ways to make the original model simpler. First, we can aggregate all species of the same trophic level into one "pseudospecies," or *trophic species*.[4] The alternative is to chose a vertical branch (a trophic chain) whose energy through-flow is much higher than the flows through all other chains, and to neglect the rest. The first way may be used when the species are similar in their trophic characteristics and the role they play in the trophic structure; the pseudospecies will then correspond to a species with some averaged values of those characteristics. The second way is preferable when there is a dominating species at each trophic level. In any case, only one trophic species remains at each trophic level after these simplifications, the trophic structure becoming a trophic chain. But when neither averaging nor the choice of a dominant species is possible in a trophic level (for instance, when there are two or more markedly different but energetically equitable ecological groups), then a number of trophic branches, or branching trophic chains, are to be considered.

Once the species community structure has been aggregated into trophic levels, the model consequently loses its species specificity, and hence, its capability of describing the dynamics of species populations. Nevertheless, stability consider-ations will still make sense: it is easy to believe that the trophic structure may be even more stable than the species composition and there is empirical evidence supporting this opinion,[7-8] although the "evidence" can be debated too.[9]

As we have already mentioned, energy input is of principal importance in making it possible for the trophic structure to persist. In the following discussion we use the notion of a single generalized "resource" entering the system at a constant rate, Q, and serving as the "food" to the first trophic level.[10] In reality it might be either energy, or a substance of vital importance (e.g., carbon, nitrogen, or phosphorus), or, more likely, a combination of vital factors. However the *limiting factor* (or *Liebig's) principle* (see, e.g., Odum[11]) allows one to stay with the hypothesis of a single resource.

While any ecosystem is open with respect to energy which flows through the system, dissipating concurrently as heat (expenditures in respiration, metabolic ac-tivity, etc.), an ecosystem may be closed (to some degree) with respect to biological substances. Closing appears due to the vital activity of the so-called "decomposers" (micro-organisms, fungi, worms), which decompose dead organic matter into min-eral components, thus providing nutrients to the primary trophic levels, and hence recycling a part of the organic matter. The same is also true for trophic chains. Both *open* and *partially closed* types of chains will be considered below. Schematically, they are shown in Figures 38a and 38b respectively.

The subsequent mathematics does not depend on whether the chain consists of trophic aggregates or represents a dominating branch of the trophic web. The model variables, therefore, will be referred to as (trophic) "species" again. The schemes in Figure 38 are supposed to display the structure of the corresponding systems of model equations, rather than the chains themselves. That is why enumeration of species, hence the main energy flow, is oriented downwards, in contrast with conventional ecological diagrams like trophic pyramids.[2] We hope this will not cause a serious conflict between the mathematical theory and ecological practice.

B. Trophic Chain of a Finite Length

In the schemes and ensuing equations, R denotes the resource to be utilized by (trophic) species 1 of the biomass density N_1. Its specific rate, $V_0(R)$, of the

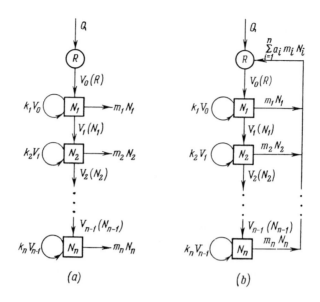

Figure 38.   Schematic representation of trophic chain models: (a) open chain; (b) partially closed chain, the degree of "closeness" being determined by parameters a_i.

resource uptake is the amount of resource consumed by a unit biomass (or a conventional "specimen") per a unit of time. Of the total amount of the resource consumed, $V_0(R)N_1$, only a fraction, k_1, contributes to reproducing new biomass of species 1, the rest being spent for maintaining vital activity. Concurrently, the species 1 biomass dies off at a rate m_1 and serves as a food to species 2, being consumed at a specific rate $V_1(N_1)$. Species 2, in turn, is consumed by species 3 at a specific rate $V_2(N_2)$, and so on. The chain ends up with species n, whose biomass is consumed by no one.

This finiteness is well-documented in the literature and well-grounded in ecological theory—not only by contrasting the famous and widely-cited infinite "flea" construction of Jonathan Swift[2−3] and the finite trophic chains observed in nature.[2−4] So, we may assume n to be great enough to cover all known cases.

The second scheme in Figure 38 differs from the first one by the existence of the $(n + 1)$th species, the *decomposers,* who utilize the dead biomass of (all or some of) the first n species and, through their vital activities, compensate, at least partially, for the expenditures at the resource level. "Dead biomass" should better be called *mortmass* and we assume that the mortmass can be decomposed at a rate high enough and the resource released by decomposition can immediately be available for uptake by species 1 again. Also, the growth rate of decomposers is assumed to be so high that they could utilize at once any amount of other species mortmass available.

Under these assumptions, it makes no sense to consider the biomass of decomposers as a state variable: one can rather consider all the mortmass (or a part of it, if inevitable losses are taken into account) to return instantaneously to the resource level.

Let the trophic graph of an ecosystem be of the types shown in Figure 38. Suppose that the whole system thus organized tends to an equilibrium state where only the first q species are nonzero. Such a stable equilibrium will be logically called a *trophic chain of length q*.[10]

Let the rate of the external resource supply to the ecosystem be constant and equal to $Q > 0$. The problem is stated as follows: given the *trophic functions* $V_0(R), V_i(N_i), i = 1, \ldots, n$, and other parameters k_i, m_i, a_i, what should the rate Q be for the model to have a stable nontrivial equilibrium with exactly q first nonzero components? In other words, what are the conditions for a q-level trophic chain to exist?

According to the schemes in Figure 38, we have the following two systems of mass-balance equations for n species.

(a) Open chain:

$$\frac{dR}{dt} = Q - V_0(R)N_1,$$

$$\frac{dN_1}{dt} = -m_1N_1 + k_1V_0(R)N_1 - V_1(N_1)N_2,$$

$$\frac{dN_i}{dt} = -m_iN_i + k_iV_{i-1}(N_{i-1})N_i - V_i(N_i)N_{i+1}, \quad i = 2, 3, \ldots, n - 1,$$

$$\frac{dN_n}{dt} = m_nN_n + k_nV_{n-1}(N_{n-1})N_n. \tag{6.1}$$

(b) Partially closed chain:

$$\frac{dR}{dt} = Q - V_0(R)N_1 + \sum_{i=1}^{n} a_im_iN_i,$$

$$\frac{dN_1}{dt} = -m_1N_1 + k_1V_0(R)N_1 - V_1(N_1)N_2,$$

$$\frac{dN_i}{dt} = -m_iN_i + k_iV_{i-1}(N_{i-1})N_i - V_i(N_i)N_{i+1}, \quad i = 2, 3, \ldots, n - 1,$$

$$\frac{dN_n}{dt} = -m_nN_n + k_nV_{n-1}(N_{n-1})N_n. \tag{6.2}$$

The physical and biological sense of parameters k_i and a_i imply that $0 < k_i < 1$ and $0 < a_i \leq 1$ for all species i.

If we assume that neither species has its trophic resource in redundancy, i.e., the trophic links are "tense,"[12] then the linear description is appropriate, so that

$$V_0(R) = \alpha_0R, \quad V_i(N_i) = \alpha_iN_i, \quad i = 1, \ldots, n, \tag{6.3}$$

and equations (6.1) and (6.2) reduce to the following equations of "almost Volterra" type (where, formally, $R \equiv N_0$ and $N_{n+1} \equiv 0$).

(a) Open chain:

$$\frac{dN_0}{dt} = Q - \alpha_0 N_0 N_1,$$

$$\frac{dN_i}{dt} = N_i(-m_i + k_i\alpha_{i-1}N_{i-1} - \alpha_i N_{i+1}), \quad i = 1, \ldots, n. \qquad (6.4a)$$

(b) Partially closed chain:

$$\frac{dN_0}{dt} = Q - \alpha_0 N_0 N_1 + \sum_{i=1}^{n} \alpha_i m_i N_i,$$

$$\frac{dN_i}{dt} = N_i(-m_i + k_i\alpha_{i-1}N_{i-1} - \alpha_i N_{i+1}), \quad i = 1, \ldots, n. \qquad (6.4b)$$

In subsequent sections the conditions are presented for the pertinent equilibria to exist and to be stable, i.e., the conditions for the q-level trophic chain to exist.

II. EQUILIBRIA IN THE TROPHIC CHAIN MODELS

Given a mathematical model (6.1) or (6.2) of a trophic chain, the definition of the trophic chain length can be quite formal.

Definition 6.1 *A trophic chain is said to be of* length q $(1 \leq q \leq n)$ *if an equilibrium of the form*

$$\mathbf{N}^q = [N_0^*, N_1^*, \ldots, N_q^*, 0, \ldots, 0], \quad N_i^* > 0, \quad i = 1, \ldots, q. \qquad (6.5)$$

exists and is stable.

It is clear theoretically that in a model there might also exist equilibria where some of N_i^* are zero for $i < q$, while N_q^* is positive. These certainly make no biological sense, so that the definition is correct when no such cases are possible in the model. Clearly, this is also a question of whether the given model is a correct representation of the trophic chain.

The "almost Volterra" type equations (6.4a) and (6.4b) make possible a finite computation of equilibria (6.5), which allows us both to verify the correctness mentioned above and to derive further constraints on model parameters, which are at least necessary for a trophic chain of length q to exist.

A. Equilibria in the Open Chain

For $Q > 0$ in system (6.4) there may exist n equilibria $\mathbf{N}^q$ of the form (6.5), where $N_i^* > 0$ can be found from the following equations:

$$
\begin{aligned}
N_1^* &= Q/(\alpha_0 N_0^*), \\
\alpha_i N_{i+1}^* &= k_1 \alpha_{i-1} N_{i-1}^* - m_i, \quad i = 1, \dots, q.
\end{aligned}
\tag{6.6}
$$

From the condition $N_{q+1}^* = 0$ it follows immediately that

$$
N_{q-1}^* = m_q/(\alpha_{q-1} k_q).
\tag{6.7}
$$

The rest of the values N_i^* can be readily obtained by induction. They are:

$$
\begin{aligned}
N_{2s}^* &= H_{2s-1}(N_0^* - f_{2s-1}), \\
N_{2s+1}^* &= H_{2s}(N_1^* - f_{2s}), \quad s = 1, 2, \dots,
\end{aligned}
\tag{6.8}
$$

where

$$
H_{2s-1} = g_1 g_3 \cdots g_{2s-1}, \quad H_{2s} = g_2 g_4 \cdots g_{2s};
\tag{6.9}
$$

$$
g_i = k_i \alpha_{i-1}/\alpha_i, \quad i = 1, \dots, n;
$$

$$
\begin{aligned}
f_{2s-1} &= \frac{\mu_1}{H_1} + \frac{\mu_3}{H_3} + \cdots + \frac{\mu_{2s-1}}{H_{2s-1}}, \\
f_{2s} &= \frac{\mu_2}{H_2} + \frac{\mu_4}{H_4} + \cdots + \frac{\mu_{2s}}{H_{2s}}, \quad \mu_i = m_i/\alpha_i, \quad i = 1, \dots, n.
\end{aligned}
\tag{6.10}
$$

These expressions show the equilibrium values to be dependent on the evenness of their numbers. Two cases therefore need to be considered.

1°. Let $q = 2S$ be *even*. Then we have

$$
N_{q-1}^* = N_{2S-1}^* = \frac{\mu_{2S}}{g_{2S}} = H_{2S-2}(N_1^* - f_{2S-2}),
$$

whereby it follows that

$$
N_1^* = f_{2S-2} + \frac{\mu_{2S}}{g_{2S} H_{2S-2}} = f_{2S-2} + \frac{\mu_{2S}}{H_{2S}} = f_{2S}.
$$

By the first of equations (6.6) we have

$$
N_0^* = \frac{Q}{\alpha_0 N_1^*} = \frac{Q}{\alpha_0 f_{2S}} .
$$

Knowing how N_0^* and N_1^* are expressed in terms of model parameters, one can now find by formulae (6.8) a finite expression for any of q steady-state values. For example,

$$N_q^* = N_{2S}^* = H_{2S-1}\left(\frac{Q}{\alpha_0 f_{2S}} - f_{2S-1}\right). \tag{6.11}$$

$2°$. Let $q = 2S + 1$ be *odd*. In a similar way, we have

$$N_{q-1}^* = N_{2S}^* = \frac{\mu_{2S+1}}{g_{2S+1}} = H_{2S-1}(N_0^* - f_{2S-1}),$$

whereby

$$N_0^* = f_{2S+1}, \qquad N_1^* = Q/(\alpha_0 f_{2S+1}),$$

and

$$N_q^* = N_{2S+1}^* = H_{2S}\left(\frac{Q}{\alpha_0 f_{2S+1}} - f_{2S}\right). \tag{6.12}$$

Obviously, the steady-state values above make sense only if they are positive. As follows from definition (6.10), functions f_{2S-1} and f_{2S} are positive and increase with s. Quantities N_0^* and N_1^* are also positive and dependent on parameter q, the length of the trophic chain. By routine algebra it can be shown[1] that if the inequalities hold

$$Q > \alpha_0 f_{2S-1} f_{2S} = Q^*(2S) \text{ for } q = 2S \tag{6.13}$$

or

$$Q > \alpha_0 f_{2S} f_{2S+1} = Q^*(2S + 1) \text{ for } q = 2S + 1, \tag{6.13'}$$

that is, if $N_q^* > 0$, then all the preceding N_i^* ($i = 1, \ldots, q - 1$) are positive too.

The above statement may seem to be trivial since a zero trophic level cannot support a nonzero biomass at the next trophic level. But in no way does this evident point follow from the mathematical model alone, which might formally allow even negative solutions to exist.

Note that inequalities (6.13) and (6.13') impose certain constraints on the rate of the external resource inflow to the system: if the chain is of length q, then the inflow rate must not exceed a threshold value $Q^*(q) = \alpha_0 f_{q-1} f_q$. However, inequality (6.13) or (6.13') holding true cannot yet guarantee that the open chain of length q is fixed in the system, as the stability of equilibrium $\mathbf{N}^q$ does not necessarily follow from the fact that it just exists. The stability conditions are therefore to be added to provide for an open chain having q trophic levels.

B. Equilibria in the Partially Closed Chain

In a model (6.4b), that describes the dynamics of an ecosystem with partial renewal of the basic resource, there also can exist n equilibrium states $\mathbf{N}^q$ of the form (6.5).

Steady-state values N_i^* can be found from the equations

$$\alpha_0 N_0^* N_1^* = Q + \sum_{i=1}^{n} \alpha_i m_i N_i^*,$$

$$\alpha_i N_{i+1}^* = k_i \alpha_{i-1} N_{i-1}^* - m_i,$$

$$N_{q+1}^* = 0, \quad i = 1, \ldots, q. \tag{6.14}$$

In a way similar to the open chain derivation, one can obtain the finite expressions for N_i^* in terms of model parameters. Once the expressions for N_0^* and N_1^* are known, the rest of N_i^* can be determined by the formulas (6.8).

1°. Let $q = 2S$ be *even*. Then we have

$$N_1^* = f_{2S},$$

$$N_0^* = \frac{Q + (f_{2S} \psi_S - \sigma_{2S})}{\alpha_0 f_{2S} - \varphi_S} \tag{6.15}$$

where

$$\varphi_S = \sum_{j=1}^{S} \alpha_{2j} m_{2j} H_{2j-1},$$

$$\psi_S = \sum_{j=1}^{S} \alpha_{2j-1} m_{2j-1} H_{2j-2},$$

$$\sigma_{2S} = \sum_{j=1}^{2S} \alpha_j m_j f_j H_{j-1} \quad (H_0 = 1, \ f_0 = 0).$$

2°. Let $q = 2S + 1$ be *odd*. In a similar way, we have

$$N_0^* = f_{2S+1},$$

$$N_1^* = \frac{Q + (f_{2S+1} \varphi_S - \sigma_{2S+1})}{\alpha_0 f_{2S+1} - \psi_{S+1}}. \tag{6.16}$$

Similar to our proof in the previous section, it can be proved that if $N_q^* > 0$, then all the preceding N_i^* ($i = 1, \ldots, q - 1$) are positive too. From the condition of N_q^* being positive a lower bound follows for the rate Q of the external resource inflow to the system. By (6.8), indeed, we have for $s = S$:

1°. $q = 2S$, $N_q^* = H_{2S-1}(N_0^* - f_{2S-1}) > 0$;

2°. $q = 2S + 1$, $N_q^* = 2H_{2S}(N_1^* - f_{2S}) > 0.$

Substituting these inequalities into the expressions for N_0^* and N_1^* of (6.15) or (6.16), then solving them with respect to Q will result in the following conditions:

$$Q > \alpha_0 f_{2S-1} f_{2S} - \{(f_{2S-1}\varphi_S + f_{2S}\psi_S) - \sigma_{2S}\} = \bar{Q}^*(q), \quad q = 2S, \tag{6.17}$$

or

$$Q > \alpha_0 f_{2S} f_{2S+1} - \{(f_{2S}\psi_{S+1} + f_{2S+1}\varphi_S) - \sigma_{2S+1}\} = \bar{Q}^*(q), \quad q = 2S + 1. \tag{6.17'}$$

As before, inequalities (6.17) and (6.17') pose the lower bounds for the resource inflow rate. It can be shown that the expressions in braces are always positive, so that a partially closed chain requires a lower inflow rate than its open counterpart to maintain the same number of trophic levels. This generally sounds quite logical since in a partially closed chain the resource undergoes, at least a partial, renewal. But still, for a finite chain to be realized in the model, it remains to be seen, as before, whether the existing equilibrium is stable under a given rate of the resource inflow.

III. FINITE CHAIN STABILITY CONDITIONS

Coupled with the existence conditions, the stability requirements for an equilibrium $\mathbf{N}^q$ in system (6.4a) or (6.4b) represent, by Definition 6.1, the conditions for a finite chain of length q to be realized in the model ecosystem. The stability problems are now reduced to investigating the eigenvalues of corresponding community matrices, i.e., the matrices of systems (6.4a) and (6.4b) linearized at equilibria $\mathbf{N}^q$. The following observation will be of use in solving these problems.

A. Sign Stability in Three-diagonal Matrices

Let a matrix A_q of size $(q + 1) \times (q + 1)$ be given in the following three-diagonal form:

$$A_q = [a_{ij}] = \begin{bmatrix} -b_0 & -d_0 & & & \\ b_1 & -h_1 & -d_1 & & \mathbf{0} \\ & \ddots & \ddots & \ddots & \\ & & b_{q-1} & -h_{q-1} & -d_{q-1} \\ \mathbf{0} & & & b_q & -h_q \end{bmatrix}, \tag{6.18}$$

where all b_i and d_i are positive and h_i nonnegative. Then the associated signed

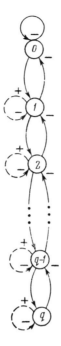

Figure 39.   The SDG associated with matrix A_q: dashed arrows vanish when $h_i = 0$.

directed graph (SDG, see Chapter 5) will take the form shown in Figure 39. In this SDG and its matrix:

(1) all entries in the principal diagonal are nonpositive while the first one is negative;

(2) all opposite arcs have opposite signs ($a_{ij}a_{ji} \leq 0$ for any $i \neq j$);

(3) there are no directed cycles longer than 2;

(4) a skeleton subgraph that indicates a nonzero term of the determinant is composed by the following 1- and 2-cycles: $\{0\}, \{1 \leftrightarrow 2\}, \ldots, \{(q-1) \leftrightarrow q\}$ when q is even, and $\{0 \leftrightarrow 1\}, \{2 \leftrightarrow 3\}, \ldots, \{(q-1) \leftrightarrow q\}$ when q is odd;

(5) there is a single predation subgraph and it fails the color test at any set of dashed arrows realized (any set of $h_i \geq 0$).

So, matrix A_q meets all conditions of the sign stability criterion and is therefore stable for any values of b_i, $d_i > 0$ and $h_i \geq 0$.

B. Stability in the Open Chain

Reverting to the stability problem for equilibrium $\mathbf{N}^q$ in system (6.4a), we can see that its community matrix (i.e., the Jacobian matrix calculated at $\mathbf{N}^q$) takes on the form of

$$F = \begin{bmatrix} A_q & \mathbf{0} \\ \mathbf{0} & D_{n-1} \end{bmatrix},$$

(6.19)

where A_q is a three-diagonal matrix (6.18) with the following entries:

$$\begin{aligned} b_0 &= \alpha_0 N_1^*, & b_i &= k_i \alpha_{i-1} N_i^*, \\ d_0 &= \alpha_0 N_0^*, & d_i &= \alpha_i N_i^*, \\ h_i &= 0, & i &= 1, \ldots, q; \end{aligned}$$

and

$$D_{n-q} = \operatorname{diag}\{-m_{q+1} + k_{q+1}\alpha_q N_q^*, -m_{q+2}, \ldots, -m_n\}.$$

Due to the block structure of matrix F (6.19) its eigenvalues are obviously the following:

$$\lambda_i = \begin{cases} \mu_i, \text{ the eigenvalues of } A_q, & i = 0, \ldots, q, \\ k_{q+1}\alpha_q N_q^* - m_q, & i = q+1, \\ -m_i, & i = q+2, \ldots, n. \end{cases}$$

For $i = 0, \ldots, q$ the eigenvalues λ_i have negative real parts as those of the sign-stable matrix A_q. For $i = q+2, \ldots, n$ the eigenvalues λ_i are clearly negative. Since $N_q^* > 0$, for λ_{q+1} to be negative it is necessary and sufficient that

$$N_q^* < \frac{m_{q+1}}{\alpha_q k_{q+1}}.$$

(6.20)

Under this condition equilibrium $\mathbf{N}^q$ is asymptotically stable. If $q = n$, then condition (6.20) disappears: the chain of length n is stable whenever equilibrium $\mathbf{N}^n$ exists.

By means of explicit expressions (6.11) and (6.12) for N_q^*, inequality (6.20) can be rewritten as follows:

$$1^\circ. \quad Q < \alpha_0 f_{2S} f_{2S+1}, \quad q = 2S;$$

(6.21)

$$2^\circ. \quad Q < \alpha_0 f_{2S+1} f_{2S+2}, \quad q = 2S + 1,$$

(6.21')

or equivalently, in terms of notations (6.13–13'), as

$$Q < \alpha_0 f_q f_{q+1} = Q^*(q+1).$$

(6.22)

This gives now the upper bound to the resource inflow rate.

Finally, with the above notation and inequalities (6.13–13'), the conclusion can be made in the form of the following

Theorem 6.1 *For an open trophic chain of length q to exist stably in model (6.4a) it is necessary and sufficient that the resource inflow rate belongs to the finite interval in the Q-axis:*

$$Q^*(q) < Q < Q^*(q+1),\tag{6.23}$$

whose bounds are defined in (6.13–13').

C. Stability in Partially Closed Chains

The stability problem for equilibrium $\mathbf{N}^q$ in system (6.4b) is reduced to investigating the spectrum of the Jacobian matrix somewhat different now from (6.19):

$$F_c = \begin{bmatrix} A_{qc} & C \\ \mathbf{0} & D_{n-q} \end{bmatrix}.\tag{6.24}$$

Here

$$A_{qc} = \begin{bmatrix} -b_0 & c_1 - d_0 & c_2 & \cdots & & c_q \\ b_1 & 0 & -d_1 & & \mathbf{0} & \\ & \ddots & \ddots & \ddots & & \\ \mathbf{0} & & b_{q-1} & 0 & -d_{q-1} \\ & & & b_q & 0 \end{bmatrix}, \quad C = \begin{bmatrix} c_{q+1} & c_{q+2} & \cdots & c_n \\ & & \mathbf{0} & \end{bmatrix},$$

$c_i = a_i m_i$, $i = 1, \ldots, n$, the rest of the notations meaning the same as before. Equilibrium $\mathbf{N}^q$ is again asymptotically stable if

$$N_q^* < \frac{m_{q+1}}{\alpha_q k_{q+1}}\tag{6.25}$$

and matrix A_{qc} is stable. The latter however does not now belong to the class of sign-stable matrices unless $c_1 < d_0$ and $c_2 = \cdots = c_q = 0$, that is, unless the resource closure be weak enough and performed from the first level only. Otherwise, simple stability conditions can hardly be given and we have to resort to more special cases of system (6.4b).

For a particular value of the chain length q, examination by pertinent stability criteria (e.g., the Hurwitz determinant criterion[13]) can be facilitated with the following recurrence relationship for the characteristic polynomial $p_q(\lambda)$ of matrix A_{qc}:

$$\begin{aligned} p_q(\lambda) &= -\lambda p_{q-1}(\lambda) + b_q d_{q-1} p_{q-2}(\lambda) - (-1)^q c_q b_1 \ldots b_q, \\ p_0(\lambda) &= -\lambda - b_0, \\ p_1(\lambda) &= \lambda^2 + b_0\lambda + b_1(d_0 - c_1). \end{aligned}\tag{6.26}$$

It can easily be obtained through expanding the characteristic determinant by its last row. If the characteristic equation $p_q(\lambda) = 0$ is written in the form

$$\lambda^{q+1} + e_q(q)\lambda^q + e_{q-1}(q)\lambda^{q-1} + \cdots + e_1(q)\lambda + e_0(q) = 0,$$

then using (6.26) one can write the recurrence equations for coefficients $e_i(q)$:

$$
\begin{aligned}
e_i(q) &= e_{i-1}(q-1) + b_q d_{q-1} e_i(q-2), & i = 1, \ldots, q \\
e_{q+1}(q) &= 1, \qquad e_i(q) = 0, & i = q+2, \ldots, \\
e_1(1) &= b_0, \\
e_0(q) &= b_q d_{q-1} e_0(q-2) - (-1)^q c_q b_1 \ldots b_q, \\
e_0(0) &= b_0, \qquad e_0(1) = b_0(d_0 - c_1).
\end{aligned}
\qquad (6.27)
$$

Consider several particular cases. Let $q = 1$. Matrix A_{1c} of size 2×2 is stable iff $c_1 < d_0$, or $m_1 a_1 < \alpha_0 N_0^*$. The latter can be rewritten as $a_1 k_1 < 1$ since $N_0^* = m_1/(\alpha_0 k_1)$. Also,

$$N_1^* = \frac{k_1 Q}{m_1(1 - a_1 k_1)} > 0,$$

only if $a_1 k_1 < 1$. Inequality (6.25) sets an upper bound to the resource inflow rate,

$$Q < \frac{m_1 m_2}{\alpha_1 k_1 k_2}(1 - a_1 k_1) = \tilde{Q}^*(2).$$

Thus, the interval

$$\frac{m_1(1 - a_1 k_1)}{k_1} < Q < \frac{m_1 m_2}{\alpha_1 k_1 k_2}(1 - a_1 k_1)$$

gives a necessary and sufficient condition for the equilibrium $\mathbf{N}^1 = [N_0^*, N_1^*, 0, \ldots, 0]$ to exist and to be stable, that is, for the existence of a trophic chain of length 1.

Let $q = 2$. Matrix A_{2c} of size 3×3 is stable iff $e_0(2), e_1(2), e_2(2) > 0$ and $e_1(2)e_2(2) > e_0(2)$, or using the expressions found by (6.27) for $e_i(2)$,

$$b_0 d_1 > b_2 c_2, \quad \text{and} \quad b_0(d_0 - c_1) + b_2 c_2 > 0.$$

The former inequality suggests that $a_2 k_1 k_2 < 1$, while the latter can be rewritten in the form

$$Q + c_2 N_2^*(1 + \alpha_2/\alpha_0) > 0,$$

that holds true whenever $N_2^* > 0$. But since we have

$$N_2^* = \frac{\alpha_1 k_1 k_2 Q - m_1 m_2(1 - a_1 k_1)}{\alpha_1 m_1(1 - a_2 k_1 k_2)} > 0$$

for $Q > \tilde{Q}^*(2)$, the stability of equilibrium $\mathbf{N}^2 = [N_0^*, N_1^*, N_2^*, 0, \ldots, 0]$ again follows from its existence.

According to their biological meaning, neither a_1k_1, nor $a_2k_1k_2$ can exceed unity. But if the parameters a_1, a_2, k_1, k_2 tend to unity, then, for the steady states to exist it is necessary that $Q \to 0$. The latter means that the system tends to be completely closed, with the total amount of resource in it remaining constant. This however brings about a degenerate case (one or several of the matrix (6.24) eigenvalues become zero or purely imaginary), so that nothing can be said of the asymptotic stability within the linearization approach.

For the case $q > 2$ it is, unfortunately, not known whether the necessary conditions for an equilibrium $\mathbf{N}^q = [N_0^*, N_1^*, \ldots, N_q, 0, \ldots, 0]$ to exist are also sufficient conditions for it to be stable. To answer the question, conditions are needed which are sufficient for the stability of matrix A_{qc}. The problem is solvable in principle for any particular value of the chain length q(for instance, the corresponding Hurwitz determinants can be considered), but it is hardly possible in this way to achieve constructive and general enough stability conditions.

By the continuity argument we can state however that, with the closure parameters c_i small enough, matrix A_{qc} will certainly be stable whenever matrix A_q of the open chain is stable. In other words, when the closing links are sufficiently weak ($a_i \ll 1$), the necessary conditions for a partially closed chain to exist are also sufficient for its stability.

Finally, the following conclusion can be drawn from the above considerations, weaker than that for the open chains.

Theorem 6.2 *If, in a model (6.4b), there exists a trophic chain of length q, then the resource inflow rate belongs to the finite interval in the Q-axis:*

$$\tilde{Q}^*(q) < Q < \tilde{Q}^*(q+1), \tag{6.28}$$

whose bounds are defined in (6.17–17').

Quantities $\tilde{Q}^*(q)$ make sense in the formulation only if they are positive. This actually takes place when

$$\alpha_0 f_{2S-1} f_{2S} + \sigma_{2S} > f_{2S-1}\varphi_S + f_{2S}\psi_S \quad \text{for } q = 2S$$

and

$$\alpha_0 f_{2S} f_{2S+1} + \sigma_{2S+1} > f_{2S}\psi_{S+1} + f_{2S+1}\varphi_S \quad \text{for } q = 2S + 1.$$

Simple, yet cumbersome, calculations can show that the conditions

$$a_i k_1 k_2 \ldots k_i < 1, \quad i = 1, \ldots, q, \tag{6.29}$$

are quite sufficient for the above inequalities to hold true. Since, by their biological meaning, $a_i \leq 1$ and $k_i < 1$, conditions (6.29) are always true in incompletely closed trophic chains.

D. Finite Chain with the First-Level Closure

Let the chain length q now be an arbitrary integer. Suppose however that the term $\sum_{i=1}^{n} a_i m_i N_i$ responsible for closing the chain at the resource level can be represented in the form

$$a_1 m_1 N_1 + \varepsilon \sum_{i=2}^{n} a_i m_i N_i,$$

where $\varepsilon > 0$ is small. The hypothesis seems to be quite plausible for natural ecosystems. In terrestrial ecosystems, for example, the major part of the dead organic matter is due to vegetation "fall-off" (dead foliage and other parts). Metabolic by-products of animals and their corpses account for quite a small part (less than 10%) of the total organic mortmass.[2]

When, traditionally, $\varepsilon \to 0$, we have

$$\sum_{i=1}^{n} a_i m_i N_i \to a_1 m_1 N_1,$$

which mean that the resource closure of the chain is performed by the link outgoing from the first, primary production, level only.

We may assume therefore all entries c_i in matrix (6.24) to be zero, excepting for $c_1 \neq 0$. In this case matrix A_{qc} is three-diagonal again and even sign-stable if $d_0 - c_1 > 0$. If, in addition, inequality (6.25) holds true, then equilibrium $\mathbf{N}^q$ is asymptotically stable, i.e., the trophic chain of length q does exist.

As now $c_i = 0$, $i = 2, \ldots, q$, we have to put $m_1 \neq 0$, $m_2 = \ldots = m_q = 0$ in the formulas for φ_S, ψ_S, and σ_{2S}. Then we have

$$\varphi_S = \sigma_{2S} = \sigma_{2S+1} = 0, \quad \psi_S = \psi_{S+1} = a_1 m_1;$$

for $q = 2S$:

$$N_0^* = \frac{Q + a_1 m_1 f_{2S}}{\alpha_0 f_{2S}}, \qquad N_1^* = f_{2S},$$

$$N_q^* = N_{2S}^* = H_{2S-1}\left(\frac{Q + a_1 m_1 f_{2S}}{\alpha_0 f_{2S}} - f_{2S-1}\right), \qquad (6.30)$$

and for $q = 2S + 1$:

$$N_0^* = f_{2S+1}, \qquad N_1^* = \frac{Q}{\alpha_0 f_{2S+1} - a_1 m_1},$$

$$N_q^* = N_{2S+1}^* = H_{2S}\left(\frac{Q}{\alpha_0 f_{2S+1} - a_1 m_1} - f_{2S}\right). \qquad (6.30')$$

Using formulas (6.30–30'), one can show (see the Appendix) that for a trophic chain of length q to exist under the first-level closure it is necessary and sufficient that the resource inflow rate belongs again to the finite interval along the Q-axis:

$$\tilde{Q}^*(q) < Q < \tilde{Q}^*(q+1),$$

whose bounds are now:

$$\tilde{Q}^*(q) = \alpha_0 m_1 f_{q-1} f_q - a_1 m_1 f_q \quad \text{for even } q,$$
$$\tilde{Q}^*(q) = \alpha_0 m_1 f_{q-1} f_q - a_1 m_1 f_{q-1} \quad \text{for odd } q,$$

and $\tilde{Q}_1^*(1) = 0$. Comparing these bounds with those given in (6.23) reveals that a partially closed trophic chain can exist under lower Q-rates than the open chain of the same length, which sounds quite logical.

Thus, the main conclusion is that the whole potential range of resource inflow rates (i.e., the Q-axis) is divided by points

$$Q^*(1) < Q^*(2) < \cdots < Q^*(q) < Q^*(q+1) < \cdots, \quad q = 1, 2, \ldots$$

(respectively by points

$$\tilde{Q}^*(1) < \tilde{Q}^*(2) < \cdots < \tilde{Q}^*(q) < \tilde{Q}^*(q+1) < \cdots$$

for partially closed chains) into consecutive intervals within which only trophic chains of a fixed length exist stably. In other words, the energy entering the system with the resource inflow undergoes "quantification"; as follows from thorough investigation of sequences $\{Q^*(q)\}$ and $\{\tilde{Q}^*(q)\}$ (see the Appendix, Statement A6.2), the higher being the number of the next trophic level, the greater being the energy "quantum" that level requires for it to be fixed in the ecosystem.

IV. STABILITY IN GENERALIZED TROPHIC CHAIN SYSTEMS

The trophic chain equations considered in foregoing sections were close to the Lotka–Volterra type of population equations. However, the assumption of "tense" trophic relations does not cover "saturation" effects in trophic functions, which exist in real nature and which mathematically mean that

$$V_i(N_i) \to \overline{V}_i < \infty \quad \text{as} \quad N_i \to \infty.$$

Neglected also were competition effects within a trophic level. These can be incorporated by adding the terms

$$-\gamma_i N_i^2, \quad \gamma_i > 0,$$

to the right-hand sides of equations (6.1) or (6.2). What might be the effects of saturation in trophic functions and of competition within trophic levels on the existence and stability of the chain is considered below.

The problem of finding equilibrium states becomes naturally more complicated in the new system. Nevertheless, it can be shown (see Svirezhev and Logofet[1]) that all those equilibria represent *trophic chains*, i.e., they take on the form $\mathbf{N}^q = [N_0^*, N_1^*, \ldots, N_q^*, 0, \ldots, 0]$, $q = 1, \ldots, n$, there being a unique chain for each fixed q.

As before, the existence problem for a trophic chain of length q is reduced to an investigation of stability for the zero solution to the linearized system, with the matrix again assuming the form

$$\bar{F}_c = \begin{bmatrix} \tilde{A}_{qc} & C \\ 0 & D_{n-q} \end{bmatrix}, \tag{6.31}$$

where

$$\tilde{A}_{qc} = \begin{bmatrix} -b_0 & c_1 - d_0 & c_2 & \cdots & & c_q \\ b_1 & -h_1 & -d_1 & & \mathbf{0} & \\ & \ddots & \ddots & \ddots & & \\ \mathbf{0} & & b_{q-1} & -h_{q-1} & -d_{q-1} \\ & & & b_q & -h_q \end{bmatrix}, \quad C = \begin{bmatrix} c_{q+1} & c_{q+2} & \cdots & c_n \\ & & \mathbf{0} & \end{bmatrix},$$

and

$$D_{n-q} = \operatorname{diag}\{-m_{q+1} + k_{q+1}V_q(N_q^*), -m_{q+2}, \ldots, -m_n\}.$$

Here we have:

$$\begin{aligned} b_0 &= N_1^* V_0'(N_0^*), & b_i &= k_i N_i^* V_{i-1}'(N_{i-1}^*), \\ d_0 &= V_0(N_0^*), & d_i &= V_i^*(N_i), \\ h_i &= \gamma_i N_i^* + N_{i+1}^*[V_i'(N_i^*) - V_i(N_i^*)/N_i^*], \\ c_i &= a_i m_i, & i &= 1, \ldots, q. \end{aligned} \tag{6.32}$$

Since the trophic functions are, though saturated but still, monotone increasing with N_i, we have $V_i'(N_i^*) > 0$ (at $N_i^* < \infty$), so that entries b_i and d_i are all positive. If now $k_{q+1}V_q(N_q^*) < m_{q+1}$, then stability of matrix $\bar{F}_c$ is equivalent to that of its submatrix $\tilde{A}_{qc}$. When all $c_i = 0$, i.e., when the trophic chain is open, the submatrix $\tilde{A}_{qc} = A_q$ is three-diagonal and, as shown in Section 6.3, sign-stable if $h_i \geq 0$. Consequently, the condition

$$h_i = \gamma_i N_i^* + N_{i+1}^*[V_i'N_i^*) - V_i(N_i^*)/N_i^*] \geq 0 \tag{6.33}$$

is sufficient for a trophic chain of length q to exist in a model with saturated trophic functions or/and self-limitation by trophic levels.

Let all $\gamma_i = 0$, so that only the saturation effects are supposed to act. Then by (6.32),

$$h_i = N_{i+1}^*[V_i'N_i^*) - V_i(N_i^*)/N_i^*]. \tag{6.34}$$

If the trophic function is of Holling type I (increasing, convex upward hyperbola with a horizontal asymptote),[14] then the expression (6.34) is negative everywhere in positive values of N_i^*. Thus matrix $\tilde{A}_{qc}$ contains positive entries in its principal diagonal and can never be sign-stable. Rather, it is likely to be unstable due to a positive trace in matrix $\tilde{A}_{qc}$, the particular examples confirming this "likelihood."[15]

But if the trophic functions are of Holling type II (piecewise linear with a horizontal asymptote) or type III (S-shaped curve with a horizontal asymptote), then the quantities (6.33)) will be negative only for great enough N_i^*. In that region submatrix $\tilde{A}_{qc}$, hence matrix $\bar{F}_c$, becomes unstable. It is typical that the faster the saturation appears, the less will be those threshold values of N_i^* above which the chain starts collapsing.[1] Moreover, the loss of stability occurs once saturation is achieved in at least one trophic level (it is sufficient that at least one of h_i becomes negative). The fact that this is actually a collapse rather than a transition into a new equilibrium state that would be also a trophic chain of the same length q, follows from the uniqueness of equilibrium $\mathbf{N}^q$ in the model.

Introducing self-limitation into the trophic chain model can hardly introduce anything else than a stabilization by the sign stability criterion. If, indeed, $\gamma_i > 0$, then an appropriate choice of its magnitude can always make the condition (6.33) true, thus promoting conservation of the chain length.

$$* \quad * \quad *$$

The general conclusion from the theory examined above that the length of a trophic chain must depend on the power of the resource flow through the ecosystem is neither new, nor surprising to any ecological scientist who appreciates the role of matter and energy flows in ecosystem organization. However, what mathematics can really contribute here is the resultant technique, given the rate of the resource inflow and other individual parameters of a trophic chain, to predict the number of trophic levels capable of being fixed in the chain, or on the other hand, given a top level of the chain, to estimate the resource input needed to provide for the stable existence of the top. Both ways of reasoning are exemplified in the next section.

V. PRACTICAL EXAMPLES

To compare the mathematical theory of trophic chains with data on a real ecosystem, one has: first, to simplify the real structure up to a simple chain; second, to assume it to be at an equilibrium state, and, last but not least, to have the equilibrium values of the chain components and flows among them estimated. What kind of conclusions may then be gained from the above mathematics is illustrated in two examples giving in the present section.

A. Silver Springs Ecosystem

Matter and energy flows in the ecosystem at warm Silver Springs, Florida, is a case study that features enough empirical evidence to apply mathematical theory.[1] In his profound monograph,[16] H. T. Odum summarized data on "standing crops" and

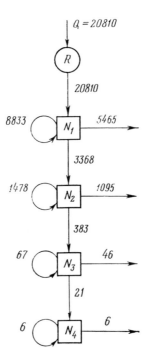

Figure 40.   Flow diagram for a simplified trophic chain in Silver Springs, Florida (the data after H. T. Odum[16]).

energy flows for five trophic levels: producers (N_1), herbivores (N_2), carnivores (N_3), top predators (N_4), and reducers closing the trophic chains. The ecosystem was considered to be at the steady state, with the variations in biomass being of seasonal origin. The equilibrium biomass densities at the main levels (expressed in grams of dry weight per square meter) were estimated respectively as follows: $N_1^* = 809, N_2^* = 37, N_3^* = 11, N_4^* = 1.5$, and the biomass of reducers as 5.

In the model, we consider energy to be the basic resource, hence, the chain to be open. By the same reasoning, reducers, which feed on dead organic matter and close the matter chains, are not included in the chain model. It should be noted however that the trophic chain equations (6.4) admit accumulating an unutilized part of the resource, whereas the solar energy does not accumulate in its original form. Nevertheless, the equations can be used in this case too, since at equilibrium, while flowing through the component R with no losses, all the incoming resource is taken up by the first trophic level. Also, the critical inflow rates under consideration, $Q^*(q)$, depend neither on N_0^*, nor upon α_0, i.e., the parameters governing the state of the resource variable.

A simplified flow diagram of the trophic chain is shown in Figure 40 representing a particular case of the diagram shown in Figure 38a. By the formulas given in

Section 6.II.A, the threshold rate values can be expressed in terms of production and mortality rates, and biomass conversion efficiencies as follows:

$$Q^*(2) = \frac{m_1 m_2}{\alpha_1 k_1 k_2},$$

$$Q^*(3) = Q^*(2)\left(1 + \frac{\alpha_1 m_3}{\alpha_2 m_1 k_3}\right),$$

$$Q^*(4) = Q^*(3)\left(1 + \frac{\alpha_2 m_4}{\alpha_3 m_2 k_4}\right),$$

$$Q^*(5) = Q^*(4)\left(1 + \frac{\alpha_1 \alpha_3 m_5}{\alpha_4 k + 5(\alpha_2 k_3 m_1 + \alpha_1 m_3)}\right). \tag{6.35}$$

It can be seen that the values of $Q^*(q)$ do not actually depend on α_0.

Since $\alpha_0 N_0^* N_1^* = Q = 20810$, while $k_1 \alpha_0 N_0^* N_1^* = 8833$, it follows that $k_1 = 0.42$. By similar argument, $k_2 = 0.44$, $k_3 = 0.175$, and $k_4 = 0.286$. The steady-state values for biomass are expressed in grams of dry weight, while those of the flows in kilocalories per time unit. If the conversion rate is denoted by ρ, so that $N_1^* = 809\rho$ [kcal/m^2], $N_2^* = \ldots$ and so on, then the particular value of ρ turns out to be irrelevant because m_i and α_i enter the algebra for $Q^*(q)$ in combinations which exclude ρ from the final expressions. Since

$$m_1 N_1^* = 5465, \quad m_2 N_2^* = 1095, \quad m_3 N_3^* = 46, \quad \text{and} \quad m_4 N_4^* = 6,$$

we have

$$m_1 = 6.755/\rho, \quad m_2 = 29.6/\rho, \quad m_3 = 4.18/\rho, \quad m_4 = 4/\rho.$$

Respectively, from the equalities

$$\alpha_1 N_1^* N_2^* = 3368, \quad \alpha_2 N_2^* N_3^* = 383, \quad \text{and} \quad \alpha_3 N_3^* N_4^* = 21$$

we find that $\alpha_1 = 0.1125/\rho^2$, $\alpha_2 = 0.956/\rho^2$, and $\alpha_3 = 1.27/\rho^2$. Substituting these values of m_i and α_i into (6.35) yields eventually

$$Q^*(2) = 9585, \quad Q^*(3) = 13611, \quad Q^*(4) = 18456.$$

The actual inflow rate Q is seen to be greater than $Q^*(4)$, i.e., the production at the primary producer level and the rest of the ecosystem parameters are such as to admit the stable existence of four trophic levels.

As concerns the fifth trophic level, let a species be introduced that feeds on the forth level and has the same parameters as those at the fourth level, i.e., $k_5 = k_4$, $m_5 = m_4$, and $\alpha_4 = \alpha_3$. Then by (6.35) $Q^*(5) = 36\,613$. This means that fixing the species at the fifth trophic level would require almost twice as much energy as the actual flow.

But even the 4-level chain appears to be highly vulnerable to changes in the ecosystem parameters. If, for example, the primary production decreases only 12%

due to an impact like, e.g., toxic pollution, then the 4-level equilibrium will lose its stability, with the top predator level being eliminated from the model.

Along similar lines of argument, other phenomena of the trophic chain ecology can be speculated and quantified as well.[1]

B. Minimal Territory Estimation for a Top Predator

Theoretical population dynamics, in general, and the mathematical theory of trophic chains, in particular, can generate approaches to obtaining minimal estimates for the size of protected areas necessary to maintain a certain level of biological diversity. The estimates make sense for those minimal bounds whose violation will result in destabilization and destruction of the system due just to the dynamic nature of interactions among and within its components, to say nothing of other, random or neglected, factors of the environment. These factors will certainly result in further expansion of the minimal bounds.[17-18]

As an illustration, a simple trophic chain can be considered in the problem of finding a minimal size of protected area for a mammal predator at the top of the chain.[17-18] The vegetation biomass is supposed to be the basic resource, so that the annual production is logically assumed to represent the resource inflow rate Q. Phytophagous animals hence occupy the first trophic levels, while their predators are at the second.

Evidently, the value of Q is closely related to area S occupied by the vegetation: if p is an average annual production rate per unit area, then $Q = pS$. The trophic chain stability theory ascertains that to maintain an open chain of length 2 (with the primary production occupying level 0), the Q-value must belong to the interval $[Q^*(2), Q^*(3)]$ defined in (6.13–13'). In particular, the minimal area size S gets bounded from below:

$$S > Q^*(2)/p.$$

Among a number of possible ways to quantify the bound, a simple one might be as follows.

The net green phytomass production (gross production minus respiration), with no regard to grazing by phytophagans, can be estimated, for instance, in the African savannah,[19] as 1200 kg/(ha × yr), the animals grazing about 10% of that production. The parameter p can consequently be estimated (in carbon units) as $p = 1200 \times 0.1 \times 0.4 = 480$ [kgC/(ha × yr)]. If now the minimal biomass size at the top trophic level is supposed to be not less than $N_2^* = 1000$ kgC (which may correspond, for instance, to an average lion pride[20]), then the area size S can be inferred from the chain stability conditions.

To solve the problem we also need: average efficiencies of biomass conversions between trophic levels, average death rates for phytophagans and predators, and parameters of trophic functions. Average efficiencies can be assumed as $k_1 = 0.01$, $k_2 = 0.02$. To estimate the death rates one may rely on the relationship $m_i = 1/L_i$, where L_i designates the average life span of organisms at a given level in a given environment. Put therefore $m_1 = 0.2 yr^{-1}$, $m_2 = 0.05 yr^{-1}$. As concerns the trophic function parameters, suppose the community to be in conditions of tense trophic relations, with the trophic functions being linearly dependent on the resource to be consumed. Also, the trophic function, V, can be assumed inversely proportional to specific expenditures, w_e, for procuring a unit food: $V w_e = \text{const.}$[21−22] Eventually, all this brings about the estimate of α_1 in the linear trophic function $V_2(N_2) = \alpha_1 N_2$ as $\alpha_1 \cong 10^{-4}$ $(\text{kgC} \times \text{yr})^{-1}$.

Now it follows from the trophic chain stability theory and the formulas of Section 6.II.A, that

$$N_2^* = Q \frac{k_1 k_2}{m_2} - \frac{m_1}{\alpha_1} = pS \frac{k_1 k_2}{m_2} - \frac{m_1}{\alpha_1},$$

whereby

$$S = \frac{m_2(N_2^* + m_1/\alpha_1)}{k_1 k_2 p}. \tag{6.36}$$

Substituting the proper estimates for the parameters in (6.36) yields $S \cong 1500ha$, which does not contradict the expert estimation.[20]

Suppose now the following situation. Let the protected area S be great enough to provide for the resource inflow rate Q, sufficient for the existence of a finite-length trophic chain (e.g., of length 2 as in the example above). If, by any reason, the size S has to be diminished, then Q will respectively decrease too and the decrease may shift the Q value down to the interval, where only a shorter trophic chain exists (interval $[Q^*(1), Q^*(2)]$ in the example). Thus, a decrease in the size of a protected area may have a consequence which is far from trivial: the loss of stability by the top trophic level and its elimination, rather than a proportional decrease in populations at each trophic level of the chain. This is a manifestation of the principal nonlinearity of ecological interactions.

Appendix

Statement 6.1. *In the Lotka–Volterra model of a first-level closure trophic chain,*

$$\frac{dN_0}{dt} = Q - \alpha_0 N_0 N_1 + a_1 m_1 N_1,$$

$$\frac{dN_i}{dt} = N_i(-m_i + k_i \alpha_{i-1} N_{i-1} - \alpha_i N_{i+1}), \quad i = 1, \ldots, n, \tag{A6.1}$$

the equilibrium $\mathbf{N}^q = [N_0^*, N_1^*, \ldots, N_q^*, 0, \ldots, 0]$, $N_i^* > 0$, $i = 1, \ldots, q$, is (locally and asymptotically) stable if the condition holds

$$\bar{Q}^*(q) < Q < \bar{Q}^*(q+1), \tag{A6.2}$$

where

$$\bar{Q}^*(q) = \alpha_0 m_1 f_{q-1} - a_1 m_1 f_q \quad \text{for even } q,$$
$$\bar{Q}^*(q) = \alpha_0 m_1 f_{q-1} f_q - a_1 m_1 f_{q-1} \quad \text{for odd } q, \tag{A6.3}$$

and quantities f_q are defined in (6.10).

Proof: The Jacobian matrix of system (A6.1) takes on the form (6.24), with the entries $c_2 = \ldots = c_q = 0$ and the submatrix A_{qc} being sign-stable if $c_1 < d_0$, or $a_1 m_1 < \alpha_0 N_0^*$. From the expressions for N_0^* of formulas (6.30–30'), it follows that the condition is equivalent to

$$a_1 m_1 f_{2S} < Q + a_1 m_1 f_{2S} \quad \text{for even } q,$$

and to

$$a_1 m_1 < \alpha_0 f_{2S+1} \quad \text{for odd } q.$$

While the former obviously holds true for any $Q > 0$, the latter can be rewritten in the form

$$\alpha_0 f_{2S} f_{2S+1} - a_1 m_1 f_{2S} > 0,$$

which is true due to the feasibility condition (6.17') for equilibrium N^q.

Now condition (6.25) remains a definitive one for stability of N^q. By means of (6.30) and (6.30') it is reduced exactly to (A6.3).   ∎

Statement A6.2. *Sequences $\{\Delta Q^*(q)\}$ and $\{\Delta \bar{Q}^*(q)\}$, where by definition*

$$\Delta Q^*(q) = Q^*(q+1) - Q^*(q), \tag{A6.4}$$

$$\Delta \bar{Q}^*(q) = \bar{Q}^*(q+1) - \bar{Q}^*(q), \tag{A6.5}$$

with $Q^*(q)$ and $\bar{Q}^*(q)$ being defined respectively by (6.13–13') and (6.17–17'), are monotone increasing with q.

Proof: The proof is hardly longer than the formulation. By the definitions, we have

$$
\begin{aligned}
\Delta Q^*(q) &= \alpha_0 f_q (f_{q+1} - f_{q-1}); \\
\Delta \bar{Q}^*(q) &= \begin{cases} \alpha_0 f_q (f_{q+1} - f_{q-1}) & \text{for even } q, \\ (\alpha_0 f_q - a_1 m_1)(f_{q+1} - f_{q-1}) & \text{for odd } q, \end{cases}
\end{aligned}
$$

whereby the statement follows from construction (6.10) of quantities f_q. To imagine the order of increasing, consider the case of equal consumption rates $\alpha_i = \alpha$ and equal efficiencies $k_i = k \ll 1$, where

$$\Delta Q^*(q) = \Delta \bar{Q}^*(q) = \frac{m^2}{\alpha k^{q+1}}[1 + O(k)].   \blacksquare$$

Additional Notes

To 6.I. Such structures as trophic chains have long invited mathematical efforts. While the cited example of a trophic web, that was published almost half a century ago by Jones[5] and, after having been canonized by E. P. Odum,[6.2] has already served to educate several generations of ecological "modelers," there was no "canonical" approach to mathematical modeling of trophic chains. The only common factor was the cognition that species-specific variables should somehow be aggregated. Alexeev and Kryshev,[23] for example, derived dynamical equations for averaged biomasses of two trophic levels by using the method of statistical mechanics (ecological characteristics of species within each level were suppose to be close enough).

A somewhat different approach was proposed in Tansky,[24] where all species were assumed to be the same within each level, although the numbers of species at each level were treated as new dynamic variables. As an ecological paradigm, the P/B-ratio (production to the total biomass) was supposed to attain its minimum at the steady state, whereby patterns were obtained that looked very much like trophic pyramids in real ecosystems.

Trophic function is the term that, after it had been coined by Svirezhev,[25] became the dominant expression in the Russian-language literature of the 70–80s for what was normally called the *functional response* (of a predator population to its prey density) after Holling.[14] While both terms may well compete in prey-predator models, the former is quite preferable in the context of aggregate trophic chains.

Equations (6.4a) for an open trophic chain are very close to those of the Lotka–Volterra model, differing only in the first, resource, equation. Modifications of this kind were applied, for instance, by V. A. Kostitzin[26] to describe some global biogeochemical cycles.

Equations (6.4b) for $Q \equiv 0$ were considered by Eman[27] as a model of a biogeocenosis, where the "biogeocenosis" was treated as a biotic community plus an "inert" component. The feasible equilibrium was shown to exist, to be unique, and to be stable under certain conditions for any fixed order of the system. In other words, the chain length was supposed to be fixed, in contrast with the variable length in the models considered above.

With only one term in the first equation corrected, systems (6.4a) and (6.4b) have gotten rid of this drawback. Their finite dimensionality is a mathematical reflection of the finiteness in trophic chains that are potentially possible in reality. However, it would be mathematically interesting to see whether the systems retain the same conclusions with respect to finite q-chains as n tends to infinity, thus resulting in countably dimensional systems of ODEs. In other words, will there be a similar dependence of the finite chain length upon the resource inflow in Jonathan Swift's infinite "flea" construction?[2,3] The answer goes beyond the mathematical scope of this book.

To 6.II. With the equilibrium sizes of trophic levels having obtained their finite expressions in terms of model parameters, some ecologically sound problems can be reduced to simple algebraic routine. In particular, conditions can be derived that will enable a "paradoxical" trophic chain to exist[1], i.e. the conditions for the size of the next level to be greater than that of the given level, or the conditions for a typical trophic pyramid[2] to become "inverted." Comparison of mathematical existence conditions for normal and "paradoxical' trophic chains of the same length showed[1] that the Q-rate had to be higher in the latter case; in nature, the "paradoxical" trophic chains were observed in eutrophying lakes: increasing nutrient load (pollution with domestic sewage, fertilizer run-off from adjacent fields) caused "normal" trophic pyramids to turn onto "paradoxical" ones.[28]

To 6.III. In some texts a three-diagonal matrix is referred to as *Jacobian matrix*, but we reserve this term for the matrix of a linearized system. Stability conditions for such matrices were, of course, known long before the sign stability criterion was developed—as long as hierarchical chain models were considered in other areas of application (see, e.g., Schwartz[29]). The criterion, however, has appeared to be the most convenient means of deducing the trophic chain stability conditions.

Several more particular cases of partially closed chains are analyzed and interpreted in Svirezhev and Logofet,[1] where, in addition, the main result on stability in the case of the first-level closure is strengthened by the Lyapunov function method.

To 6.IV. Saturation and self-limitation effects on the trophic chain dynamics were naturally the object of further mathematical efforts, although with less contribution from matrix analysis.[1,30−31] What are the dynamical consequences of the distinction between the shapes of type I and type III functional response curves, is considered in more detail by Svirezhev.[30] Both speculations and some analytical evidence were also proposed[1] in the "ecological stability" problem: whether the extension of the Lyapunov stability concept will expand the range of trophic chain existence. Some more "evidence" was obtained recently,[31−32] although the problem is quite far from being solved.

Another approach to stability in trophic chains deals with the so-called "food-web resilience," understood as the rate at which the system returns to steady state following a perturbation.[33−34] Unfortunately, no simple or general relationship has yet been revealed between stability and nutrient input. For example, in a partially closed, 3-level chain model $(0 \to 1 \to 2 \to 3 + \text{detritus})$ with a type-I trophic function for the nutrient uptake and type-III functions for other trophics, "resilience may not always increase as nutrient input is increased and resilience may not always decrease as the number of levels in a food chain increases" (p. 803 in Ref. 33).

For branching trophic chains and for those of a completely closed matter cycle, see, e.g., Svirezhev and Logofet[1] and Alexeev.[35] In the latter case, there exist interesting dynamical regimes like "automodel" solutions (i.e., self-oscillations) and oscillations of the trigger, or "flip-flop," type.

To 6.V. Minimal territory estimation is one of problems intensively discussed in the context of nature conservation,[36−37] to which the trophic chain stability considerations may add just one more theoretical aspect. For experiments on a laboratory flow-through microcosm, designed to test the conclusions drawn from theoretical food-chain models, see DeAngelis et al.[34]

REFERENCES

1. Svirezhev, Yu. M. and Logofet, D. O. *Stability of Biological Communities* (revised from the 1978 Russian edition), Mir Publishers, Moscow, 1983, 319 pp., Chap. 5.

2. Odum, E. P. *Basic Ecology*, Saunders College Publishing, Philadelphia, 1983, Chap. 3.

3. Pimm, S. L. *Food Webs*, Chapman and Hall, London, 1982, 219 pp., Chap. 6.

4. Cohen, J. E., Briand, F. and Newman, C. M. Community food webs: Data and theory, *Biomathematics*, Springer-Verlag, Berlin, 20, 1990, 308 pp., Chap. 1.

5. Jones, J. R. E. A further ecological study of a calcareous stream in the "Black Mountain" district of South Wales, *J. Anim. Ecol.*, 18, 142–159, 1949.

6. Odum, E. P. *Fundamentals of Ecology*, Saunders, Philadelphia, 1953, Chap. 4.

7. Simberloff, D. S. and Wilson, O. E. Experimental zoogeography of islands: The colonization of empty islands, *Ecology*, 50, 278–296, 1969.

8. Heatwale, H. and Levins, K. Trophic structure, stability and faunal change during recolonization, *Ecology*, 53, 531–534, 1972.

9. Simberloff, D. S. Trophic structure determination and equilibrium in an arthropod community, *Ecology*, 57, 395–398, 1976.

10. Svirezhev, Yu. M. On the trophic chain length, *Zhurnal Obshchei Biologii* (*Journal of General Biology*), 39, 373–379, 1978 (in Russian).

11. Odum, E. P. *Basic Ecology*, Saunders College Publishing, Philadelphia, 1983, Chap. 5.

12. Svirezhev, Yu. M. and Logofet, D. O. *Stability of Biological Communities*, Nauka, Moscow, 1978, Chap. 5 (in Russian).

13. Gantmacher, F. R. *The Theory of Matrices*, Vol. 2. Chelsea, New York, 1960, Chap. 16.

14. Holling, C. S. The functional response of predator to prey density and its role in mimicry and population regulation, *Mem. Entomol. Soc. Canada*, 45, 1–60, 1965.

15. Svirezhev, Yu. M. and Logofet, D. O. *Stability of Biological Communities* (revised from the 1978 Russian edition), Mir Publishers, Moscow, 1983, Chap. 3.

16. Odum, H. T. Trophic structure and productivity of Silver Springs, Florida, *Ecological Monographs*, 27, 55–112, 1957.

17. Logofet, D. O. and Svirezhev, Y. M. Modeling of population and ecosystem dynamics under reserve conditions. In: *Conservation, Science and Society. Contributions to the First International Biosphere Reserve Congress, Minsk, Byelorussia/USSR, 26 Sep.– 2 Oct. 1983*, Vol. 2. UNESCO-UNEP, 331–339, 1984.

18. Logofet, D. O. and Svirezhev, Yu. M. Modeling the dynamics of biological populations and communities under reserve conditions. In: *Mathematical Modeling of Biogeocenotic Processes*, Svirezhev, Yu. M., ed., Nauka, Moscow, 1985, 25–37 (in Russian).

19. Bazilevich, N. I. Personal communication, 1983.

20. Boitani, L., Bartoli, S. T. and Anderson, S. *Guide to Mammals*, Simon and Schuster, 1983, 512 pp.

21. Dolnik, V. R., ed. *Time and Energy Budgets in Wild Birds*, (*Transactions of Zoological Institute of the USSR Acad. Sci.*, Vol. 113) Leningrad Sate University, Leningrad, 1982, 158 pp.

22. Karelin, D. V. and Gilmanov, T. G. Simulation model for population density dynamics in insectivorous forest passerines: An energy budget approach, *Zhurnal Obshchei Biologii* (*Journal of General Biology*), 53, 31–46 (in Russian).

23. Alexeev, V. V. and Kryshev, I. I. Kinetic equations to describe ecosystem dynamics, *Biophysics*, 19, 754–759, 1974 (in Russian).

24. Tansky, M. Structure, stability and efficiency of ecosystem. In: *Progress in Theoretical Biology*, Rosen, R. and Snell, F. M., eds., Academic Press, New York, 1976, pp. 206–262.

25. Svirezhev, Yu. M. On mathematical model of biological communities and related control and optimization problems. In: *Mathematical Modeling in Bioligy*, Molchanov, A. B., ed., Nauka, Moscow, 1975, 30–52 (in Russian).

26. Kostitzin, V. A. *L'evolution de l'atmosphere*, Hermann, Paris, 1935.

27. Eman, T. I. On some mathematical models of biogeocenoses. In: *Problemy Kibernetiki (Problems of Cybernetics)*, Vol. 16, Lyapunov, A. A., ed., Nauka, Moscow, 1966, pp. 191–202 (in Russian).

28. Loucks, O. L., Prentki, R. T., Watson, V. J., Reynolds, B. J., Weiler, P. R., Bartell, S. M., and D'Alessio, A. B. Studies of the Lake Wingra Watershed: an Interim Report. IED Report 78. Institute for Environmental Studies, University of Wisconsin, Madison, 1977.

29. Schwartz, H. R. Ein Verfahren zur Stabilitatsfrage von Matrizen-Eigenwert-Problem, *Z. Agnew. Math. Phys.*, 7, 473–500, 1956.

30. Svirezhev, Yu. M. *Nonlinear Waves, Dissipative Structures and Catastrophes in Ecology*, Nauka, Moscow, 1987, 368 pp., Chap. 9 (in Russian).

31. Adzhabyan, N. A. and Logofet, D. O. Population size dynamics in trophic chains. In: *Problems of Ecological Monitoring and Ecosystem Modelling*, Vol. 14, Izrael, Yu. A., ed., Gidrometeoizdat, St. Petersburg, 135–153, 1992 (in Russian).

32. Adzhabyan, N. A. and Logofet, D. O. Extention of the stability concept in a mathematical theory of trophic chains, *Zhurnal Obshchei Biologii (Journal of General Biology)*, 53, 81–87, 1992 (in Russian).

33. DeAngelis, D. L., Bartell, S. M. and Brenkert, A. L. Effects of nutrient recycling and food-chain length on resilience, *American Naturalist*, 134, 778–805, 1989.

34. DeAngelis, D. L., Post, W. M., Mulholland, P. J., Steinman, A. L. and Palumbo, A. L. Nutrient limitation and the structure of food webs, *Ecology International Bulletin*, 19, 15–28, 1991.

35. Alexeev, V. V. Biogeocenoses: Autogenerators and triggers, *Zhurnal Obshchei Biologii* (Journal of General Biology), 37, 738–744, 1976 (in Russian).

36. Diamond, J. D. and May, R. M. Island biogeography and the design of natural reserves. In: *Theoretical Ecology: Principles and Applications*, Blackwell, Oxford, 1976, 163–186.

37. Pimm, S. L. and Gilpin, M. E. Theoretical issues in conservation biology. In *Perspectives in Ecological Theory*, Rougharden, J., May, R. M. and Levin, S. A., eds., Princeton University Press, Princeton, NJ, 287–305, 1989.

CHAPTER 7

Ecological Niche Overlap and Stability in Horizontal-Structured Communities

While, in the previous chapter, we considered vertical-structured communities, with interactions between trophic levels, our concern in the present chapter is with interactions between species within a trophic level, or, in other words, with horizontal-structured communities. Prey-predator or resource-consumer type of relations between a pair of species should place the species into different trophic levels; therefore, only mutualism (+ +) or competition (− −) remain as possible types of "bilateral" relations among species of the same trophic level. Mutualism is touched upon in relation with *M*-matrices (Section 4.V), and the present chapter concentrates on competition models.

Because of (and perhaps, due to) the fact that competition models can no longer be sign-stable, their mathematics is fairly far developed, particularly in conjunction with theoretical insight into the role that competition plays in community organization. Of fundamental importance for the theory are the concepts of the *ecological niche* and the *competition exclusion principle*. Such phenomena as the introduction or invasion of new species can also be studied in terms of stability conditions for pertinent species compositions.

I. ECOLOGICAL NICHE AND THE DYNAMIC EQUATIONS FOR COMPETITIVE SPECIES

Since the models in this chapter are deliberately confined to a single trophic level, they do not focus on the resource dynamics explicitly, but rely instead upon the notion of ecological niche, which treats the resources implicitly and in static rather than dynamic terms. This notion follows logically from the simple and straightforward idea that organisms of any biological species are characterized by a certain range of physical and ecological conditions within which they survive and successfully reproduce. *Ecological niche* is therefore that domain in the space of vitally important environmental factors (for example, the specific composition and size of food, site conditions, etc.) within which the existence of the species is ensured and beyond which this existence is impossible or unlikely.[1]

Despite this concept being intuitively clear, there is still some discussion in the literature[2,3] about what should be an unambiguous definition of the niche. However, there is no doubt that each species occupies its own ecological niche, while the niches of species coexisting in real communities are typically observed to intersect, or to *overlap*, with each other. Species may coexist, not only by cohabiting the same or similar physical and geographical conditions, but also by consuming some resources in common. Resource limitation imposes natural constraints on the total population sizes of species sharing the resource, so that the population growth rates are mutually restricted. Ecological niche overlap thus naturally results in relations of competition, and the ecological niche naturally determines the place and role of a given species in the structure of a competitive community.

Since competitive communities cannot be qualitatively stable (cf. Section 5.II.C), the problem arises of determining stability domains for the parameters of a competition model, or in other words, to determine a structure of niche overlap which provides for the stable coexistence of competitors, the so-called *species packing*.

The ecological niche concept is closely related to the well-known *competitive exclusion principle* of G. F. Gause,[4] according to which two species with similar ecological requirements cannot coexist in one habitat. This is the limit to the niche overlap in a stable community which indicates the limits to the ecological similarity of species that persist together. Species packing analysis must therefore bring about those theoretical conditions which account for the elimination of particular species from the competitive community, as well as the conditions for new species to invade or to be successfully introduced into the community.

A. Resource Spectrum and Utilization Function

Definitions of the ecological niche are fairly abundant in the literature, but from the mathematician's viewpoint, the most appropriate among them is the one that, while presenting the essence, allows formalization too. A successful compromise of this kind is the niche definition based upon the notions of a *resource spectrum* and a *utilization function*[5-8].

Suppose the resource consumed by a species is characterized by a parameter $\mathbf{x}$ (size of the food, its specific composition, space coordinates of the habitat, etc.), which is multi-dimensional in the general case; and suppose the amount of the resource with characteristics $\mathbf{x}$ which is available for consumption is determined by a function $K(\mathbf{x})$. Then the set of values $\mathbf{x}$ with the function $K(\mathbf{x})$ defined on it is called a *resource spectrum* (or *resource space*).

Moreover, let the resource consumption by a given species be described by a probability distribution, whose density function, $f(\mathbf{x})$, with the mean $\mathbf{x}_0$ and a finite variance σ^2 is termed *utilization function*. Then the *ecological niche* is defined by the point $\mathbf{x}_0$ on the resource spectrum and the function $f(\mathbf{x})$, the density of a random distribution around $\mathbf{x}_0$. For example, organisms of some species prefer the food of size $\mathbf{x}_0$, while the consumption of food of other sizes is governed by a— theoretical or empirical—probability law with the density $f(\mathbf{x})$. The preferable size $\mathbf{x}_0$ is logically thought of as the *center of the niche*, and the mean square deviation σ as the *niche breadth* characterizing to what extent the species is specialized

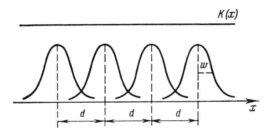

Figure 41.   One-dimensional resource spectrum $K(x)$ and the overlap of ecological niches determined by normal utilization functions.

in uptaking resource **x**: small σ values suggest a high specialization, large ones suggest a low specialization.

It is clear that one might attain a theoretically complete and adequate definition of the niche that accommodates the living conditions of a biological species by properly expanding the dimensionality of vector **x**. However, to make the quantitative analysis of models clearer and simpler we assume hereafter that the resource spectrum is one-dimensional. In many cases the results obtained can be generalized in a simple and natural way.

For a community of several species competing for common resources, the points x_0 corresponding to different species are naturally thought of (and this is in accord with observations) as finitely distant from each other. In this case, the competition phenomenon itself, caused by the overlap of ecological niches for competing species, appears within the formalism adopted as a result of intersecting domains of definition of the appropriate utilization functions $f_i(x)$ in the resource space. The most simple and idealized example[6,7] of this kind is shown in Figure 41, where $f_i(x)$ represent normal distribution density curves, of the same shape, uniformly spaced along the resource spectrum with the same distance d between adjacent mean values; $K(x) \equiv$ const. If w is the mean square deviation in common for all the curves, then the ratio w/d can be considered a measure of closeness of the niches, or a measure of how densely species are "packed" in the spectrum. Should this measure enter any condition of stability, one could judge about the limits to niche overlap.

B. Lotka–Volterra Competition Equations

Below is an argument to explain the form of dynamic model equations for an n-species competitive community, where the variables $N_i(t)$ are expressed in units of the resource taken up. The product $f_i(x)N_i(t)$ makes sense of a fraction of the overall resource that the ith species consumes at point x of the resource spectrum.

Furthermore, the quantity

$$K(x) - \sum_{i=1}^{n} f_i(x)N_i(t)$$

is nothing else than the local (in the resource spectrum) difference between the amount of resource which is potentially available and that which is actually necessary for species composition $[N_1(t), \ldots, N_n(t)]$ to coexist. As far as the resource does not limit the species population growth (for small populations), the population can be considered to grow exponentially with its intrinsic growth rate r:

$$\frac{d(f_i(x)N_i)}{dt} = r_i f_i(x)N_i, \quad i = 1, \ldots, n. \tag{7.1}$$

The limiting effect of competition can be expressed by means of a relative measure of the resource depletion at point x:

$$[K(x) - \sum_{i=1}^{n} f_i(x)N_i(t)]/K(x), \tag{7.2}$$

whereby the ith species dynamics at point x of its ecological niche obeys the equation

$$f_i(x)\frac{dN_i}{dt} = \frac{r_i N_i}{K(x)}f_i(x)\left[K(x) - \sum_{j=1}^{n} f_j(x)N_j(t)\right], \tag{7.3}$$

Integrating (7.3) throughout the resource spectrum yields

$$K_i\frac{dN_i}{dt} = r_i N_i\left(K_i - \sum_{j=1}^{n} \alpha_{ij}N_j\right), \quad i = 1, \ldots, n, \tag{7.4}$$

where

$$K_i = \int K(x)f_i(x)\,dx \tag{7.5}$$

represents the total amount of resource consumed by the ith species, the *niche carrying capacity*, while the *coefficient of competition* between species i and j,

$$\alpha_{ij} = \int f_i(x)f_j(x)\,dx, \tag{7.6}$$

is in proportion to the total probability (throughout the spectrum) that the two species, when utilizing the resource, collide at the same point of the spectrum, and thus characterizes their niche overlap.

Dividing both sides of (7.4) by K_i, we come to the following system of dynamic equations:

$$\frac{dN_i}{dt} = r_i N_i - \frac{r_i}{K_i}\sum_{j=1}^{n} \alpha_{ij}N_iN_j, \quad i = 1, \ldots, n, \tag{7.7}$$

which coincides, to within notations, with the general-type Lotka–Volterra system [cf. (4.4)]. Notice that the scaling change of variables $\tilde{N}_i = K_i N_i / r_i$ makes (7.7) take the form

$$\frac{d\tilde{N}_i}{dt} = r_i \tilde{N}_i - \sum_{j=1}^{n} \gamma_{ij} \tilde{N} \tilde{N}_j, \tag{7.8}$$

where

$$\gamma_{ij} = \alpha_{ij} \frac{r_i r_j}{K_i K_j} , \tag{7.9}$$

so that the interaction matrix $[\gamma_{ij}]$ retains the symmetry property of the competition matrix $A = [\alpha_{ij}]$.

System (7.7) is dissipative in the Volterra sense (Section 4.IV.C) since

$$F(\mathbf{y}) = \sum_{i,j=1}^{n} \alpha_{ij} y_i y_j = \int \sum_{i,j} f_i(x) y_i f_j(x) y_j \, dx = \int \left[\sum_{i=1}^{n} f_i(x) y_i \right]^2 dx > 0$$

for all $\mathbf{y} \neq \mathbf{0}$, i.e., the quadratic form is positive definite. Dissipativeness implies that if (7.7) has a positive equilibrium, then it is globally asymptotically stable throughout the positive orthant $\mathbb{R}_+^n$ and, what is more, this equilibrium stability appears to be not only a sufficient condition for stable functioning of an n-species competitive community, but a necessary one as well.

Therefore, a stability criterion for a competitive community is equivalent to the existence conditions of a positive equilibrium, that is a positive solution to the linear system

$$A N = K, \quad K = [K_1, \ldots, K_n]^T > 0 \tag{7.10}$$

A "geometric" interpretation for positiveness of solutions to (7.10) is given in Section 7.III.

C. Normal Utilization Functions and the Limits to Similarity

MacArthur and May[5,6] analyzed in detail a particular case of (7.7) with Gaussian utilization functions. For each species i,

$$f_i(x) = (2\pi \omega_i^2)^{-1/2} \exp \left\{ \frac{-(x - x_i)^2}{2\omega_i^2} \right\}, \tag{7.11}$$

where x_i is the mean value, or the niche center, and ω_i^2 is the variance of the normal distribution. If the niche centers for the ith and jth species are distant by d_{ij}, then

$$\begin{aligned} \alpha_{ij} &= \frac{1}{2\pi \omega_i \omega_j} \int_{-\infty}^{\infty} \exp \left\{ -\frac{x^2}{2\omega_i^2} - \frac{(x - d_{ij})^2}{2\omega_j^2} \right\} dx \\ &= \frac{1}{\sqrt{2\pi(\omega_i^2 + \omega_j^2)}} \exp \left\{ -\frac{d_{ij}^2}{2(\omega_i^2 + \omega_j^2)} \right\}. \end{aligned} \tag{7.12}$$

If all $f_i(x)$ have equal variances $\omega_i^2 = \omega^2$ and the niches are labeled sequentially, then $d_{ij} = |i - j|d$ $(d > 0)$ and

$$\alpha_{ij} = \frac{1}{2\omega\sqrt{\pi}} \left\{ -\frac{(i-j)^2 d^2}{4\omega^2} \right\}. \tag{7.13}$$

Normalizing the coefficients α_{ij} in a way such that the measure of niche "self-overlap" equals unity, yields

$$\alpha_{ij} = a^{(i-j)^2}, \quad a = \exp\{-d^2/4\omega^2\}; \tag{7.14}$$

and the corresponding *competition matrix* $A = [\alpha_{ij}]$ takes on the form

$$A = \begin{bmatrix} 1 & a & a^4 & \cdots & a^{(n-1)^2} \\ a & 1 & a & \cdots & a^{(n-2)^2} \\ a^4 & a & 1 & \cdots & \\ \cdots & \cdots & \cdots & \cdots & \cdots \\ a^{(n-1)^2} & & & & 1 \end{bmatrix}. \tag{7.15}$$

Dissipativeness of matrix (7.15), resulting from the integral representation (7.6) of the competition coefficients, is thus preserved for all values of the ratio d/ω, that is a measure of how closely the species niches are packed in the spectrum. Hence, equilibrium $\mathbf{N}^*$ will formally be stable for any arbitrarily close packing (however small d/ω).

If translated into ecological terms, the conclusion would mean that there are no limits to niche similarity and it would apparently contradict reality (see, e.g., May[9] and references therein). It can be shown however[9] that for $n \gg 1$ the smallest eigenvalue of A, which specifies how fast a perturbed trajectory returns to equilibrium, approximates

$$\lambda_{\min} \approx 1 - 2a + 2a^4 - 2a^9 + 2a^{16} - \cdots,$$

which gives, for small d/ω,

$$\lambda_{\min} \approx 4\sqrt{\pi}\frac{\omega}{d} \exp\{-\pi^2\omega^2/d^2\}.$$

It follows that for substantial niche overlap, i.e., for $d/\omega \to 0$, although $\lambda_{\min}$ is still positive, providing formally for equilibrium asymptotic stability, it tends to zero faster than any finite power of d/ω, thus having an essential singularity at zero. This results in extremely long return times for the perturbed trajectory.

To find a quantitative estimate for the "limits to niche similarity," MacArthur and May[6] resorted to a stochastic counterpart of competition equations (7.7), with "white-noise" perturbations interpreted as variability of the environment. Stable coexistence then required the value of d/w to be bounded from below, so that the "limits to similarity" were related to environmental stochasticity.[6,10]

But there is another argument in favor of deterministic theory, which reverts to the equilibrium existence conditions. Direct computer runs for different n within the range of $n = 100$ indicate[11] that for small d/ω, that is, for values of a sufficiently close to 1, negative components occur in the solution of (7.10) (when the right-hand sides are all identical). This implies that in order to preserve stability an increase in niche overlap necessarily requires that the function $K(x)$ be changed in a quite specific way, such that the set of niche capacities would keep the solutions of (7.10) positive. Further comments on the positiveness of equilibrium solutions are given below in Section 7.III.

D. Other Special Cases

Some special, pure theoretical, constructions of the resource spectrum are worthy of being mentioned.[7-9] Suppose each of n species competes only with its closest neighbors but does not at all with the other species. If $a < 1$ denotes the measure of niche overlap for two adjacent niches, then the competition matrix takes on the form

$$A = \begin{bmatrix} 1 & a & & & \\ a & 1 & a & & \mathbf{0} \\ & \ddots & \ddots & \ddots & \\ \mathbf{0} & & a & 1 & a \\ & & & a & 1 \end{bmatrix}. \tag{7.16}$$

Clearly, a cannot be arbitrarily close to 1 in this case since, otherwise, if the 1st species has more overlap with the 2nd, while the 2nd overlaps to the same extent with the 3rd, then the 1st and the 3rd species must have an overlap too.

In another particular case, where every species competes equally with all the rest, the competition matrix has the form

$$A = \begin{bmatrix} 1 & a & a & \cdots & a \\ a & 1 & a & \cdots & a \\ & & \cdots & & \\ a & a & a & \cdots & 1 \end{bmatrix}. \tag{7.17}$$

This case is realized, for instance, for three species with identical niches centered at vertices of an equilateral triangle on a two-dimensional resource spectrum; or for 4 species with their niche centers at vertices of a regular tetrahedron in a three-dimensional resource space; or, in the general case, by n species with niche centers at vertices of a regular n-hedron, the simplex spanning $n - 1$ equal vectors, in an $(n - 1)$-dimensional resource space.

Since both these matrices can be constructed in terms of an especially chosen resource spectrum and utilization functions, one may assert a priori that they are positive definite, that is, the corresponding competitive communities are dissipative.

Nevertheless, it is still of interest to compute the eigenvalues (see the Appendix) to know how fast the perturbed trajectories converge to equilibrium or to estimate a limit to niche overlap.

II. ANOTHER REPRESENTATION OF COMPETITION COEFFICIENTS

If utilization functions of all species in the competitive community are known, the calculation of competition coefficients presents no principal difficulties. However, an exact form of $f_i(x)$ is generally unknown in practice and occasionally, when the limiting resource is essentially multivariate, it can hardly be specified for sure.[12-14] Neither are the methods themselves unique in calculating a measure of niche overlap.[3,15-16] Therefore, it would be of interest to study the properties of system (7.7) with no reference to the integral representation (7.6) for the competition coefficients α_{ij}.[8,17-18]

A. Niche Partition as a Metric in the Resource Space

Any attempt to construct the resource spectrum from field or laboratory observations will require the application of statistical techniques which are generically called *descriptive models* and which are generally based upon representing the data as points in a multi-dimensional space of relevant parameters (see, e.g., Jeffers[19]). Quite logically the idea then arises to use a metric on that space as the basis for quantitative methods of similarity and cluster analysis.

Arguing along these lines and refusing the integration form (7.6) of the competition coefficients α_{ij}, consider now ecological niches i and j as points of a metric space and assume the *metric*, $\rho(i,j)$, or a generalized *distance* in the resource space, to be a measure of niche separation in the resource continuum. The assumption is grounded as far as the sampling data are representative for those species-specific parameters which are relevant to (and sufficient to represent) the competition among species under study. The simple argument that the farther the niches are from each other in the space, the lesser must be the competition between the species, results in a representation like

$$\alpha_{ij} = \exp\{-\rho(i,j)\}. \tag{7.18}$$

Then, two standard axioms of the metric $\rho(i,j)$, namely, $\rho(i,j) = \rho(j,i)$ and $\rho(i,i) = 0$, correspond respectively to the symmetry in effects of competitors upon each other and to the convention adopted earlier that the measure of niche self-overlap should equal unity. (The third axiom, the so-called *triangle inequality*, unfortunately, finds no meaningful interpretations apart from banal allusions to the "eternal triangle" of competition for a breeding-mate.)

Suppose further that niches of n competitor species can be ordered (and enumerated) in such a way as to form a regular sequence, the "regularity" implying that the metric is now a function of the "normal" distance between niches, i.e., the difference in their numerals, $|i - j|$. Then, with no further loss of generality, the competition coefficients (7.18) are transformed into

$$\alpha_{ij} = \alpha(|i - j|), \tag{7.19}$$

where $\alpha(z) > 0$ is a decreasing function of a nonnegative integral argument.

Note that the Gaussian competition coefficients (7.14) is a particular case of the above relationship for $\alpha(z) = a^{z^2}$, with the niche pattern of Figure 41 being a particular case of the regular niche sequence.

So, the target for the further analysis is a symmetric matrix A with positive entries depending only upon the difference between the row and column numbers. The locus where the difference is constant represents a matrix diagonal, i.e., the matrix is *diagonal-specific*.

Recall that in the case of a symmetric matrix the properties of stability, D-stability, total stability, and dissipativeness are all equivalent (Section 4.IV.E), so that, for system (7.7) with competition coefficients (7.19) to be dissipative, it is necessary and sufficient that the spectrum of A be positive. As can be seen from examples, the restriction (7.19) alone is still insufficient to answer the question of whether or not a positive equilibrium $\mathbf{N}^*$ is stable. It is, however, sufficient to establish an intuitively evident property of the solution to the steady-state system of equations (7.10), namely, that the solution be symmetric about the "center" when it is so for the vector on the right-hand side of (7.10).

Theorem 7.1 *Let $A = [\alpha_{ij}]$ be an $n \times n$ matrix of entries $\alpha_{ij} = \alpha(|i-j|)$, where $\alpha(z)$ is a function of integral argument such that $\det[\alpha_{ij}] \neq 0$; let vector $\mathbf{K} = [K_1, \ldots, K_n]^T$ be symmetric, i.e., $K_i = K_{n-i}$. Then the solution to the system of linear equations $A\mathbf{x} = \mathbf{K}$ is symmetric, too, i.e., $x_i = x_{n-i}$.*

The proof is given in the Appendix.

Consider now a few particular cases where the niche overlap decreases steadily with species distance in the resource space.

Example 7.1 An analytical investigation can be performed for the case of

$$\alpha_{ij} = a^{|i-j|}, \quad 0 < a < 1, \tag{7.20}$$

where a still characterizes the overlap of two adjacent niches. Then

$$A = \begin{bmatrix} 1 & a & a^2 & \cdots & a^{n-1} \\ a & 1 & a & \cdots & a^{n-2} \\ a^2 & a & 1 & \cdots & a^{n-3} \\ \cdots & \cdots & \cdots & \cdots & \cdots \\ a^{n-1} & a^{n-2} & a^{n-3} & \cdots & 1 \end{bmatrix} \tag{7.21}$$

and expanding the $n \times n$ determinant of A in elements of its first row leads to the recurrence equation

$$\det A = D_n = (1 - a^2)D_{n-1},$$

whereby

$$D_n = (1 - a^2)^{n-1} > 0 \tag{7.22}$$

for all a, $0 \le a < 1$. Inequality (7.22) for $n = 1, 2, \ldots$ implies that the Sylvester criterion for A to be positive definite holds true, hence the competitive community with matrix (7.21) is again dissipative.

The solution to the steady-state system takes on the form

$$N_1^* = \frac{K_1 - aK_2}{1 - a^2}, \qquad N_n^* = \frac{K_n - aK_{n-1}}{1 - a^2},$$

$$N_i^* = \frac{K_i(1 + a^2) - a(K_{i-1} + K_{i+1})}{1 - a^2}, \qquad i = 2, 3, \ldots, n - 1, \tag{7.23}$$

which can be easily verified, for instance, by induction in n. It is peculiar to this pattern that, although each of the species competes with all of the rest, its population at the steady state is dependent only on the carrying capacities of its own niche and of the two adjacent ones. Ensuing from (7.23), the conditions for niche capacities K_i, which ensure N^* being positive, are as follows:

$$
\begin{aligned}
K_1 - aK_2 &> 0, \\
K_i(1 + a^2) - a(K_{i-1} + K_{i+1}) &> 0, \quad i = 2, \ldots, n - 2, \\
K_n - aK_{n-1} &> 0. \tag{7.24}
\end{aligned}
$$

A consequence of inequalities (7.24) is the condition that

$$aK_i < K_{i+1} < K_i/a, \quad i = 1, \ldots, n - 1. \tag{7.25}$$

Clearly, if the equilibrium were required to stay positive for any value of $a \to 1$, i.e., for niches packed arbitrary close, then necessarily

$$K_1 = K_2 = \cdots = K_n = K > 0. \tag{7.26}$$

In that case it follows from (7.23) that the populations of the "boundary" species, N_1^* and N_n^*, approach $K/2$ as $a \to 1$, whereas those of the "interior" species, N_i^*, approach zero. The total population of n species,

$$K \frac{n(1 - a) + 2a}{1 + a},$$

declines steadily and approaches K as $a \to 1$. □

Example 7.2 Another special case of decrease in competition coefficients (7.19), namely,

$$\alpha_{ij} = \left[1 + |i + j|^p\right]^{-1}, \quad p > 0, \tag{7.27}$$

fails analytical treatment. Computer runs for different p and n values indicate,[11] however, that both the solution to the steady-state system [with the right-hand

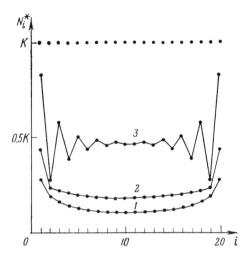

Figure 42.   Positive equilibria in the competition model (7.27) for $n = 20$ species and particular values of p: 0.5 (1); 1.0 (2); 4.0 (3).

sides (7.26)] and the spectrum of matrix A are positive, thus illustrating that a competitive community with matrix (7.27) can be dissipative and have a globally stable equilibrium if, for instance, all the niche capacities are equal.

Some qualitative conclusions can be inferred from the computer-aided studies[11] on how the population sizes are distributed among competitors at the steady state (see Figure 42). It appears that the higher the values of p, i.e., the faster the decline in competition at the passage from one niche to the next one, the higher the mean equilibrium populations. For parameter values $p \geq 2$, the solution exhibits "damped oscillations," and the lower the p values, the faster they are damped (when approaching the middle). For $p \leq 1$ the solution takes on an intuitively clearer U-shaped form (monotone decreasing from the margins to the center). It is still a question whether these peculiarities make real sense in the organization of a competitive community, or are they, rather, an artifact of a peculiar mathematical description?   □

B. Convex Metrics and Stable Matrices

Whether or not the equilibrium distributions were stable in the examples we considered above depended on the pattern of decrease in function $\alpha(|i-j|)$ with increasing $|i-j|$, i.e., on the pattern of how the competition pressure attenuates with increasing distance in the resource space. Special structure (7.19) alone has proved to be quite insufficient for the competition matrix to be positive definite, hence, for

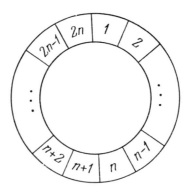

Figure 43.   Annular pattern of niche overlap resulting in matrix (7.29).

the competitive community to be dissipative. The eigenvalues of matrix A are not yet known in a general form, nor it is possible to prove their positiveness, without loss of generality, by using general localization criteria such as Geršgorin discs or ovals of Cassini.[20]

The heuristics however prompts the following restriction on the pattern of decrease in competition coefficients with the distance between species in the resource space. Suppose that a declining function of integral argument $\alpha(z) > 0$ meets the conditions of strict convexity:

$$\alpha(m) < \frac{1}{2}[\alpha(m-1) + \alpha(m+1)], \quad m = 1, \ldots, n-1. \tag{7.28}$$

This condition turns out to be quite sufficient for stability, which is established by the following theorem.[18]

Theorem 7.2 *Matrix $A = [\alpha_{ij}]$ of entries $\alpha_{ij} = \alpha(|\,i-j\,|)$ that satisfy conditions (7.28) is positive definite.*

In the proof (given in the Appendix) a method happened to be useful that is quite "distant," mapped in the space of mathematical disciplines, from what has been already used in the book, namely, the *linear programming* method. The idea relies upon the special case of niche arrangement illustrated in Figure 43—somewhat artificial, but useful for the purposes of proof. If the individual niche geometry is invariant, then the coefficients of competition between a species and its neighbors first decline to a certain extent and then grow in a reverse manner.

With the notation $\alpha(m) = \alpha_m$ and assumption that $\alpha(0) = 1$, the competition matrix of an annular community assumes the form

$$
C(A)=\left[\begin{array}{cccccc|ccc}
1 & \alpha_1 & \alpha_2 & \cdots & \alpha_{n-1} & \alpha_n & \alpha_{n-1} & \cdots & \alpha_1 \\
\alpha_1 & 1 & \alpha_1 & \cdots & \alpha_{n-2} & \alpha_{n-1} & \alpha_n & \cdots & \alpha_2 \\
\cdots & \cdots & \cdots & \cdots & \cdots & \cdots & \cdots & \cdots & \cdots \\
\alpha_{n-1} & \alpha_{n-2} & \alpha_{n-3} & \cdots & 1 & \alpha_1 & \alpha_2 & \cdots & \alpha_n \\
\alpha_n & \alpha_{n-1} & \alpha_{n-2} & \cdots & \alpha_1 & 1 & \alpha_1 & \cdots & \alpha_{n-1} \\
\alpha_{n-1} & \alpha_{n-2} & \alpha_{n-3} & \cdots & \alpha_2 & \alpha_1 & 1 & \cdots & \alpha_{n-2} \\
\cdots & \cdots & \cdots & \cdots & \cdots & \cdots & \cdots & \cdots & \cdots \\
\alpha_1 & \alpha_2 & \alpha_3 & \cdots & \alpha_n & \alpha_{n-1} & \alpha_{n-2} & \cdots & 1
\end{array}\right] \left.\begin{array}{c} \\ \\ \\ \\ \\ \\ \\ \\ \end{array}\right\} (n+1) \text{ rows}
$$

$$(7.29)$$

Matrix $C(A)$ is cyclic and symmetric, and its submatrix A made up of the first $n+1$ rows and columns corresponds to an ordinary horizontal-structured community. Clearly, matrix A is symmetric if and only if the circulant $C(A)$ is symmetric. Notice, by the way, that any symmetric circulant of size $2n \times 2n$ is uniquely determined by the set of the first $n + 1$ entries in its first row and has the pattern of (7.29).

Since the sum of entries is the same in each row of $C(A)$, the positive steady state $\mathbf{N}^*$, as a solution to the system of equations

$$C(A)\mathbf{N}^* = \mathbf{K}, \quad \mathbf{K} = [K, \ldots, K]^T,$$

will have all the components equal to

$$N_i^* = \frac{K}{1 + 2(\alpha_1 + \alpha_2 + \ldots + \alpha_{n-1}) + \alpha_n}, \quad i = 1, 2, \ldots, 2n,$$

and positive for any $\alpha_i \geq 0$. Equal competition pressure on species in the pattern of Figure 43 generates a uniform equilibrium distribution. It turns out stable due to the stability of matrix (7.29) under conditions (7.28) (Lemma A7.1 in the Appendix).

The statement is proved by using the main result of linear programming theory: the minimum of a linear function $\mu(\mathbf{x})$ over a simplex of $\mathbf{x}$ is attained at one or several vertices of the simplex.[18] Each eigenvalue of the circulant matrix $C(A)$ has proved to be a linear function of competition coefficients, while the simplex emerges from the monotone decrease and convexity constraints on the coefficients. Examination at a finite number of the vertices has shown the minimum to be positive, thereafter the positive definiteness of the original matrix A follows from how the structure of A is related to that of $C(A)$.

Thus, if a competition structure (7.19) has a convex pattern of decrease, i.e., if the decrease is smooth enough, then matrix A is positive definite and the community is dissipative. "Convex" metrics of the resource space thus generate stable matrices of the competition coefficients. In particular, it is this decline pattern which is inherent in the above competitive structures (7.21) and (7.27), as well as in (7.15) for values of a not too close to 1. The spectrum being positive for these matrices illustrates Theorem 7.2.

Also, as follows from Lemma A7.1, a choice of α_js close to the boundary of the constraints may result in arbitrarily small, yet positive, eigenvalues, which may cause some technical problems in computer calculation of the eigenvalues.

As a fairly simple sufficient condition, Theorem 7.2 can also be useful in stability studies of statistical ensembles of competition matrices. For example, a decrease in stability by increasing the number of competitor species, revealed by Lawlor,[21] can now be explained by the decrease in a statistical portion of those matrices which meet the convexity condition (7.28). Also, the convexity of the competition function $\alpha(m)$ turns out to be connected with the effect of predation on species packing in a stable community where competing species serve as food for predatory species.[22]

III. POSITIVE AND PARTLY POSITIVE EQUILIBRIA

In the examples considered above, the competition matrices appeared to be dissipative, so that the stability problem amounted to finding whether or not, given a set of niche capacities, system (7.7) has an equilibrium $\mathbf{N}^*$ with all components positive. But what can be said about *partly positive* equilibrium states, i.e., those with some of their components being zero? Will they be stable or at least *partly stable* with respect to the nonzero part of their components?

A. When "Partly Positive" Means "Partly Stable"

Stability analysis of partly positive equilibria is of importance in the context of the following ecological problem. Suppose a steadily existing community is invaded by a small group of organisms of some new species. Will the new species become fixed in the community or should we expect it to decline to extinction? Clearly, the first outcome is possible only if the pertinent partly positive equilibrium is unstable in the expanded community, while the positive equilibrium stable. If, on the contrary, the partly positive point is stable under perturbations of any of its components, then the new species will become extinct in the community.

In Lotka–Volterra population equations of a general type, the sufficient conditions for a partly positive equilibrium to be locally stable can be readily formulated on the basis of the theorem on Lyapunov stability by linear approximation: the subsystem of nonzero components should form a stable community matrix, while deviations from zero components of the equilibrium $\mathbf{N}^\circ = [N_1^\circ, \ldots, N_n^\circ]$ will vanish if

$$\varepsilon_k - \sum_{j=1}^{n} \gamma_{kj} N_k^\circ < 0 \tag{7.30}$$

for all k with zero N_k° (the so-called *transversality* condition[23]).

For dissipative communities the answer is more straightforward. As long as any subset of species in a dissipative community again makes up a dissipative system, any partly positive point is stable unless its zero components are perturbed, and loses stability if those zero components are affected by perturbation which can generate a higher-dimensional partly positive equilibrium; in case the new species composition has an equilibrium solution, the perturbed trajectory approaches exactly this equilibrium, while otherwise it goes away, to subspaces of other partly positive points.

Thus, a variety of species compositions may exist in a dissipative community (if only the niche capacities provide for an appropriate partly positive equilibrium to exist), and invading species may fix in the community (under the same condition). So, dissipative communities bear a simple answer to the question posed in the heading: a partly positive equilibrium is always partly stable.

Normally, the positiveness of a partly nonzero equilibrium can be verified by direct algebraic calculations and there also exist some sufficient conditions for a solution to system (7.10) to be positive. Abrosov et al.[24] proposed a survey of such algebraic conditions for the case of unitary right-hand sides. A general case can be reduced to the unitary one through multiplying matrix A by a diagonal matrix with appropriate positive entries. This does not lead A beyond the class of D-stable, totally stable, dissipative, or quasi-dominant matrices (see Sections 4.IV and 4.V), thus retaining the category of stability in the original matrix.

A "geometric" interpretation to an equilibrium being positive or partly positive is considered below.

B. Geometric Measures of "Equilibriumness"

Let $A = [a_{ij}]$ be a nonnegative nonsingular $n \times n$ matrix, $\mathbf{e}_1, \ldots, \mathbf{e}_n$ be vectors of the standard orthonormal basis. Consider the image of $\mathbb{R}^n_+$, the positive orthant of the n-dimensional space, after having been transformed by the linear operator with matrix A. The orthant $\mathbb{R}^n_+$ consists of vectors taking the form

$$\mathbf{x} = x_1\mathbf{e}_1 + \ldots + x_n\mathbf{e}_n,$$

with $x_i \geq 0$. Operator A transforms basis vectors $\mathbf{e}_j$ into the columns of matrix $A : A\mathbf{e}_j = [a_{1j}, a_{2j}, \ldots, a_{nj}]^T$ (which are linearly independent due to nonsingularity of A), while any vector $\mathbf{x} \in \mathbb{R}^n_+$ is transformed into their linear combination with nonnegative coefficients:

$$A\mathbf{x} = \sum_{j=1}^{n} x_j A\mathbf{e}_j. \tag{7.31}$$

Thus, the orthant $\mathbb{R}^n_+$ is contracted by operator A into an n-hedral angle, $A\mathbb{R}^n_+$, with the generatrices $A\mathbf{e}_j$. The existence of a positive solution to system

$$A\mathbf{x} = \mathbf{K}, \quad \mathbf{K} > 0, \tag{7.32}$$

is equivalent to vector $\mathbf{K}$ belonging to the interior of the angle since the continuous operator A^{-1} transfers an interior point of $A\mathbb{R}^n_+$ into an interior point of $\mathbb{R}^n_+$ (all components x_i are positive).

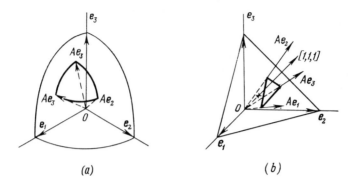

Figure 44.   Illustration of a measure of "equilibriumness" for a 3×3 competition matrix A: (a) in the Euclidean norm; (b) in the sum norm (direction $\mathbf{K} = [1, 1, 1]$ goes outside the simplex $\sigma^3(A)$).

Since A is linear, vectors $\mathbf{K}$ which give a positive solution to (7.32), may be determined up to a positive constant factor, i.e., they generate whole directions of positive solutions. Therefore, the variety of all vectors of this kind can be restricted by the condition that, for instance, the Euclidean norm $\|\mathbf{K}\|$ be equal to 1. These vectors now terminate at the sphere of a unit radius, S^n, and the n-hedral angle $A\mathbb{R}_+^n$ cuts out of its surface a spherical n-hedron, $s^n(A)$, comprised by "arcs" of $(n-1)$-dimensional "circles," whose hyperplanes pass through the origin (Figure 44a).

The volume of $s^n(A)$ (which is actually the surface area in the three-dimensional case) characterizes the number of directions in $\mathbb{R}_+^n$ which generate positive equilibria; if related to the $(1/2^n)$th fraction of the "surface area" confined within S^n, it can serve as a measure of how many different sets of carrying capacities K_i furnish a positive equilibrium to the competition structure given by matrix A, or in shorter terms, a measure of "equilibriumness" in the competition matrix. Clearly, $\mu(A) = 1$ only if $A = D$, a positive diagonal matrix of a scaling transformation. It can be interpreted as the matrix of a "community" of n noninteracting, self-limited species, which is stable for any values $K_i > 0$. In general, the smaller the value of $\mu(A)$, the fewer the number of sets of K_i which provide for the equilibrium to be positive.

For $n = 3$ this measure is calculated as follows:

$$\mu(A) = s^3(A)/\left(\frac{1}{8}4\pi\right) = 2s^3(A)/\pi, \tag{7.33}$$

where the area of the spherical triangle with angles $A, B,$ and C equals

$$s^3 = A + B + C - \pi \tag{7.34}$$

if the angles are measured in radians.[25] If a, b, and c are the sides (arcs) of the triangle which are opposite to those angles, then by the formulas of analytical geometry one can calculate

$$\cos a = \frac{a_{11}a_{12} + a_{21}a_{22} + a_{31}a_{32}}{\sqrt{a_{11}^2 + a_{21}^2 + a_{31}^2}\sqrt{a_{12}^2 + a_{22}^2 + a_{32}^2}},$$

$$\cos b = \dots, \quad \cos c = \dots$$

By the spherical triangle formulas[25] the angles A, B and C can be found, thereafter the area s^3 can be calculated by (7.34). The reader may generalize the scheme to the case of an arbitrary n.

The calculation becomes simpler if the so-called *sum*, or *Manhattan*, norm[20] is used instead of the Euclidean norm of vectors:

$$\|\mathbf{x}\|_\Sigma = |x_1| + |x_2| + \cdots + |x_n|.$$

Now the condition $\|\mathbf{K}\|_\Sigma = 1$, in place of the sphere, yields the $(n-1)$-dimensional simplex, $\sum^n$, spanning the standard basis vectors $\mathbf{e}_1, \dots, \mathbf{e}_n$ (Figure 44b), which represents the set of all possible convex linear combinations, or *convex hull*, of those vectors:

$$\mathbf{x} = \lambda_1\mathbf{e}_1 + \lambda_2\mathbf{e}_2 + \cdots + \lambda_n\mathbf{e}_n, \quad \lambda_j \geq 0, \quad \lambda_1 + \lambda_2 + \cdots + \lambda_n = 1.$$

The n-hedral angle $A\mathbb{R}^n$ cuts another simplex out of the former one, $\sigma^n \subset \Sigma^n$, which spans normalized vectors $A\mathbf{e}_1, \dots, A\mathbf{e}_n$, while the ratio of their "surface areas" serves as a measure of "equilibriumness" (or of stability in the dissipative case) in the competition structure:

$$\mu_\Sigma(A) = \sigma^n(A)/\Sigma^n, \quad 0 < \mu \leq 1. \tag{7.35}$$

For $n = 3$ the simplex area is equal, to within a sign, to

$$\sigma^3(A) = \frac{1}{2\rho} \begin{bmatrix} a_{11} & a_{12} & a_{13} \\ a_{21} & a_{22} & a_{23} \\ a_{31} & a_{32} & a_{33} \end{bmatrix} \left(\|A\mathbf{e}_1\|_\Sigma\|A\mathbf{e}_2\|_\Sigma\|A\mathbf{e}_3\|_\Sigma\right)^{-1},$$

where $\rho = 1/\sqrt{3}$ is the distance from the simplex plane to the origin and $\|A\mathbf{e}_j\|_\Sigma$ is, by definition, the sum of absolute values of entries in the jth column of A. Hence,

$$\mu_\Sigma(A) = \frac{\sigma^3(A)}{\Sigma^3} = \frac{|\det A|}{\|A\mathbf{e}_1\|_\Sigma \|A\mathbf{e}_2\|_\Sigma \|A\mathbf{e}_3\|_\Sigma} . \tag{7.36}$$

For any regular niche sequence (any competition matrix of pattern (7.19)) this gives

$$\mu_\Sigma\left(\begin{bmatrix} 1 & \alpha & \beta \\ \alpha & 1 & \alpha \\ \beta & \alpha & 1 \end{bmatrix}\right) = \frac{(1-\beta)(1+\beta-2\alpha^2)}{(1+\alpha+\beta)^2(1+2\alpha)} . \tag{7.37}$$

In the n-dimensional case, as simplexes σ^n and Σ^n belong to the same hyperplane, the ratio of their "surfaces" is equal to that of the volumes of the supporting n-hedral "pyramids," each making up a $(1/n!)$th part of the parallelepiped constructed on the corresponding vectors. The parallelepiped of Σ^n is just a unitary n-dimensional cube, so that

$$\mu_\Sigma(A) = \frac{\sigma^n(A)}{\Sigma^n} = \frac{|\det A|}{\|A\mathbf{e}_1\|_\Sigma \|A\mathbf{e}_2\|_\Sigma \ldots \|A\mathbf{e}_n\|_\Sigma} . \tag{7.38}$$

In view of these interpretations, it becomes clear that statements like "If there exists a positive equilibrium, then it has to be stable," may not be true in a more general than dissipative case. In fact, one can easily imagine a community matrix which is not stable. Yet any matrix A of nonnegative entries produces a nonempty angle $A\mathbb{R}^n_+ \subset \mathbb{R}^n_+$, i.e., there exist vectors $\mathbf{K}$ such that they give a positive equilibrium point, which is unstable by the choice of matrix A.

It is also clear how system (7.7) with a fixed collection of K_i may loose equilibrium when the species packing becomes more dense, which has been mentioned in Section 7.I.C for a particular competition structure (7.14). For values of a sufficiently close to 1, the simplex $\sigma^n(A)$ (as well as the spherical n-hedron $s^n(A)$) shrinks so rapidly as vector $\mathbf{K} = [1, 1, \ldots, 1]$ falls beyond its boundary (Figure 44b), the stability measures μ and μ_Σ (7.37–38) approaching zero in this case.

By means of these measures one can also explain the above-mentioned effect which the convexity in competition coefficients renders to the existence of the positive equilibrium. Formula (7.37) shows, for example, that if the convexity is disturbed strongly enough, i.e., if α exceeds $(1 + \beta)/2$ by as much as $(1 + \beta - 2\alpha^2)$ becomes close to zero, the measure μ_Σ appears quite small. Mentioned in Section 7.I.C, the direct computer solutions to system (7.10) in particular cases of normal utilization functions (7.14) for values of a close enough to 1 (specifically, for those violating the convexity condition $a < (1 + a^4)/2$) indicate[11] that positive solutions are possible under a highly special choice of the right-hand sides K_i only; in other words, the corresponding n-hedral angle is quite narrow (see Figure 44b) and, in particular, it does not contain the direction of $\mathbf{K} = [1, 1, \ldots, 1]^T$.

Calculated[26] by formula (7.38) for particular cases of the competition matrix $n \times n$, the measure μ_Σ assumes the following forms:

$$\mu_\Sigma^{(n)}(7.17) = \frac{|\det A|}{(1 + a^2)(1 + 2a)^{n-2}} \, ;$$

$$\mu_\Sigma^{(n)}(7.18) = \left[\frac{1 - a}{1 + (n - 1)a}\right]^{n-1} \, ;$$

$$\mu_\Sigma^{(n)}(7.21) = \frac{(1 - a)(1 - a^2)^{n-1}}{(1 - a^n)} \, . \tag{7.39}$$

For particular values of n and a one can compare various competition structures in their degrees of "equilibriumness." It can be shown, for example, that all the above-cited measures μ_Σ tend to zero as $n \to \infty$ and a fixed, the measure $\mu_\Sigma^{(n)}(7.21)$ decreasing slower than $\mu_\Sigma^{(n)}(7.17)$ and $\mu_\Sigma^{(n)}(7.18)$, whose order of decrease is equivalent to $(1/n)^n$. As $a \to 1$ and n fixed, the measures tend to zero at the order of $\varepsilon^n = (1 - a)^n$.[26] Thus, if species packing becomes more and more dense, either by increase in the number of species or by strengthening the overlap between adjacent niches, the positive equilibrium will only be retained under a very special choice of right-hand sides in steady-state equations (7.10), i.e., for highly specific patterns of the niche capacity distribution among species.

C. Whether "Positive" Implies "Partly Positive"?

Developing further the "geometric" interpretation for the existence of the positive equilibrium, we come below to general conditions for the existence of partly positive equilibria.

Obeying the system of equations

$$N_i \sum_{j=1}^{n} \left(\alpha_{ij} N_j - K_i\right) = 0, \quad i = 1, \ldots, n, \tag{7.40}$$

partly positive equilibrium points may have zeros in any of their coordinate places, while the nonzero coordinates are solutions to the linear system with matrix A', which is derived from matrix A by deleting the rows and columns corresponding to zero coordinates (a "truncated" matrix). Clearly, the maximum number of all possible points is 2^n (including the trivial point $\mathbf{0}$ and the positive equilibrium $\mathbf{N^*}$). A nontrivial question is whether there always exists a positive solution to the "truncated" system of equations

$$A'\mathbf{N}' = \mathbf{K}' \tag{7.41}$$

and whether it is an implication from the existence of the nontrivial equilibrium $\mathbf{N^*}$.

Deletion of any m rows and the corresponding columns from the original matrix ($m = 1, \ldots, n-1$), or switching to the truncated system (7.41), means that the linear operator A is constricted to the lower-dimensional subspace $\mathbb{R}^{n-m}$, into which the right-hand side vector $\mathbf{K}$ is also projected. If the projecting operator is denoted by p_m, then the projection of the n-hedral angle can be written as $p_m A \mathbb{R}^n_+$, while $A' \mathbb{R}^{n-m}_+$ designates the $(n - m)$-hedral angle we get when the orthant $\mathbb{R}^{n-m}_+$ is contracted by the operator A' (when A' is nonsingular). It can be seen that, in the general case,

$$A' \mathbb{R}^{n-m}_+ \subset p_m A \mathbb{R}^n_+, \tag{7.42}$$

and $p_m A \mathbb{R}^n_+ \setminus A' \mathbb{R}^{n-m}_+$ is nonempty in general. Indeed, for any vector $\mathbf{x}'$ of the subspace, $\mathbf{x}' = [x_{j_1}, \ldots, x_{j_{n-m}}]^T \in \mathbb{R}^{n-m}_+$, we have

$$A' \mathbf{x}' = \sum_{k=1}^{n-m} x_{j_k} A \mathbf{e}_{j_k},$$

and if vector $\mathbf{x}'$ is extended by zero coordinates up to the vector $\mathbf{x}'^\circ \in \mathbb{R}^n_+$, then

$$A' \mathbf{x}' = p_m A \mathbf{x}'^\circ \in p_m A \mathbb{R}^n_+.$$

Consonant with the general reason mentioned in point **B** of this section, for a positive solution of the truncated system (7.41) to exist, it is necessary and sufficient that the projection of vector $\mathbf{K}$ falls within the $(n - m)$-hedral angle:

$$\mathbf{K}' = p_m \mathbf{K} \in A' \mathbb{R}^{n-m}_+, \tag{7.43}$$

the inner set of implication (7.42). For matrices of horizontal-structured communities there may well be cases where $p_m A \mathbb{R}^n_+$, the outer set, is actually wider than $A' \mathbb{R}^{n-m}_+$ (cf. Example 7.3 later in this section); in other words, there exist vectors $\mathbf{K} \in A \mathbb{R}^n_+$ (yielding a positive equilibrium) for which condition (7.43) is violated, the truncated system thus having no positive solution. In this sense, the "positive" may well not imply the "partly positive," even in dissipative systems.

In terms of dynamic behavior of the corresponding model, this means that, after a species has been eliminated from the community (for instance, after $N_4(t)$ has been set to zero in Example 7.3), the remaining composition of $(n - 1)$ species is unstable and degrades further, losing other species too (species N_1 in Example 7.3) until a truncated system reveals a positive solution (composition N_2, N_3 in Example 7.3). At this partly positive equilibrium, which is stable to perturbations of its nonzero components, a perturbation of zero components can cause a return to the original composition of the complete set of species only if all zero components (species N_1 and N_4) have been simultaneously shifted to positive values.

In ecological terms, a consecutive introduction of eliminated species is condemned to failure in such a model, whereas the species can only be fixed altogether.

In contrast to the situation mentioned above, vectors $\mathbf{K}$ are also possible such that the solutions to (7.41) are all positive, no matter how the original matrix A

has been truncated, i.e., condition (7.43) holds true for any operator p_m. This is observed in Example 7.4 given below for matrix A of the size 4×4 and type (7.19) and vector $\mathbf{K} = [1, 1, 1, 1]^T$: the model has all the $2^4 = 16$ steady-state points, each being stable within the subspace of its nonzero components.

In ecological terms, any subset of the total 4-species composition can coexist stably in such a model, or in other words, species can colonize the living space in any sequence. Obviously, such a freedom can only exist in such a highly specific and artificial construction as the competition model (7.53).

Note that in the examples mentioned vectors $\mathbf{K}$ are supposed to be invariant, i.e., the spectrum of available resource is assumed independent of the presence or absence of species in the community under scrutiny. But if we require that a positive solution to the truncated system exists there for any vector $\mathbf{K} > \mathbf{0}$ which gives a positive solution to the original system (7.10), then, in "geometric" terms, the projection of the n-hedral angle $A\mathbb{R}_+^n$ must coincide with the $(n-m)$-hedral angle $A'\mathbb{R}_+^{n-m}$, the implication in (7.42) thus becoming the identity:

$$A'\mathbb{R}_+^{n-m} = p_m A \mathbb{R}_+^n. \tag{7.44}$$

If the transformation of A is nonsingular, then the generatrices of the angle $A'\mathbb{R}_+^{n-m}$ (the columns of matrix A') make up a basis in $\mathbb{R}_+^{n-m}$. Condition (7.44) then indicates that, when projected into $\mathbb{R}_+^{n-m}$, those generatrices of angle $A\mathbb{R}_+^n$ which correspond to the coordinates to be truncated, namely $p_m A\mathbf{e}_{i_1}, \ldots, p_m A\mathbf{e}_{i_m}$, can be expressed linearly (with nonnegative coefficients) through the generatrices of angle $A'\mathbb{R}_+^{n-m}$. This, in turn, is equivalent to the condition that each of the m systems

$$A'\mathbf{x}' = p_m A\mathbf{e}_{i_k}, \quad k = 1, \ldots, m, \tag{7.45}$$

has a nonnegative solution. The latter, by Cramer's rule for solution of a linear system, requires that the signs of all nonzero determinants we get from matrix A' by substituting each of vectors $p_m A\mathbf{e}_{i_k}$ for any one of matrix columns, coincide with the sign of $\det A'$.

This is what guarantees that the "positive" implies "partly positive" in our models like in common logic, and this is what actually takes place for matrix A of type (7.21) in Example 7.5, where $n = 4$ and $m = 1, 2, 3$: once the model has a positive equilibrium $\mathbf{N}^*$, there exist all the 14 partly positive equilibria, too. The model thus carries a variety of coexisting compositions out of the four competing species, stable under wide variations in the niche carrying capacities—as wide as are projections of the 4-hedral angle into the proper coordinate subspaces.

Example 7.3: To see that condition (7.44) may well be violated, consider a 4×4 competition matrix of pattern (7.19):

$$A = \begin{bmatrix} 1 & \alpha & \beta & \gamma \\ \alpha & 1 & \alpha & \beta \\ \beta & \alpha & 1 & \alpha \\ \gamma & \beta & \alpha & 1 \end{bmatrix}, \quad 0 < \alpha, \beta, \gamma < 1, \quad \det A \neq 0. \tag{7.46}$$

Suppose the last coordinate of $\mathbf{N}^*$ is annulled, that is, the last row and column have to be deleted. For vector $p_1 A e_4 = [\gamma, \beta, \alpha]^T$ to be a linear nonnegative combination of angle $A'\mathbb{R}_+^3$'s generatrices, which are the columns of matrix A', it is necessary and sufficient that the system of equations

$$\begin{bmatrix} 1 & \alpha & \beta \\ \alpha & 1 & \alpha \\ \beta & \alpha & 1 \end{bmatrix} \begin{bmatrix} x_1 \\ x_2 \\ x_3 \end{bmatrix} = \begin{bmatrix} \gamma \\ \beta \\ \alpha \end{bmatrix} \tag{7.47}$$

has a nonnegative solution. The system determinant is

$$D' = (1 - \beta)(1 + \beta + 2\alpha^2) > 0,$$

and the determinant to determine x_1,

$$D^{(1)} = \alpha^3 + \alpha\beta + \gamma - \alpha^2\gamma - 2\alpha\beta,$$

is negative, say, for $\alpha = 0.7$, $\beta = 0.5$, and $\gamma = 0.31$, where D' is positive. Hence, $x_1 < 0$ and condition (7.44) is consequently violated. It might be possible, therefore, that even though there be a positive equilibrium in a model with matrix (7.46), the partly positive point pertinent to matrix $A'(7.47)$ will be absent. This case is realized, for instance, for the vector $\mathbf{K} = [\gamma, \beta, \alpha, K_4]^T$, $0 < K_4 < 1$.   □

Example 7.4: Consider now all sorts of truncations in system

$$\begin{bmatrix} 1 & \alpha & \beta & \gamma \\ \alpha & 1 & \alpha & \beta \\ \beta & \alpha & 1 & \alpha \\ \gamma & \beta & \alpha & 1 \end{bmatrix} \begin{bmatrix} x_1 \\ x_2 \\ x_3 \\ x_4 \end{bmatrix} = \begin{bmatrix} 1 \\ 1 \\ 1 \\ 1 \end{bmatrix}, \tag{7.48}$$

assuming that

$$1 > \alpha > \beta > \gamma > 0. \tag{7.49}$$

If the first or fourth coordinate is annulled, the truncated matrix takes on the form

$$A' = \begin{bmatrix} 1 & \alpha & \beta \\ \alpha & 1 & \alpha \\ \beta & \alpha & 1 \end{bmatrix},$$

and the condition that guarantees the solution of $A'\mathbf{x}' = [1, 1, 1]^T$ to be positive is

$$\alpha < (1 + \beta)/2. \tag{7.50}$$

When the second coordinate is zeroed, the truncated system appears to have a somewhat different form:

$$\begin{bmatrix} 1 & \beta & \gamma \\ \beta & 1 & \alpha \\ \gamma & \alpha & 1 \end{bmatrix} \begin{bmatrix} x_1 \\ x_3 \\ x_4 \end{bmatrix} = \begin{bmatrix} 1 \\ 1 \\ 1 \end{bmatrix}.$$

For the determinant of this system,

$$D' = 1 + 2\alpha\beta\gamma - \alpha^2 - \beta^2 - \gamma^2,$$

to be positive it is sufficient, for instance, that

$$\alpha^2 < (1 + \gamma)/2. \tag{7.51}$$

The determinants pertinent to x_1, x_3, x_4 are respectively:

$$
\begin{aligned}
D^{(1)} &= 1 + \alpha\beta + \alpha\gamma - \gamma - \alpha^2 - \beta, \\
D^{(3)} &= 1 + \alpha\gamma + \beta\gamma - \gamma^2 - \alpha - \beta, \\
D^{(4)} &= 1 + \alpha\beta + \beta\gamma - \gamma - \alpha - \beta^2.
\end{aligned}
$$

Since

$$\frac{\partial D^{(1)}}{\partial \alpha} = \beta + \gamma - 2\alpha, \quad \frac{\partial D^{(1)}}{\partial \beta} = \alpha - 1, \quad \frac{\partial D^{(1)}}{\partial \gamma} = \alpha - 1,$$

it can be easily seen that the set of extremal points for determinant $D^{(1)}$ in the space of parameters (α, β, γ) has no intersection with the bounded domain Ω defined by conditions (7.49). Yet $D^{(1)} \geq 0$ on the boundary of Ω and one can find points where $D^{(1)} > 0$ within Ω. Hence, due to $D^{(1)}$ being continuously differentiable with respect to α, β, γ, it follows that $D^{(1)} > 0$ everywhere within Ω. Under the additional constraint,

$$\beta < (\alpha + \gamma)/2, \tag{7.52}$$

a similar argument shows that $D^{(3)}$ and $D^{(4)}$ are positive, too. Thus, in the intersection of Ω with the domains defined by (7.50)–(7.52)—denote it by $\tilde{\Omega}$—the solution to the truncated system is positive.

The constraints defining domain $\tilde{\Omega}$ are already sufficient for the truncation in the $3rd$ component to result in a positive solution as well. When any two or three of components are truncated, the solution is obviously positive in domain $\tilde{\Omega}$. Whether $\tilde{\Omega}$ is nonempty can be readily seen by the same particular example of α, β, γ as was proposed for subsystem (7.47).

Thus, matrix (7.48) with competition coefficients lying in domain $\tilde{\Omega}$ and vector of carrying capacities $\mathbf{K} = [1, 1, 1, 1]^T$ (or proportional to $\mathbf{K}$) possess partly positive solutions for any kind of matrix truncations. In model terms, there exist all the $2^4 - 2 = 14$ partly positive equilibria, each being globally stable, by Theorem 7.2, in its own subspace. □

Example 7.5: In the special case (7.20) of a regular niche sequence, matrix (7.46) is reduced to the positive definite matrix

$$A = \begin{bmatrix} 1 & a & a^2 & a^3 \\ a & 1 & a & a^3 \\ a^2 & a & 1 & a \\ a^3 & a^2 & a & 1 \end{bmatrix}, \quad 0 < a < 1. \tag{7.53}$$

Zeroing the 1st component leaves

$$A' = \begin{bmatrix} 1 & a & a^2 \\ a & 1 & a \\ a^2 & a & 1 \end{bmatrix}$$

with $D' = \det A' > 0$. The determinants for the unknowns are

$$D^{(2)} = aD' > 0, \quad D^{(3)} = 0, \quad D^{(4)} = 0.$$

Similarly, $D^{(j)} \geq 0$ also for any other mode of truncating (7.53).

This means that for matrix (7.53) the implication (7.42) is in fact the identity (7.44), i.e., once there exists a positive equilibrium $\mathbf{N}^*$, there also exist all the rest of the 14 partly positive equilibrium points. Note the difference from the statement in Example 7.4, where only some of vectors $\mathbf{K}$ could provide for this property.   □

To summarize from the standpoint of stability in the competition community dynamics, the limits to similarity among competitor species arise not only from the stability conditions of the competition matrix but also—and quite essentially!—from the existence conditions of the positive equilibrium. Both conditions have to be taken into account in studying both deterministic models of community dynamics and statistical ensembles of community matrices (see, e.g., Svirezhev and Logofet[27] and the references therein).

IV. EXTREMAL PROPERTIES IN HORIZONTAL-STRUCTURED COMMUNITY MODELS

In mathematical ecology, as in other sciences with longer histories (mechanics, optics, etc.), there has long existed a tendency to formulate basic regularities relevant in population and community dynamics in the form of some *variational*, or *extremal*, principles, i.e., to think of trajectories as solutions that can furnish an extremum to an interpretable functional. Perhaps, the reason for this tendency is the quest for more concise and elegant description, on the one hand, and the perennial teleological bend of our thinking, on the other. Leaving aside the issue of whether and to what an extent the teleological approach is legitimate in studying population and community dynamics,[28-29] we now consider only some of the extremal properties which ensue from the form of population equations for competition communities.

A. Lotka–Volterra Dynamics as a Steepest Ascent Movement

While extremal principles are generally concerned with local properties of trajectories, the global stability in Volterra dissipative systems gives a good reason to anticipate a global pattern in the extremal properties too, i.e., the principle is expected to be true for any trajectory initiating at an interior point of the positive orthant. Consider, therefore, the classic form of Lotka–Volterra equations with an interaction matrix Γ of the competition type (i.e., nonnegative and symmetric):

$$\frac{dN_i}{dt} = N_i \left(\varepsilon_i - \sum_{j=1}^{n} \gamma_{ij} N_j \right), \quad i = 1, \ldots, n$$

$$(\varepsilon_i, \gamma_{ij} \geq 0, \quad \gamma_{ij} = \gamma_{ji}). \tag{7.54}$$

Consider also the mapping $\eta_i = \pm 2\sqrt{N_i}$ $(i = 1, \ldots, n)$, which transforms the positive orthant $\mathbb{R}_+^n$ into the entire coordinate space $\mathbb{R}_\eta^n$. It appears[29] that in $\mathbb{R}_\eta^n$ the trajectories of (7.54) represent the steepest ascent trajectories for the function

$$W(\boldsymbol{\eta}) = \frac{1}{4} \sum_{i=1}^{n} \varepsilon_i \eta_i^2 - \frac{1}{32} \sum_{i,j=1}^{n} \gamma_{ij} \eta_i^2 \eta_j^2 = \sum_{i=1}^{n} \varepsilon_i N_i - \frac{1}{2} \sum_{i,j=1}^{n} \gamma_{ij} N_i N_j, \tag{7.55}$$

i.e., the equations hold true

$$\frac{d\eta_i}{dt} = \frac{\partial W}{\partial \eta_i}, \quad i = 1, \ldots, n, \tag{7.56}$$

and

$$\frac{dW}{dt} = \sum_{i=1}^{n} \frac{\partial W}{\partial \eta_i} \frac{d\eta_i}{dt} = \sum_{i=1}^{n} \left(\frac{\partial W}{\partial \eta_i} \right)^2. \tag{7.57}$$

Obviously, $dW/dt = 0$ only at points where all $d\eta_i/dt$ are zero, i.e., at stationary points of system (7.56). Note also that all these points meet the necessary extremum conditions for the function $W(\boldsymbol{\eta})$. If (7.54) has the maximal number $(2^n - 1)$ of all possible stationary points with $n, n - 1, \ldots, 1$ nonzero coordinates, then any point which differs from these only in signs of nonzero components, will be stationary for system (7.56). The number of such points is evidently equal to $3^n - 1$. Thus, the stationary points of system (7.56) are derived from the those of system (7.54) as their symmetric reflections with respect to all sorts of coordinate hyperplanes and the origin, the values of function $W(\eta_1, \ldots, \eta_n)$ being identical at symmetric points.

It follows from (7.57) that the function $W[\boldsymbol{\eta}(t)]$ always increases along the trajectories of (7.56) and attains its local maximum at a stationary point $\boldsymbol{\eta}^*$, if the latter is asymptotically stable. Indeed, since trajectories of the dynamical system (7.56) are everywhere dense in a domain G containing point $\boldsymbol{\eta}^*$, for any point $\boldsymbol{\eta}$ from any neighborhood of $\boldsymbol{\eta}^*$ there is a trajectory $\boldsymbol{\eta}(t) \to \boldsymbol{\eta}^*$ that goes through $\boldsymbol{\eta}$. Then, by the continuity of function $W(\boldsymbol{\eta})$ and its monotone increasing along $\boldsymbol{\eta}(t)$, it follows that $W(\boldsymbol{\eta}^*) > W(\boldsymbol{\eta})$. Hence,

$$W(\boldsymbol{\eta}^*) = \max_t W[\boldsymbol{\eta}(t)] = \max_{\boldsymbol{\eta} \in G} W(\boldsymbol{\eta}). \tag{7.58}$$

As the inverse mapping from $\mathbb{R}_\eta^n$ to $\mathbb{R}_+^n$ merges all the symmetric prototypes into one image—the stationary point of (7.54)—with the value of W unchanged, an assertion similar to (7.58) must hold for trajectories in $\mathbb{R}_+^n$.

On the other hand, if the function $W(N_1, \ldots, N_n)$ has an isolated maximum at a stationary point $\mathbf{N}^\circ = [N_1^\circ, \ldots, N_n^\circ]$, either positive or partly positive, then $\mathbf{N}^o$ will be asymptotically stable since, in this case, the function

$$L(\mathbf{N}) = W(\mathbf{N}_n^\circ) - W(\mathbf{N}) \qquad (7.59)$$

is a Lyapunov function for system (7.54) at $\mathbf{N}^\circ$. This is so because $L(\mathbf{N}) \geq 0$ in a domain which contains $\mathbf{N}_n^\circ$ ($L = 0$ only at $\mathbf{N}_n^\circ$) and the derivative by virtue of the system is $dL/dt = -dW/dt \leq 0$ with $dL/dt = 0$ only at $\mathbf{N}^\circ$.

The sufficient maximum condition for $W(\boldsymbol{\eta})$ is that the quadratic form

$$\sum_{i,j} \sum \gamma_{ij} N_i N_j$$

be positive definite, and in this case $W(\boldsymbol{\eta})$ is strictly concave (convex upwards). As far as the orthant $\mathbb{R}_+^n$ also represents a convex set, $W(\boldsymbol{\eta})$ has a unique isolated maximum in $\mathbb{R}_+^n$ (inside or on the boundary).[30] Hence, the local maximum of W is also the global one and system (7.54) has a unique stable equilibrium. Moreover, since $L(\mathbf{N}) \to \infty$ as $\|\mathbf{N}\| \to \infty$, this equilibrium, by the theorem about stability on the whole,[31] is stable under any initial trajectory deviation within $\mathbb{R}_+^n$ (globally stable, absolutely stable), thus suggesting that any system trajectory starting inside $\mathbb{R}_+^n$ approaches equilibrium as $t \to \infty$.

In case max $W(\boldsymbol{\eta})$ is not attained inside the positive orthant $\mathbb{R}_+^n$, it will be attained on its boundary, in the appropriate coordinate hyperplane. As such an equilibrium again becomes stable, it means that one or several species have to be eliminated in the course of community dynamics. This happens when, for instance, there are some negative components N_i° among those of the solution $\mathbf{N}^\circ = [N_1^\circ, \ldots, N_n^\circ]$ to the system of equations $\sum_j \gamma_{ij} N_j^\circ = \varepsilon_i$, $i = 1, \ldots, n$.

The squared velocity of movement along trajectories of steepest assent can be found as

$$v^2 = \sum_{i=1}^n \left(\frac{d\eta_i}{dt}\right)^2 = \sum_{i=1}^n \left(\frac{\partial W}{\partial \eta_i}\right)^2 = \frac{dW}{dt} = \sum_{i=1}^n N_i \left(\varepsilon_i - \sum_{i=1}^n \gamma_{ij} N_j\right)^2.$$

From this expression it is clear that the movement becomes more and more decelerated while approaching the steady state, although it may be fast enough when far from this state.

Together with similar results for the frequency form of Lotka–Volterra equations,[29] the extremal property of system (7.54) not only corroborates the qualitative description given earlier[32] (see also Section 4.III) for dynamic behavior of dissipative systems, but also invites a meaningful interpretation.[29] The value of $V(\mathbf{N}) = \sum_{i=1}^n \varepsilon_i N_i$ characterizes, in essence, the rate of biomass increase in case neither resource competition, nor any other kind of limitation regulates the community but the growth is only determined by the physiological potential of organisms to reproduce and by their natural mortality. It will be logical, therefore, to call V

the *reproductive potential* of the community, an analog to the notion of ecosystem *gross production*. The expression

$$G(\mathbf{N}) = \frac{1}{2} \sum_{i,j=1}^{n} \gamma_{ij} N_i N_j$$

can further be regarded as a measure for the rate of energy dissipation due to intra- and interspecies competition, i.e., the total community biomass expenditures for competition. Therefore, we shall refer to $G(\mathbf{N})$ as the *total competition expenditures*, or, in ecological terms again, the ecosystem *respiration losses*. Hence, the increase in $W(\mathbf{N})$ in the course of community evolution may be interpreted as the tendency of the community to maximize the difference between its reproductive potential and the total competition expenditures. This can be done in several ways: either the reproductive potential is maximized under fixed competition expenditures, or the competition expenditures are minimized under limited reproduction potential (there may be some mixed cases as well). Empirical evidence indicates that all these cases are realized in nature.

Consider, for example, the so-called *r-competition* and *K-competition*, which refer to two opposite mechanisms of population density regulation.[33]

The origin of these terms becomes clear if we turn to the form (7.7) of the competition equations, which is related to (7.54) by a scaling change of variables and notation:

$$N_i = \bar{N}_i \frac{r_j}{K_i}, \quad r_i = \varepsilon_i, \quad \alpha_{ij} = \gamma_{ij} \frac{K_i K_j}{r_i r_j}, \tag{7.60}$$

and obviously does not affect the main properties of trajectory behavior.

As $r_i = \varepsilon_i$ is nothing but an intrinsic rate of natural increase, those species which win the competition by increasing their own fertility are said to use the *r-strategy*. Similarly, as K_i is the environment carrying capacity for the ith species, those species which win the competition by optimizing their resource consumption (uptake in a wide resource spectrum, reduction of competition expenditures by means of effective resource niche partition, etc.) are said to use the *K-strategy*. Thus, the *r-strategy* of a community can be said to consist in maximization of the reproductive potential, whereas the *K-strategy* to consist in minimization of the total competition expenditures.

B. Towards the Closest Species Packing

If the previous "extremal" formulation concerns, in essence, how trajectories move in a model, one more formulation can be proposed with regard to where the movement arrives. To see "where," we revert to interpreting the coefficients of population equations in terms of Section 7.I, i.e., in terms of the resource spectrum and utilization functions. If the resource is considered to be the whole living space, including all environmental factors of vital importance such as food, habitat, temperature, etc., then $K(\mathbf{x})$ can be interpreted as a volume of the vital space available at a given state $\mathbf{x}$ of the environment and $f_i(\mathbf{x})$ as an elementary volume of the space needed for a "unit" of species i to survive at the same state $\mathbf{x}$ in the environment.

Hence, the function

$$\mathcal{D}(\tilde{N}_1, \ldots, \tilde{N}_n) = \int \left[\mathbf{K}(\mathbf{x}) - \sum_{i=1}^{n} f_i(\mathbf{x}) \tilde{N}_i \right]^2 d\mathbf{x}, \qquad (7.61)$$

where the integral is taken over the entire space of environmental factors, is interpreted as the mean square difference between the really existing vital space and the space needed for a community to exist in species composition $\tilde{N}_1, \ldots, \tilde{N}_n$. If every species is considered to occupy a certain volume in this space, then the function $\mathcal{D}$ can be regarded as a measure of how densely the species are "packed" in the given environment, the less being $\mathcal{D}$, the closer being the species packing.

If an equilibrium state, say $\mathbf{N}^\circ$, is stable, then the function $\mathcal{D}(N_1, \ldots, N_n)$ attains a minimum, which is global on the positive orthant $\mathbb{R}_+^n$, and, moreover, $d\mathcal{D}/dt \geq 0$, along all the trajectories $\mathbf{N}(t)$, with $d\mathcal{D}/dt = 0$ only at the equilibrium point. Indeed, using the expressions for K_i and α_{ij} in terms of $K(\mathbf{x})$ and $f_i(\mathbf{x})$, i.e., formulas (7.5) and (7.6), the function $\mathcal{D}$ can be written in the form

$$\mathcal{D} = \int \mathbf{K}^2(\mathbf{x}) \, d\mathbf{x} - 2 \sum_{i=1}^{n} K_i \tilde{N}_i + \sum_{i,j=1}^{n} \alpha_{ij} \tilde{N}_i \tilde{N}_j, \qquad (7.62)$$

the integral being assumed convergent. If we denote by $\mathcal{D}_0$ the first term in (7.62), which is independent both of $\mathbf{x}$ and i, and make use of the formulae (7.60) to return to the original variables N_i, then function $\mathcal{D}$ assumes the form

$$\mathcal{D} = \mathcal{D}_0 - 2 \sum_{i=1}^{n} \varepsilon_i N_i + \sum_{i,j=1}^{n} \gamma_{ij} N_i N_j = \mathcal{D}_0 - 2W, \qquad (7.63)$$

where $W(N_1, \ldots, N_n)$ is the function introduced in (7.55). Since W increases everywhere along trajectories leading to the stable equilibrium point and attains an isolated maximum at that point, function $\mathcal{D}$ will decrease accordingly along the trajectories, reaching a minimum at that equilibrium.

The extremal points of $\mathcal{D}$ obviously coincide with those of W and, consequently, they can be found among equilibrium points of system (7.7) or (7.54). Competition matrix $[\alpha_{ij}]$ being positive definite, which is equivalent to matrix $[\gamma_{ij}]$ being positive definite, indicates that the function $\mathcal{D}$ has a unique minimum over the positive orthant $\mathbb{R}_+^n$ (in the interior or on the boundary); this minimum coincides with the only stable equilibrium point, and $\mathcal{D}$ declines along any trajectory initiating inside $\mathbb{R}_+^n$.

Thus, if there is a positive equilibrium $\mathbf{N}^*$ in a system functioning in an ecological space of "volume" $K(\mathbf{x})$, then the equilibrium is stable, i.e., all the n species coexist and function $\mathcal{D}$ attains its minimum on $\mathbb{R}_+^n$ at point $\mathbf{N}^*$. If there is no such equilibrium, then one of the partly stable equilibria $\mathbf{N}^\circ$, which gives $\mathcal{D}$ the minimum on $\mathbb{R}_+^n$, proves to be stable and the species vanish corresponding to the zero components of the equilibrium point.

By examining the function $\mathcal{D}(\mathbf{N})$ we may formally determine the largest number of new species that may consolidate the community after it has been invaded by a small number of individuals belonging to those species. If we assume that the new species do not affect the properties of the ecological space [the function $K(\mathbf{x})$], then we have to consider an expanding sequence of state spaces for systems of increasing dimensionality, i.e., the sequence of orthants

$$\mathbb{R}_+^n \subset \mathbb{R}_+^{n+1} \subset \mathbb{R}_+^{n+2} \subset \cdots,$$

and the associated sequence of minimal values for function $\mathcal{D}(\mathbf{N})$:

$$\min_{\mathbf{N} \in \mathbb{R}_+^n} \mathcal{D}(\mathbf{N}) \geq \min_{\mathbf{N} \in \mathbb{R}_+^{n+1}} \mathcal{D}(\mathbf{N}) \geq \min_{\mathbf{N} \in \mathbb{R}_+^{n+2}} \mathcal{D}(\mathbf{N}) \geq \cdots \tag{7.64}$$

By Definition (7.61) $\mathcal{D}(\mathbf{N})$ is never negative, and so the sequence (7.64) is bounded from below; by the classic Weierstrass Lemma of Real Analysis it has a limit which is greater than or equal to zero.

If this limit is attained at a member of the sequence with a finite number, then the dimensionality of the corresponding orthant is just the largest possible number of species in the community. This is the case, for instance, when the function $K(\mathbf{x})$ is such that, for some finite n, $K(\mathbf{x}) = k_1 f_1(\mathbf{x}) + k_2 f_2(\mathbf{x}) + k_n f_n(\mathbf{x})$, where all $k_i > 0$. It can be seen that in this case $\mathbf{N}^* = [k_1, \ldots, k_n]$ represents a nontrivial equilibrium point and $\mathcal{D}(k_1, \ldots, k_n) = 0$.

It is possible of course [for some highly specific constructions of $K(\mathbf{x})$ and $f_i(\mathbf{x})$] that sequence (7.64) proceeds indefinitely. In view of the previous section, this may suggest, for instance, that, given a spectrum $K(\mathbf{x})$, there exists a positive equilibrium in a system of any dimension n, thereby being stable and giving a minimum to function $\mathcal{D}(\mathbf{N})$. This could be traced in the competition pattern (7.21) with a spectrum $K(\mathbf{x})$ such that $K_i = \text{const} > 0$ for all i.

The results above can be formulated as the following principle of MacArthur: *a community of species competing for the vital space evolves towards the state of the closest species packing (the minimum of $\mathcal{D}$), the density of packing always increasing in the course of community evolution and attaining, at the equilibrium, the maximal possible closeness for the given environment.*

As $\mathcal{D}$ is minimal at a stable equilibrium, this means that there is no free living space in the community, needed, for instance, for introduction and establishment of a new species with characteristics similar to those of a species already present in the community. Such an introduction becomes possible only if the new species requires the types of resources (or occupies the regions of the environment-specific living space) which have been left without use by the former species. But, in terms of the formal scheme, this now suggests a change in properties of the ecological space , i.e., the change in the form of function $K(\mathbf{x})$ and, accordingly, of function $\mathcal{D}(N_1, \ldots, N_n)$.

Note that the most dense packing may also be attained when one or several species are eliminated from the community. This does not contradict common sense, as the packing density is a function of not only the species set but also of the species population sizes: function $\mathcal{D}(N_1,\ldots,N_n)$ is to reach the minimum in one of the coordinate hyperplanes in that case.

This is a more general view of what we considered in more detail in the previous section, and this is where "extremal" formulations may help, when a dissipative model is investigated in terms of stable species compositions.

Appendix

Calculating the Spectrum of Matrices (7.16) and (7.17). The characteristic polynomial, $D_n(\lambda)$, of the $n \times n$ matrix (7.16) can be shown to possess the recurrence property:

$$D_n(\lambda) = (1 - \lambda)D_{n-1}(\lambda) - a^2 D_{n-2}(\lambda).$$

The solution to this difference equation can be written down as

$$D_n(\lambda) = a^n \frac{\sin(n + 1)\theta}{\sin \theta}, \qquad \cos \theta = \frac{1 - \lambda}{2a}. \qquad (A7.1)$$

Tackling now the equation $D_n(\lambda) = 0$, we obtain

$$\lambda_k = 1 - 2a \cos \frac{\pi k}{n + 1}, \qquad k = 1, 2, \ldots, n, \qquad (A7.2)$$

whereby it follows, in particular, that for a competitive structure (7.16) to be stable the overlap of adjacent niches should measure less than

$$a_{\max} = 1 \Big/ \left(2 \cos \frac{\pi}{n + 1}\right),$$

the latter decreasing to $1/2$ for large n.   □

Matrix (7.17) is a special case of the matrix whose every row, starting from the second one, is derived from the preceding row by the cyclic shift of entries by one position. A matrix of this kind is termed a *cyclic matrix*, or *circulant*.[34] If the entries of its first row are denoted by $c_0, c_1, \ldots, c_{n-1}$, then the eigenvalues are:[34]

$$\lambda_k = c_0 + c_1 \varepsilon_k + c_2 \varepsilon_k^2 + \cdots + c_{n-1} \varepsilon_k^{n-1}, \qquad k = 0, 1, \ldots, n - 1, \qquad (A7.3)$$

where ε_k is the kth root of the equation $\varepsilon^n = 1$. If a cyclic matrix is also symmetric, then λ_k are real, hence only cosine terms are to remain in the sum (A7.3) when ε_k are expressed in trigonometrical terms:

$$\begin{aligned} \lambda_k &= c_0 + c_1 \cos \varphi_k + c_2 \cos 2\varphi_k + \cdots + c_{n-1} \cos(n - 1)\varphi_k, \\ \varphi_k &= 2\pi k/n; \quad k = 0, 1, \ldots, n - 1. \end{aligned} \qquad (A7.4)$$

For a particular set of c_i defining matrix (7.17) it follows that the spectrum consists of eigenvalues

$\lambda = 1 - a$ of multiplicity $(n - 1)$,
$\lambda = 1 + (n - 1)a$ of multiplicity 1.   □

Theorem 7.1 *Let $A = [\alpha_{ij}]$ be an $n \times n$ matrix of entries $\alpha_{ij} = \alpha(|i - j|)$, where $\alpha(z)$ is a function of the integral argument such that $\det[\alpha_{ij}] \neq 0$; let vector $\mathbf{K} = [K_1, \ldots, K_n]^T$ be symmetric, i.e., $K_i = K_{n-i}$. Then the solution to the system of linear equations $A\mathbf{x} = \mathbf{K}$ is also symmetric, i.e., $x_i = x_{n-i}$.*

Proof: For the sake of clearness and simplicity we consider the case of $n = 6$ (for an arbitrary n the argument is similar). With no loss of generality, we assume that

$$\alpha(0) = 1, \quad \alpha(1) = \alpha, \quad \alpha(2) = \beta, \quad \alpha(3) = \gamma, \quad \alpha(4) = \delta, \quad \alpha(5) = \varepsilon,$$

whereby the system of equations takes on the form

$$
\begin{bmatrix}
1 & \alpha & \beta & \gamma & \delta & \varepsilon \\
\alpha & 1 & \alpha & \beta & \gamma & \beta \\
\beta & \alpha & 1 & \alpha & \beta & \gamma \\
\gamma & \beta & \alpha & 1 & \alpha & \beta \\
\delta & \gamma & \beta & \alpha & 1 & \alpha \\
\varepsilon & \delta & \gamma & \beta & \alpha & 1
\end{bmatrix}
\begin{bmatrix}
x_1 \\ x_2 \\ x_3 \\ x_4 \\ x_5 \\ x_6
\end{bmatrix}
=
\begin{bmatrix}
K_1 \\ K_2 \\ K_3 \\ K_4 \\ K_5 \\ K_6
\end{bmatrix}.
\tag{A7.5}
$$

Rearrange (A7.5) in the following way: subtract the last equation from the first one, the last but one equation from the second one, and so on. This results in

$$
\begin{bmatrix}
1-\varepsilon & \alpha-\delta & \beta-\gamma & \gamma-\beta & \delta-\alpha & \varepsilon-1 \\
\alpha-\delta & 1-\gamma & \alpha-\beta & \beta-\alpha & \gamma-1 & \beta-\alpha \\
\beta-\gamma & \alpha-\beta & 1-\alpha & \alpha-1 & \beta-\alpha & \gamma-\beta \\
\gamma & \beta & \alpha & 1 & \alpha & \beta \\
\delta & \gamma & \beta & \alpha & 1 & \alpha \\
\varepsilon & \delta & \gamma & \beta & \alpha & 1
\end{bmatrix}
\begin{bmatrix}
x_1 \\ x_2 \\ x_3 \\ x_4 \\ x_5 \\ x_6
\end{bmatrix}
=
\begin{bmatrix}
0 \\ 0 \\ 0 \\ K_4 \\ K_5 \\ K_6
\end{bmatrix}.
\tag{A7.6}
$$

In the first three equations of (A7.6), group together the variables x_1 and x_6, x_2 and x_5, x_3 and x_4 to yield

$$
\begin{bmatrix}
1-\varepsilon & \alpha-\delta & \beta-\gamma \\
\alpha-\delta & 1-\gamma & \alpha-\beta \\
\beta-\gamma & \alpha-\beta & 1-\alpha
\end{bmatrix}
\begin{bmatrix}
x_1 - x_6 \\ x_2 - x_5 \\ x_3 - x_4
\end{bmatrix}
= \mathbf{0}.
\tag{A7.7}
$$

If the determinant of (A7.7) $D' \neq 0$, the only solution is zero, that is, $x_1 = x_6$, $x_2 = x_5$, $x_3 = x_4$, exactly what the theorem states. To prove that $D' \neq 0$, notice that the passage from (A7.5) to (A7.6) keeps the determinant D of (A7.5) unchanged, while further rearrangement of its columns gives

$$D = \begin{bmatrix} 1-\varepsilon & \alpha-\delta & \beta-\gamma & 0 & 0 & 0 \\ \alpha-\delta & 1-\gamma & \alpha-\beta & 0 & 0 & 0 \\ \beta-\gamma & \alpha-\beta & 1-\alpha & 0 & 0 & 0 \\ \gamma & \beta & \alpha & 1+\alpha & \alpha+\beta & \beta+\gamma \\ \delta & \gamma & \beta & \alpha+\beta & 1+\gamma & \alpha+\delta \\ \varepsilon & \delta & \gamma & \beta+\gamma & \alpha+\delta & 1+\varepsilon \end{bmatrix}$$

$$= D' \begin{bmatrix} 1+\alpha & \alpha+\beta & \beta+\gamma \\ \alpha+\beta & 1+\gamma & \alpha+\delta \\ \beta+\gamma & \alpha+\delta & 1+\varepsilon \end{bmatrix}.$$

In accordance with the condition of the theorem, we have $D \neq 0$ and consequently $D' \neq 0$, thus proving the theorem. For an odd n the argument is similar, with the only difference being that the "unpaired" component of the solution, $x_{(n+1)/2}$, takes no part in system (A7.7). But neither value of $x_{(n+1)/2}$ can disturb the symmetry of $\mathbf{x}$. ∎

Lemma A7.1. *Given the conditions of Theorem 7.2, all the eigenvalues of the circulant matrix $C(A)$ (7.29) are positive, although they may be arbitrarily close to zero.*

Proof: According to (A7.4) the eigenvalues of $C(A)$ are

$$\mu_k = 1 + 2\sum_{j=1}^{n} \alpha_j \cos j\varphi_k + (-1)^k \alpha_n, \quad \varphi_k = \pi k/n;$$

$$k = 0, 1, \ldots, 2n - 1,$$

and it can be easily seen that

$$\mu_k = \mu_{2n-k}, \quad k = 1, \ldots, n - 1.$$

Thus, in order to prove that $C(A)$ is positive definite we need to verify whether the conditions

$$\mu_k = 1 + 2\sum_{j=1}^{n} \alpha_j \cos j\varphi_k + (-1)^k \alpha_n > 0, \tag{A7.8}$$

hold true for any $k = 0, 1, \ldots, n$, under the restrictions

$$1 > \alpha_1 > \alpha_2 > \cdots > \alpha_n > 0, \tag{A7.9}$$

$$\alpha_m < (\alpha_{m-1} + \alpha_{m+1})/2, \quad m = 1, \ldots, n - 1. \tag{A7.10}$$

Note that the left-hand side of inequality (A7.8) represents a linear function of variables α_j, while the restrictions (A7.9) and (A7.10) are of the same kind as those in problems of linear programming, with the only difference being that those problems deal with nonstrict inequalities. Consider the following sets of parameters α_j:

$$\Omega = \Big\{(\alpha_1,\ldots,\alpha_n) : 1 > \alpha_1 > \cdots > \alpha_n > 0, \ \alpha_m < (\alpha_{m-1} + \alpha_{m+1})/2,$$

$$m = 1, 2, \ldots, n-1\Big\},$$

$$\overline{\Omega} = \Big\{(\alpha_1,\ldots,\alpha_n) : 1 \geq \alpha_1 \geq \cdots \geq \alpha_n \geq 0, \ \alpha_m \leq (\alpha_{m-1} + \alpha_{m+1})/2,$$

$$m = 1, 2, \ldots, n-1\Big\}.$$

Clearly, $\overline{\Omega} \supset \Omega$ and

$$\inf_{\Omega} \mu_k \geq \min_{\overline{\Omega}} \mu_k, \quad k = 0, 1, \ldots, n. \tag{A7.11}$$

The problem to minimize a linear function μ_k over the set $\overline{\Omega}$ represents a classic problem of linear programming, and by the well-known result of this theory it will be shown further on that, first,

$$\min_{\overline{\Omega},k} \mu_k = 0,$$

second, this minimum is attained at neither interior point of set $\overline{\Omega}$, and yet, third,

$$\inf_{\Omega}\Big\{\min_k \mu_k\Big\} = 0.$$

To reduce the minimization of μ_k to a linear programming problem in its standard form, we introduce new variables, x_j, related with the original α_js by the following formulas:

$$
\begin{aligned}
1 &= \alpha_0 = x_0 + 2x_1 + \cdots + nx_{n-1} + x_n, \\
&\quad\ \alpha_1 = x_1 + 2x_2 + \cdots + (n-1)x_{n-1} + x_n, \\
&\quad\ \cdots\cdots\cdots\cdots\cdots\cdots\cdots\cdots\cdots\cdots \\
&\quad\ \alpha_m = x_m + 2x_{m+1} + \cdots + (n-m)x_{n-1} + x_n, \\
&\quad\ \cdots\cdots\cdots\cdots\cdots\cdots\cdots\cdots\cdots\cdots \\
&\quad\ \alpha_{n-1} = x_{n-1} + x_n, \\
&\quad\ \alpha_n = x_n.
\end{aligned}
\tag{A7.12}
$$

Then it follows from conditions (A7.10) that

$$x_{m-1} \geq 0, \quad m = 1, 2, \ldots, n-1,$$

while the condition $\alpha_{n-1} \geq \alpha_n \geq 0$ leads to $x_{n-1} \geq 0$, $x_n \geq 0$. Thus, we have come to a linear programming problem in the standard form: find the minimum to the linear function $\mu_k(\mathbf{x})$ under one equality constraint:

$$x_0 + 2x_1 + \cdots + nx_{n-1} + x_n = 1 \tag{A7.13}$$

and $n + 1$ inequality constraints:

$$x_m \geq 0, \quad m = 0, 1, \ldots, n. \tag{A7.14}$$

It is known[35] that the solution to a linear programming problem is attained at one or several vertices of simplex $\overline{\Omega}$ defined by conditions (A7.13)–(A7.14); if attained at several vertices, the solution being so at any point of the convex hull spanning those vertices.

In the space of variables x_j the vertices of simplex $\overline{\Omega}$ have the following coordinates:

$$
\begin{aligned}
\mathbf{V}_0 &= [1, \quad 0, \quad 0, \quad \ldots, \quad 0], \\
\mathbf{V}_1 &= [0, \quad 1/2, \quad 0, \quad \ldots, \quad 0], \\
&\cdots \\
\mathbf{V}_{n-1} &= [0, \quad 0, \quad 0, \quad \ldots \quad 1/n, \quad 0], \\
\mathbf{V}_n &= [0, \quad 0, \quad 0, \quad \ldots, \quad 0, \quad 1],
\end{aligned}
$$

whereas in the space of variables $(\alpha_1, \ldots, \alpha_n)$, with regard to formulae (A7.12), the coordinates are:

$$
\begin{aligned}
\mathbf{V}_0 &= [0, 0, \ldots, 0], \\
\mathbf{V}_1 &= [1/2, 0, \ldots, 0], \\
&\cdots \\
\mathbf{V}_m &= \left[\frac{m}{m+1}, \frac{m-1}{m+1}, \ldots, \frac{1}{m+1}, 0, \ldots, 0 \right], \quad m = 1, \ldots, n-1, \\
&\cdots \\
\mathbf{V}_n &= [1, 1, \ldots, 1].
\end{aligned}
$$

The value of μ_k at vertex $\mathbf{V}_0$ is apparently

$$\mu_k(\mathbf{V}_0) = 1, \tag{A7.15}$$

and those at vertices $\mathbf{V}_1, \ldots, \mathbf{V}_{n-1}$ are

$$\mu_k(\mathbf{V}_m) = 1 + 2 \sum_{l=1}^{m} \frac{m+1-l}{m+1} \cos l\varphi_k = 1 + \frac{2}{m+1} Y_m(\varphi_k), \tag{A7.16}$$

where the following notation is used:

$$Y_m(\varphi) = m\cos\varphi + (m-1)\cos 2\varphi + \cdots + \cos m\varphi.$$

By mathematical induction one can show that the following identity:

$$Y_m(\varphi) = \frac{1}{2}\left[\frac{\cos\varphi - \cos(m+1)\varphi}{1-\cos\varphi} - m\right], \tag{A7.17}$$

holds true for all admissible values of $\varphi \neq 0, \pm 2\pi, \ldots$, whereby the expression (A7.16) can be rearranged into

$$\mu_k(\mathbf{V}_m) = \frac{1}{m+1}\frac{1-\cos(m+1)\varphi_k}{1-\cos\varphi_k}, \quad m = 1, \ldots, n-1. \tag{A7.18}$$

(Note that the denominator in (A7.18) is nonzero for each $\varphi_k, k = 1, \ldots, n$.) From (A7.18) it is clear that

$$\mu_k(\mathbf{V}_m) > 0 \tag{A7.19}$$

for $m = 1, \ldots, n-2$, whereas for $m = n-1$ we have

$$\mu_k(\mathbf{V}_{n-1}) = \frac{1}{n}\frac{1-(-1)^k}{1-\cos\varphi_k} = \begin{cases} 0, & \text{if } k \text{ is even,} \\ \frac{2}{n(1-\cos\varphi_k)} > 0, & \text{if } k \text{ is odd.} \end{cases} \tag{A7.20}$$

Further, for $m = n$, the identity

$$\cos\alpha + \cos 2\alpha + \cdots + \cos N\alpha = \frac{1}{2}\left[\frac{\sin(N+1/2)\alpha}{\sin\alpha/2} - 1\right], \quad \alpha \neq 0, \pm 2\pi, \ldots$$

guarantees that

$$\mu_k(\mathbf{V}_n) = 1 + 2\sum_{j=1}^{n-1}\cos j\varphi_k(-1)^k = \frac{\sin(n-1/2)\varphi_k}{\sin\varphi_k/2} + (-1)^k = (-1)^{k-1} + (-1)^k = 0 \tag{A7.21}$$

for all $k = 1, \ldots, n$.

Summarizing (A7.15) and (A7.19)–(A7.21), we see that

$$\min\mu_k = 0, \quad k = 1, 2, \ldots, n,$$

the minimum being attained at the only point $\mathbf{V}_n$ for odd k, while for even k it is attained at the vertex $\mathbf{V}_{n-1}$ as well. In the latter case, by linear programming theory, $\mu_k = 0$ also at any point of the interval $\mathcal{I}$ spanning vertices $\mathbf{V}_{n-1}$ and $\mathbf{V}_n$. Thus, we have

$$\min_k\min_{\overline{\Omega}}\mu_k = 0$$

and the minimum is attained on the segment $\overline{\mathcal{I}} \subset \overline{\Omega}$.

Any point V of the segment,

$$V = aV_{n-1} + bV_n, \quad a, b \geq 0, \quad a + b = 1,$$

can be represented as a vector in the space of α_j,

$$V = \left[b + a \left(1 - \frac{1}{n} \right), b + a \left(1 - \frac{2}{n} \right), \dots, b + \frac{a}{n} \right],$$

whose coordinates decrease linearly with the ordinal number. Hence, everywhere in the interval $\mathcal{I}$ conditions (A7.10) turn into equalities, so that $\mathcal{I}$ belongs entirely to the boundary of $\overline{\Omega}$ and $\mathcal{I} \cap \Omega = \phi$.

Thus, for any point of the open set Ω, that is, for any set of α_j which obeys strict inequalities (A7.9)–(A7.10), the eigenvalues μ_k of the circulant $C(A)$ are positive because $\Omega \subset \overline{\Omega}$ and all the points of $\overline{\Omega}$ at which $\mu_k = 0$ belong to $\mathcal{I}$. This proves the first part of the lemma statement.

As to the second, from μ_k being continuous with respect to α_j it follows that

$$\inf_{\Omega} \left\{ \min_k \mu_k \right\} = 0.$$

Therefore, in Ω there exists a collection of α_js that gives an arbitrary small, yet positive, eigenvalue to matrix $C(A)$. ∎

Theorem 7.2 *Matrix $A = [\alpha_{ij}]$ with entries $\alpha_{ij} = \alpha(|i - j|)$ that satisfy conditions (7.28) is positive definite.*

Proof: Denote by λ_i the eigenvalues of matrix A, which are real due to symmetry of A. Note that matrices A and $C(A)$ satisfy the terms of the Sturm theorem[36] on separability of eigenvalues in a finite sequence $\{A_r\}$ of "nested" symmetric matrices $A_r = [a_{ij}]$, $i, j = 1, \dots, r$ ($r = 1, 2, \dots, M$): if $\lambda_k(A_r)$ denotes the k-th eigenvalue of A_r, where $\lambda_1(A_r) \geq \lambda_2(A_r) \geq \cdots \geq \lambda_r(A_r)$, then

$$\lambda_{k+1}(A_{i+1}) \leq \lambda_k(A_i) \leq \lambda_k(A_{i+1}), \quad i = 1, \dots, M - 1. \tag{A7.22}$$

As seen from (7.29), matrices $A = A_1$ and $C(A) = A_M$ represent the extreme terms in such a sequence of $M = n$ "nested" submatrices. Hence, by Sturm's theorem it follows that, at least,

$$\min \mu_k \leq \lambda_i \leq \max \mu_k, \quad i = 0, 1, \dots, n. \tag{A7.23}$$

Since all the μ_k are proved to be positive by Lemma A7.1, it follows from (A7.23) that the spectrum of A is positive, too. ∎

Additional Notes

To 7.I. Evoked by set theory, the idea of an *ecological niche* of a species was proposed by Hutchinson[1] in 1957, with the thesis that the concept "requires to be stated formally in an unambiguous way to prevent further confusion."[1] Developed afterwards in several directions (for a survey see, e.g., Levins[37] and Williamson[2]), the notion has not however avoided a bit of "confusion" and "ambiguity," apparently attributable to a certain complexity and ambiguity relevant to what is necessary and sufficient both for a species to exist in nature and to be defined in theory, rather than to a deficiency in the fundamental idea of the niche.

If Hutchinson's article is to be regarded as a benchmark paper in the theory of the ecological niche, then the paper by MacArthur[5] should be regarded as the benchmark in introducing the theory into population dynamic models. It is to this type of models that both the major part of the criticism of the Gause *exclusion principle* and the constructive methods fostering its generalization have been addressed.[24,38−39]

Derivation of the Lotka–Volterra equations for a competition community over a one-dimensional resource spectrum follows the ideas of MacArthur.[40−41] A method to evaluate the rate of convergence to equilibrium and also particular estimates to show that the rate is too low can be found in May.[9] The effects of nonuniform niche patterns and multidimensional resource spectra were intensively discussed in the literature of the 1970s.[42−47]

The lack of sign stability in the SDG of any competition community matrix was probably why graph-theoretic considerations were not too popular in the contexts of competition. Recent studies by J. Cohen and coauthors[48−49] have however revealed at least one more, though negative again, contribution from graph theory. The contribution relies upon the notion of *intervality* in the *predator overlap graph*, which is defined as an ability of the graph to be mapped into a number of overlapping intervals of a straight line and which is interpreted as unidimensionality of the food niche organization. Real food webs have revealed a low probability of being *interval*: the greater the number of species in a web, the less is the chance that a single dimension will be sufficient to describe the niche pattern.[48]

In conjunction with and parallel to models of competition for a resource in common, models of competition due to direct conflicts between specimens, or *interference* models, were also developed (see Case and Gilpin,[50] Shigesada et al.,[51−52] and their references). This approach may avoid considerations of niche overlap, yet the stability problem of the competition matrix still remains to be solved.

For example, the stable competitive community considered by Shigesada et al.[52] appears to fall in between dominant and recessive patterns, since the $(N - 1)$ of N ordered species are assumed to be "auto-competitors" (with diagonally dominated rows of the community matrix), while the N-th species is a "hetero-competitor" (with diagonally recessive row of the matrix). Kawasaki et al.[53] have proved that if an equilibrium composition of this kind is locally stable then it is also globally stable, the characteristic statement for dissipative systems. In addition to the general reasons, this specific example justifies the structure of the "dissipative petal" in the "matrix flower" of Figure 17: besides quasi-dominant and quasi-recessive subsets of the petal, there are parts which belong to neither of these two types.

To 7.II. The definition of competition coefficients relying on a discrete analog to the integration form (7.6) can be found in Levins,[37] where competition matrix (7.17) was also studied, or in Schoener.[54] The particular case (7.20) was studied by Logofet and Svirezhev.[55] The scheme to prove that convexity is sufficient for stability in the competition matrix of generic form (7.19) was proposed by Logofet.[17]

To 7.III. The "geometric" interpretation for positive equilibrium to exist in a competition model, as well as the measure of "equilibriumness," can obviously be extended to other community types too, although they are most apparent for nonnegative interaction matrices.

To 7.IV. Mathematical ecologists could hardly neglect such a treasure of analytical mechanics as variational methods (Svirezhev and Yelizarov,[28] Svirezhev and Logofet[29]), nor could they miss such a crowded place to display the treasure as in the discussions on the role that teleological principles play in theoretical biology (see, e.g., Waddington,[56] Svirezhev and Logofet,[29] Semevsky and Semenov,[57] and their references).

"Extremal" formulation of the close packing principle in terms of the "vital space" was proposed in Svirezhev and Logofet.[58] The ideas underlying the principle had been repeatedly expressed in the works of R. H. MacArthur,[40−41,59] a godfather of the formal principle, and developed by later authors. Gatto,[60] for example, by reasoning similar to that of theoretical mechanics, generalized the principle to a less stringent set of assumptions, in particular, with a semi-definite rather than positive definite competition matrix. However, what remains unclear is whether the less stringent situations may actually bring about non-exotic new examples of community models.

REFERENCES

1. Hutchinson, G. E. Population studies: Animal ecology and demography. In: *Cold Spring Harbor Symposia on Quantitative Biology*, Vol. 22, Long Island Biological Association, New York, 1957, pp. 415–427. Reprinted in *Bulletin of Mathematical Biology*, 55, 193–213, 1991.

2. Williamson, M. H. *The Analysis of Biological Populations*, Edward Arnold, London, 1972, Chap. 9.3.

3. Petraitis, P. S. Algebraic and graphical relationships among niche breadth measures, *Ecology*, 62, 545–548, 1981.

4. Gause, G. F. *The Struggle for Existence*, Williams and Wilkins, Baltimore, 1934.

5. MacArthur, R. H. The theory of niche. In: *Population Biology and Evolution*, Lewontin, R., ed., Syracuse University Press, Syracuse, 1968, 159–176.

6. May, R. M. and MacArthur, R. H. Niche overlap as a function of environmental variability, *Proc. Natl. Acad. Sci. U.S.A.*, 69, 1109–1113, 1972.

7. May, R. M. On the theory of niche overlap, *Theor. Pop. Biol.*, 5, 3, 297–332, 1974.

8. Svirezhev, Yu. M. and Logofet, D. O. *Stability of Biological Communities* (revised from the 1978 Russian edition), Mir Publishers, Moscow, 1983, 319 pp., Chap. 6.

9. May, R. M. *Stability and Complexity in Model Ecosystems*, Princeton University Press, Princeton, NJ, 1973, Chap. 6.

10. Abrams, P. Niche overlap and environmental variability, *Math. Biosci.*, 28, 357–372, 1976.

11. Logofet, D. O. *Stability of Biological Communities (Mathematical Models)*. Candidate of Sciences Thesis, Computer Center of the USSR Academy of Sciences, Moscow, 1976.

12. Yoshiama, R. M. and Roughgarden, J. Species packing in two dimensions, *Amer. Natur.*, 111, 107–121, 1976.

13. Morowitz, H. J., The dimensionality of niche space, *J. Theor. Biol.*, 86, 259–263, 1980.

14. Abrosov, N. S., Kovrov, B. G. and Cherepanov, O. A. *Ecological Mechanisms of Co-existence and Species Regulation*, Nauka, Novosibirsk, 1982, 301 pp. (in Russian).

15. Harner, H. J. and Whitmore, C. Multivariate measures of niche overlap using discriminant analysis, *Theor. Pop. Biol.*, 12, 21–36, 1977.

16. Abrams, P. Some comments on measuring niche overlap, *Ecology*, 61, 44–49, 1980.

17. Logofet, D. O. On the stability of a class of matrices arising in the mathematical theory of biological associations, *Soviet Math. Dokl.*, 16, 523–527, 1975.

18. Logofet, D. O. and Svirezhev, Yu. M. Stability in models of interacting populations. In: *Problemy Kibernetiki (Problems of Cybernetics)*, Vol. 32, Yablonsky, S. V., ed., Nauka, Moscow, 1977, 187–202 (in Russian).

19. Jeffers, J. N. R. *An Introduction to Systems Analysis: with Ecological Applications*, Edward Arnold, London, 1978, 198 pp. Chap. 6.

20. Horn, R. A. and Johnson, C. R. *Matrix Analysis*, Cambridge University Press, Cambridge, 1990, Chaps. 5, 6.

21. Lawlor, L. R. Structure and stability in natural and randomly constructed competitive communities, *Amer. Natur.*, 116, 394–408, 1980.

22. Roughgarden, J. and Feldman, M. Species packing and predation pressure, *Ecology*, 56, 489–492, 1975.

23. Hofbauer, J. and Sigmund, K. *The Theory of Evolution and Dynamical Systems*, Cambridge University Press, Cambridge, 1988, 341 pp., Chap. 19.

24. Abrosov, N. S., Kovrov, B. G., and Cherepanov, O. A. *Ecological Mechanisms of Co-existence and Species Regulation*, Nauka, Novosibirsk, 301 pp., Appendix (in Russian).

25. Korn, G. A. and Korn, T. M. *Mathematical Handbook for Scientists and Engineers*, McGraw-Hill, New York, 1968, Chap. 1.12.

26. Logofet, D. O. *Matrices and Graphs: The Stability Problem in Mathematical Ecology*. Doctoral dissertation, Computer Center of the USSR Academy of Sciences, Moscow, 1985 (in Russian).

27. Svirezhev, Yu. M. and Logofet, D. O. *Stability of Biological Communities* (revised from the 1978 Russian edition), Mir Publishers, Moscow, 1983, 319 pp., Chap. 9.

28. Svirezhev, Yu. M. and Elizarov, E. Ya. *Mathematical Modeling of Biological Systems* (*Problems of Space Biology*, Vol. 20), Nauka, Moscow, 1972, 159 pp. (in Russian).

29. Svirezhev, Yu. M. and Logofet, D. O. *Stability of Biological Communities* (revised from the 1978 Russian edition), Mir Publishers, Moscow, 1983, 319 pp., Chap. 7.

30. Carr, C. R. and Howe, C. W. *Quantitative Decision Procedures in Management and Economics*, McGraw-Hill, New York, 1964.

31. Barbashin, E. A. *Introduction to Stability Theory*, Nauka, Moscow, 1967, Chap. 1.12 (in Russian).

32. Volterra, V. *Lecons sur la Théorie Mathématique de la Lutte pour la Vie*, Gauthier-Villars, Paris, 1931, 214 pp., Chap. 3.3.

33. MacArthur, R. H. and Conell, J. H. *The Biology of Populations*, John Wiley, New York, 1966.

34. Marcus, M. and Minc, A. *A Survey of Matrix Theory and Matrix Inequalities*, Allyn and Bacon, Boston, 1964, Chap. 4.9.

35. Korn, G. A. and Korn, T. M. *Mathematical Handbook*, McGraw-Hill, New York, 1968, Chap. 11.4-1.

36. Bellman, R. *Introduction to Matrix Analysis*, McGraw-Hill, New York, 1960, Chap. 7.8.

37. Levins, R. *Evolution in Changing Environments*, Princeton University Press, Princeton, 1968.

38. Armstrong, R. A. and MacGehee, R. Coexistence of species competing for shared resources, *Theor. Pop. Biol.*, 9, 317–328, 1976.

39. Armstrong, R. A. and MacGehee, R. Coexistence of two competitors on one resource, *J. Theor. Biol.*, 56, 499–502, 1976.

40. MacArthur, R. H. Species packing, or what competition minimizes, *Proc. Natl. Acad. Sci. U.S.A.*, 64, 1369–1375, 1969.

41. MacArthur, R. H. Species packing and competitive equilibrium for many species, *Theor. Pop. Biol.*, 1, 1–11, 1970.

42. Roughgarden, J. Evolution of niche width, *Amer. Natur.*, 106, 683–718, 1972.

43. May, R. M. Some notes on estimating the competition matrix, α, *Ecology*, 56, 3, 737–741, 1975.

44. Emlen, J. M. Niches and genes: Some further thoughts, *Amer. Natur.*, 109, 474–476, 1975.

45. Abrams, P. Limiting similarity and the form of the competition coefficient, *Theor. Pop. Biol.*, 8, 356–375, 1975.

46. MacMurtrie, R. On the limit to niche overlap for nonuniform niches, *Theor. Pop. Biol.*, 10, 96–107, 1976.

47. Yodzis, P. *Competition for Space and the Structure of Ecological Communities* (*Lecture Notes in Biomathematics*, Vol. 25), Springer-Verlag, Berlin, 1978, 191 pp.

48. Cohen, J. E. and Palka, Z. J. A stochastic theory of community food webs. V. Intervality and triangulation in the trophic-niche overlap graph, *The American Naturalist*, 135, 435–463, 1990.

49. Pimm, S. L., Lawton, J. H., and Cohen, J. E. Food web patterns and their consequences, *Nature*, 350, 669–674, 1991.

50. Case, T. J. and Gilpin, M. E. Interference competition and niche theory, *Proc. Natl. Sci. U.S.A.*, 71, 3073–3077, 1974.

51. Shigesada, N., Kawasaki, K., and Teramoto, E. The effects of interference competition on stability, structure and invasion of multi-species system, *J. Math. Biol.*, 21, 97–113, 1984.

52. Shigesada, N., Kawasaki, K., and Teramoto, E. Direct and indirect effects of invasions of predators on a multiple-species community, *Theor. Pop. Biol.*, 36, 311–338, 1989.

53. Kawasaki K., Nakajima, H., Shigesada, N., and Teramoto, E. Structure, stability and succession in model ecosystems. In: *Theoretical Studies of Ecosystems*, Nigashi, M. and Burns, T., eds., Cambridge University Press, London, 1991, pp. 179–210.

54. Schoener, T. W. Some methods for calculating competition coefficients from resource-utilization spectra, *Amer. Natur.*, 108, 332–340, 1974.

55. Logofet, D. O. and Svirezhev, Yu. M. On Volterra models for communities of coexisting species with overlapping ecological niches. In: *Theoretical and Experimental Biophysics*, Kaliningrad State University, Kaliningrad, 1975, 180–184 (in Russian).

56. Waddington, C. H., ed. *Towards a Theoretical Biology. I. Prolegomena*, Aldine, Birmingham, 1968, 181 pp.

57. Semevsky, F. N. and Semenov, S. M. *Mathematical Modeling of Ecological Processes*, Gidrometeoizdat, Leningrad, 1982, 280 pp. (in Russian).

58. Svirezhev, Yu. M. and Logofet, D. O. On stability and optimality in models of biological communities. In: *Optimization Problems in Ecology*, Novik, I. B., ed., Nauka, Moscow, 1978, 271–291 (in Russian).

59. MacArthur, R. H. *Geographical Ecology*, Harper and Row, New York, 1972.

60. Gatto, M. A general minimum principle for competing populations: some ecological and evolutionary consequences, *Theor. Pop. Biol.*, 37, 369–388, 1990.

CHAPTER 8

Stability in "Box" Models of Spatial Distribution

It is strongly believed, at least by theoretically minded ecologists, that the ability of a community to persist in spite of perturbations, is strongly promoted by its being distributed in space. If the conditions of vital importance are identical throughout the space, then the dynamics at any point gives an idea of the global picture. This plain assumption is inherent in all the models considered in the previous chapters, and it is precisely because of this aspect that those models are sometimes referred to as *lumped* or *point* models: the *local* description is assumed to be sufficient for the global characterization. This is obviously not the case when the environment is apparently inhomogeneous, or there are explicit *gradients* in the environmental conditions. For example, resource distributions are typically *mosaic*, or *patchy* in nature, inducing patchy distributions of consumers; or the amount of the resource available varies gradually in the direction of the gradient, for instance, the sunlight available to oceanic phytoplankton decreases with depth.

A direct way to account for these effects—and perhaps the most natural course for a mathematician—would be to include the spatial dimension(s) x into the model, thus considering the space densities, $N_i(x, t)$, rather than the population sizes, $N_i(t)$, as model variables, while studying the stability of pertinent solutions to the corresponding integral-differential or partial derivative equations. Unfortunately, the relevant mathematics is more complicated here than in difference or ordinary differential equations, so that the idea arises to represent the spatial model as a combination of a number of point models, united accordingly to the biology of the species under study.

For motile organisms, this usually involves migrations which combine the local *habitats*, or "boxes" into the global spatial system; then the question arises about the effects the migrations exert on stability in the combined system. What and how matrix theory can contribute to answering this question is considered in the present chapter.

I. CAN MIGRATION STABILIZE A COMMUNITY?

When the environment is thought to be inhomogeneous in space, so that the model parameters are essentially different in various points of the space, it is easy to believe by intuition and to show mathematically in model examples[1] that the in-homogeneity may drastically affect the dynamic behavior of the whole system, bringing about, in particular, the loss of stability by some equilibrium states of the model. But is there any effect on stability by the space dimension itself, or, in other words, we ask whether a homogeneous space can cause any new effects in stability of a spatial system as compared with its lumped counterpart? An answer is given below in the framework of a general model for several identical communities combined together by migration flows.[2]

A. Migration Flows and United Systems

Let there exist in the space a certain number, n, of identical, relatively isolated habitats, each inhabited by a community of the same collection of p species and of the same pattern of interaction among them for all the habitats. Suppose also that individuals of the species may pass, in accordance with their radii of individual activity, from some habitats to other ones, i.e., there are migration flows among species of different habitats. It is natural to think that the intensity of migration is not indifferent to the population sizes of migrating species in the habitats which are sources and sinks for the migration flows.

Clearly, the aggregate of n communities now makes up a whole *system*—let us call it *united*—consisting of n *subsystems*. Within a subsystem, species are linked by biological interactions, while the migration flows link the subsystems. Should one expect the united system to be stable if all the isolated subsystems exhibit stability? Can the migration links stabilize a behavior that is unstable within the isolated subsystem? Ecologists are prone to give positive answers to these questions—both in experiments[3-7] and theoretical works.[8-9]

Let the dynamics of an isolated community of p interacting species be modeled by a system of equations,

$$\frac{dN_i}{dt} = f_i(N_1, \ldots, N_p), \quad i = 1, \ldots, p, \tag{8.1}$$

where the right-hand sides be restricted by the only qualitative requirements: that an equilibrium $\mathbf{N}^* > \mathbf{0}$ exist and that linearization be possible at point $\mathbf{N}^*$. In the united system of n subsystems linked by migration flows, let the superscript at variables N_i refer to the subsystem. Then, for any kth system, we have

$$\frac{dN_i^k}{dt} = f_i(N_1^k, \ldots, N_p^k) + M_{ik}(N_i^1, \ldots, N_i^n),$$
$$k = 1, \ldots, n; \quad i = 1, \ldots, p, \tag{8.2}$$

where the set of functions f_i specifies the structure of the biological interactions within the kth subsystem (arbitrary enough, but identical for all the subsystems), and the functions M_{ik} characterize the effect of immigration flows from all the other subsystems to the kth one.

The simplest form of these functions M_{ik} is linear and is associated with migration flows of constant intensities—fixed for each ith species or specific to the *route* of migration. Obviously, the linear relationship is by no means the only type of relationship by which the migration may depend on the population sizes of migrating species, but it is quite sufficient for the purpose of stability analysis as far as the analysis is to be done by the linearization method.

The scheme of migration flows (the routes of migrations among subsystems) is first determined by the geometry of how the habitats are located in the space relative to each other (or in other words, by the geography of habitats). If the scheme is supposed to be the same for all the migrating species, then it can be represented by a single *matrix of migration structure*,

$$M_n = [m_{ks}], \quad k, s = 1, \ldots, n,$$

where the entry $m_{ks} \geq 0$ ($k \neq s$) specifies the *migration intensity* on the route $s \to k$. The system being closed with respect to migration implies that

$$m_{kk} = -\sum_{\substack{s=1 \\ s \neq k}}^{n} m_{sk}, \quad k = 1, \ldots, n, \tag{8.3}$$

i.e., all the column sums are zero.

Second, the migratory behavior should certainly be species-specific. If $m_i \geq 0$ denotes the factor of migration intensity that is responsible for the species-specificity and independent of the migration route, then the system of equations for the united community takes on the following form:

$$\frac{dN_i^k}{dt} = f_i(N_1^k, \ldots, N_p^k) + m_i \sum_{s=1}^{n} m_{ks} N_i^s, \tag{8.4}$$

with $i = 1, 2, \ldots, p$ denoting a species number and $k = 1, 2, \ldots, n$ a subsystem number. The equality $m_i = 0$ will indicate that the ith species (in all the subsystems) takes no part in migration.

B. Migrations in Isotropic Media

Suppose now that the migration intensities are entirely determined by the biology of the species but are independent of any particular direction that the migration may take, that is, the medium is *isotropic* with respect to migration. Then all of the nonzero off-diagonal entries in matrix M_n must be identical, just indicating those particular pairs of habitats which are linked by migration flows (of the same intensity in both directions). In this case one can put all these entries equal to unit; matrix M_n is symmetric, and by (8.3) all its row sums are also equal to zero:

$$m_{kk} + \sum_{s=1, s \neq k}^{n} m_{ks} = 0, \quad k = 1, \ldots, n. \tag{8.5}$$

It can be readily seen that the steady-state equations for the system (8.4),

$$f_i(N_1^k, \ldots, N_p^k) + m_i \sum_{s=1}^n m_{ks}N_i^s = 0, \quad k = 1, \ldots, n; \ i = 1, \ldots, p, \qquad (8.6)$$

admit a solution in the form of an np-vector

$$\mathcal{N}^* = [\mathbf{N}^*, \ldots, \mathbf{N}^*],$$

which preserves all the previous equilibrium population sizes for isolated subsystems.

Clearly, in the general case there may be other, additional, solutions to (8.6) with all the components being positive. It can be proved, for instance, that there are no such solutions when two second-order conservative Volterra subsystems are united. From the continuity argument it follows that no such solutions exist, at least in a small vicinity of $\mathcal{N}^*$.

Linearizing (8.4) at this point yields the matrix $\mathcal{A}$ of the block-diagonal structure

$$\mathcal{A} = I_n \otimes A + M_n \otimes D \qquad (8.7)$$

where

$$A = \left[\frac{\partial f_i}{\partial N_i} \bigg|_{\mathbf{N}^*} \right]$$

is a $p \times p$ matrix of the linearized isolated system, I_n is the identity matrix,

$$D = \text{diag}\{m_1, \ldots, m_p\}$$

is the diagonal matrix of species-specific migration intensities, and symbol $\otimes$ stands for the Kronecker product of matrices. The following lemma allows one to judge about stability of the equilibrium $\mathcal{N}^*$:

Lemma 8.1 *Let a matrix $M_n = [m_{ks}]$ have a zero eigenvalue and let $\mathbf{a} = [a_1, \ldots, a_n]^T$ be an eigenvector associated with the zero eigenvalue. If $\mathbf{x} \in \mathbb{R}^p$ is a column eigenvector of matrix A with an eigenvalue λ, then the same value λ is associated with the column eigenvector $\mathcal{X} = [a_1\mathbf{x}^T, \ldots, a_n\mathbf{x}^T]^T \in \mathbb{R}^{np}$ of matrix $\mathcal{A}$ (8.7), too.*

The proof is given in the Appendix.

Matrix M_n of a migration-closed structure in a migration-isotropic environment has, by (8.5), the eigenvector $\mathbf{a} = [1, 1, \ldots, 1]^T$ with $\lambda = 0$, i.e., satisfies the conditions of Lemma 8.1. Thus, the spectrum of matrix $\mathcal{A}$ for the united system necessarily contains the set of all eigenvalues λ_i, which determine the stability of equilibrium $\mathbf{N}^*$ in the isolated subsystem. This means that if $\mathbf{N}^*$ was unstable in the subsystem (i.e. if there existed eigenvalues of A with Re $\lambda(A) > 0$), then the instability is also preserved in system (8.4); if matrix A was stable (all Re $\lambda(A) < 0$), then for the equilibrium $\mathcal{N}^*$ to be stable in the united system the real parts of all the remaining eigenvalues of the block matrix $\mathcal{A}$ are required to be negative, hence constricting further (or making empty) the stability domain in the space of the model parameters.

If the spectrum of A is real, while there are complex eigenvalues among the additional values arising in the spectrum of $\mathcal{A}$, then the monotonic pattern of trajectory behavior in the neighborhood of the isolated subsystem equilibrium becomes oscillatory. The opposite change is obviously impossible, since the spectrum of A is preserved.

Example 8.1: Let $n = p = 2$ (see Figure 45a). The united system is then of the fourth order:

$$
\begin{aligned}
dN_1^1/dt &= f_1(N_1^1, N_2^1) + m_1(-N_1^1 + N_1^2), \\
dN_2^1/dt &= f_2(N_1^1, N_2^1) + m_2(-N_2^1 + N_2^2), \\
dN_1^2/dt &= f_1(N_1^2, N_2^2) + m_1(N_1^1 - N_1^2), \\
dN_2^2/dt &= f_2(N_1^2, N_2^2) + m_2(N_2^1 - N_2^2),
\end{aligned}
$$

$$
M_2 = \begin{bmatrix} -1 & 1 \\ 1 & -1 \end{bmatrix}, \qquad D = \begin{bmatrix} m_1 & 0 \\ 0 & m_2 \end{bmatrix}.
$$

Denoting the entries of matrix

$$
A = \left[\partial f_i / \partial N_j |_{\mathbf{N}\cdot} \right] \quad \text{by} \quad A = \begin{bmatrix} a & b \\ c & d \end{bmatrix},
$$

we have the united system matrix in the following form:

$$
\mathcal{A} = \left[\begin{array}{cc|cc} a - m_1 & c & m_1 & 0 \\ b & d - m_2 & 0 & m_2 \\ \hline m_1 & 0 & a - m_1 & c \\ 0 & m_2 & b & d - m_2 \end{array} \right].
$$

Its characteristic equation,

$$
[(\lambda - a)(\lambda - d) - bc]\{[\lambda - (a - 2m_1)][\lambda - d - 2m_2)] - bc\} = 0,
$$

along with the roots of the isolated subsystem,

$$
\lambda_{1,2} = \left[a + d \pm \sqrt{(a - d)^2 + 4bc} \right] / 2,
$$

has a pair of new roots:

$$
\lambda_{3,4} = \left[\tilde{a} + \tilde{d} \pm \sqrt{(\tilde{a} - \tilde{d})^2 + 4bc} \right] / 2
$$

where $\tilde{a} = a - 2m_1$, $\tilde{d} = d - 2m_2$.

The first approximation stability criterion gives the conditions

$$a + d < 0, \quad ad - bc > 0, \tag{8.8}$$

$$\tilde{a} + \tilde{d} < 0, \quad \tilde{a}\tilde{d} - bc > 0, \tag{8.8'}$$

and if they are satisfied, then all the Re $\lambda_i(\mathcal{A})$ are negative.

It is clear that if inequalities (8.8), the conditions for an isolated subsystem to be stable, hold true, then for sufficiently small m_1 and m_2 the inequalities (8.8') are also satisfied, i.e., the united system is stable too. If trajectories of the isolated subsystem are monotonically stable in the neighborhood of $\mathbf{N}^*$, which is the case when

$$(a - d)^2 + 4bc \geq 0, \quad ad - bc > 0, \tag{8.9}$$

then those of the united system will be is monotonically stable too if

$$(\tilde{a} - \tilde{d})^2 + 4bc \geq 0, \quad \tilde{a}\tilde{d} - bc > 0. \tag{8.9'}$$

When conditions (8.9) hold true, the inequalities (8.9') are also satisfied for sufficiently small m_1 and m_2. But if m_1 and m_2 are such that the first of conditions (8.9') fails, i.e., there appear the complex conjugate roots $\lambda_{3,4}$, and the condition $(\tilde{a} - \tilde{d}) < 0$ holds true, then the monotonic behavior of stable trajectories in the isolated subsystem alters into an oscillatory stable one in the united system. The opposite alteration cannot take place, as the pair of roots $\lambda_{1,2}$ from the isolated subsystem spectrum is present in the spectrum of $\mathcal{A}$, i.e., if the subsystems demonstrated oscillations when approaching the equilibrium $\mathbf{N}^*$, then the oscillations are preserved everywhere after the integration into the united system.

In a like manner one may consider the conditions under which a monotonic instability turns into an oscillatory one after integration.

But if under the terms of (8.8) the second inequality in (8.8') alters into the opposite (obviously, the first one always remains true), then the equilibrium of the united system turns out to be unstable, that is, migration destabilizes the system. □

Example 8.2: Another example of a migration structure is shown in Figure 45b, where $n = 3$. Migrations proceed here with the same intensity between any two of the n habitats, which may be explained, for instance, by habitats being mutually equidistant in the plane. In a three-dimensional space, a similar geometric argument results in four habitats placed at the vertices of a regular tetrahedron. For $n \geq 4$ habitats such geometric constructions are no longer possible in a 3-dimensional space.

The matrix of migration structure (in the isotropic case) takes on the form

$$M_n = \begin{bmatrix} -(n-1) & 1 & 1 & 1 \\ 1 & -(n-1) & 1 & 1 \\ \cdots & \cdots & \cdots & \cdots \\ 1 & 1 & 1 & -(n-1) \end{bmatrix} \tag{8.10}$$

and satisfies the conditions of Lemma 8.1.   □

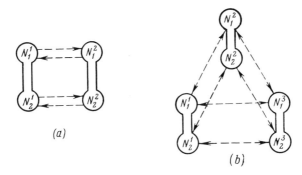

Figure 45.   Examples of migration structures: dashed arrows indicate migration flows; interactions within habitats are not specified.

Generally, one may think of even more complex schemes of linking the subsystems than those illustrated in Figure 45a,b. Lemma 8.1 enables us to state that the original eigenvalues are preserved in the spectrum of the united system for any migration structure M_n with zero row sums. Since the condition of zero row sums is also a criterion for the previous equilibrium values to be retained in the steady-state solution to the united system (8.6), this results in the following conclusion: if uniting by a migration scheme preserves the equilibrium values of the isolated subsystems, then the united system spectrum carries all the previous eigenvalues and all the above statements are true as to the behavior of trajectories in the neighborhood of equilibrium.

C. Unrealistic Scheme with Real Eigenvalues

In a highly hypothetical case where all the "biological" migration intensities are identical, i.e.,

$$m_i = m, \quad i = 1, \ldots, p,$$

the technique of Kronecker matrix products yields the explicit forms for all the eigenvalues of the united system. In this case, indeed, the matrix is represented as

$$\mathcal{A} = I_n \otimes A + m M_n \otimes D \tag{8.11}$$

and, given $\lambda(A)$ and $\lambda(M_n)$, one can find $\lambda(\mathcal{A})$ by making use of the following lemma.

Lemma 8.2 *Let X and Y be $p \times p$ and $n \times n$ matrices, respectively, with eigenvalues λ_i ($i = 1, \ldots, p$) and μ_j ($j = 1, \ldots, n$). Then the eigenvalues of matrix $I_n \otimes X + Y \otimes I_p$ are given by all possible sums of the form $\lambda_i + \mu_j$.*

The proof is given in the Appendix.

If applied to matrix $\mathcal{A}$ (8.11), Lemma 8.2 states that the eigenvalues of $\mathcal{A}$ represent all possible pairwise sums

$$\lambda(\mathcal{A}) = \lambda_i + m\,\mu_j,$$

where λ_i and μ_j are the eigenvalues of A and M_n, respectively.

Example 8.3: For a migration structure resulting from the annular arrangement of habitats, similar to the annular arrangement of ecological niches for competing species (cf. Figure 43 in Chapter 7), the matrix of migration structure in an isotropic medium is the circulant matrix:

$$M_n = \begin{bmatrix} -2 & 1 & 0 & \cdots & 0 & 1 \\ 1 & -2 & 1 & \cdots & 0 & 0 \\ \cdots & \cdots & \cdots & \cdots & \cdots & \cdots \\ 1 & 0 & 0 & \cdots & 1 & -2 \end{bmatrix}$$

with the eigenvalues (see formula (A7.4))

$$\mu_j = -2 + \cos j\frac{2\pi}{n} + \cos j\frac{2\pi(n-1)}{n} \;=\; -2\left(1 - \cos j\frac{2\pi}{n}\right),$$

$$j \;=\; 0, 1, \ldots, n-1.$$

In the special case of equal $m_i = m$ the eigenvalues of $\mathcal{A}$ are

$$\lambda(\mathcal{A}) = \lambda_i + m\mu_j.$$

The presence of $\mu_0 = 0$ illustrates Lemma 8.1 and the conclusion that all of the original eigenvalues λ_i are preserved in the spectrum of $\mathcal{A}$. Since all the other μ_js are real and negative, one can say that the integration by no means affects stability, as compared with the isolated case, and the qualitative picture of trajectory behavior in the neighborhood of equilibrium remains unchanged.   □

Hence, for any pattern of biological interactions in an isolated subsystem, uniting by means of migration in an isotropic medium fails to improve the system stability properties; rather it may only change the monotonic behavior of trajectories in the neighborhood of equilibrium (either stable or unstable) into oscillations, but not vice versa. In some cases such an integration may even destabilize stable subsystems (cf. Example 8.1).

As shown above, the ground for these mathematical results lies in the singularity of the migration structure matrix M_n for an isotropic medium. But it is hard to believe that any real environment, with some prevailing directions of migration practically always present, could ensure strict isotropy in all migration routes. Observed in experiments, the stabilizing effects of migration may therefore be explained by distinctions in intensities of migration in different directions, and within the framework of mathematical models such effects are to be sought in schemes with the intensities essentially depending upon the direction, i.e., in media which are *anisotropic* with respect to migration. Two further sections consider the schemes of this kind.

II. MIGRATIONS IN ANISOTROPIC MEDIA

If a migration intensity depends not only upon the biology of the migrating species, but also upon the migration route, or the direction, then the environment, or medium, is said to be *anisotropic* with respect to migration.[2] As seen from the preceding section, this is also a "medium" in which to look for stabilization effects of migration in models.

A. Migration Exchange between Two Habitats

In its simplest case the influence of environmental anisotropy can be studied in a system with $n = 2$ habitats, within which $p = 2$ biological species of population sizes $x(t)$ and $y(t)$ interact in the same general way as before:

$$
\begin{aligned}
dx/dt &= f_x(x, y), \\
dy/dt &= f_y(x, y).
\end{aligned}
\tag{8.12}
$$

To simplify the notation of the preceding section, assume that

$$
\begin{aligned}
(m_{12})_x &= r + \rho, \quad (m_{21})_x = r - \rho, \\
(m_{12})_y &= s + \sigma, \quad (m_{21})_y = s - \sigma,
\end{aligned}
\tag{8.13}
$$

and notice that parameters ρ and σ specify the distinctions in the intensities of migrations in the opposite directions, respectively, for the first and second species, the case of $\rho = \sigma = 0$ corresponding to the scheme of the preceding section. Let, for certainty,

$$(m_{12})_x > (m_{21})_x \quad \text{and} \quad (m_{12})_y > (m_{21})_y,$$

hence, in the notation adopted, we have

$$0 < \rho < r \quad \text{and} \quad 0 < \sigma < s.$$

Dynamics of the united community is modeled by the fourth-order system (small superscripts indicate the habitat number rather than a power exponent):

$$
\begin{aligned}
\dot{x}^1 &= f_x(x^1, y^1) - (r - \rho)x^1 + (r + \rho)x^2, \\
\dot{y}^1 &= f_y(x^1, y^1) - (s - \sigma)y^1 + (s + \sigma)y^2, \\
\dot{x}^2 &= f_x(x^2, y^2) + (r - \rho)x^1 - (r + \rho)x^2, \\
\dot{y}^2 &= f_y(x^2, y^2) + (s - \sigma)y^1 - (s + \sigma)y^2.
\end{aligned}
\tag{8.14}
$$

The fact that migration actually takes place for at least one of the species and with direction-specific intensities is expressed by the formal condition that

$$r + s > 0, \quad \rho^2 + \sigma^2 \neq 0.$$

If there is a nontrivial equilibrium $\mathbf{z}^* = [\overset{*}{x}{}^1, \overset{*}{y}{}^1, \overset{*}{x}{}^2, \overset{*}{y}{}^2]$ in system (8.14), then linearization at this equilibrium yields a matrix of the following block structure:

$$\mathcal{A}(\mathbf{z}^*) = \left[\begin{array}{c|c} A(\overset{*}{x}{}^1, \overset{*}{y}{}^1) - \mathrm{diag}\{r - \rho, s - \sigma\} & \mathrm{diag}\{r + \rho, s + \sigma\} \\ \hline \mathrm{diag}\{r - \rho, s - \sigma\} & A(\overset{*}{x}{}^2, \overset{*}{y}{}^2) - \mathrm{diag}\{r + \rho, s + \sigma\} \end{array} \right],$$

(8.15)

where $A(\overset{*}{x}{}^i, \overset{*}{y}{}^i)$ are matrices of the subsystems linearized at points $[\overset{*}{x}{}^i, \overset{*}{y}{}^i]$ ($i = 1, 2$). By the spectrum of matrix $\mathcal{A}(\mathbf{z}^*)$ we shall judge, as before, about the stability of equilibrium $\mathbf{z}^*$, and by comparing this spectrum with the eigenvalues of matrix $A(x^*, y^*)$, about the effect of migration on the system stability.

An additional problem which arises in the anisotropic set-up is that of finding new equilibrium points, since the previous equilibrium values, x^* and y^*, for isolated population sizes—such that $f_x(x^*, y^*) = f_y(x^*, y^*) = 0$—are no longer at equilibrium in the united system. Since the system is still migration-closed, it is legitimate to assume that, due to the nonsymmetric migration exchange, the equilibrium populations are redistributed, so that new equilibria take on the following form:

$$\begin{aligned} \overset{*}{x}{}^1 &= x^* + \Delta x, & \overset{*}{x}{}^2 &= x^* - \Delta x, \\ \overset{*}{y}{}^1 &= y^* + \Delta y; & \overset{*}{y}{}^2 &= y^* - \Delta y. \end{aligned}$$

(8.16)

For such a stationary solution to exist in system (8.14) it is necessary that the functions $f_x(x, y)$ and $f_y(x, y)$ be *skew-symmetric* at point $[x^*, y^*]$ with respect to increments in x and y, that is,

$$f_{x(y)}(x^* + \Delta x, y^* + \Delta y) = -f_{x(y)}(x^* - \Delta x, y^* - \Delta y).$$

(8.17)

If, indeed, we substitute (8.16) into the steady-state equations for system (8.14), we will come to the following equations:

$$\begin{aligned} 2\rho x^* - 2r\Delta x + f_x(x^* + \Delta x, y^* + \Delta y) &= 0, \\ -2\rho x^* + 2r\Delta x + f_x(x^* - \Delta x, y^* - \Delta y) &= 0, \\ 2\rho y^* - 2s\Delta y + f_y(x^* + \Delta x, y^* + \Delta y) &= 0, \\ -2\rho y^* + 2s\Delta y + f_y(x^* - \Delta x, y^* - \Delta y) &= 0, \end{aligned}$$

which, when added pairwise, give the relationships needed.

Even to see if these conditions hold true, before speculating on the stability of the new equilibrium, we need now to specify the pattern of interaction within isolated subsystems.

B. Prey-Predator Models: Who Migrates after Whom

Prey-predator models, especially those with a neutral pattern of stability, are probably the most interesting target for investigating the stabilization effects of migration, which are the most likely to be certain in this case.

In the Volterra description of the prey-predator interactions with no self-regulation we have (see Chapter 4):

$$f_x(x, y) = \alpha x - \beta xy, \qquad (\alpha, \beta, \gamma, \delta > 0) \qquad x^* = \gamma/\sigma,$$
$$f_y(x, y) = -\gamma y + \delta xy, \qquad \qquad y^* = \alpha/\beta.$$

The skew-symmetry conditions (8.17) require that

$$2\beta \Delta x \Delta y = 0, \qquad 2\delta \, \Delta x \, \Delta y = 0,$$

this being possible only if either $\Delta x = 0$ or $\Delta y = 0$ (the case of $\Delta x = \Delta y = 0$ refers to an isotropic medium).

If $\Delta x = 0$, the system of equations $f_x = f_y = 0$ and the condition that $y^* \pm \Delta y > 0$ will result in a solution of the form

$$\Delta x = 0, \, \Delta y = 2\rho/\beta, \begin{cases} \alpha > 2\rho & \text{if } s = \sigma = 0, \\ \alpha\sigma/s = 2\rho & \text{if } s, \sigma, \rho > 0. \end{cases} \qquad (8.18a)$$

In case $\Delta y = 0$, the condition that $x^* \pm \Delta x > 0$ generates similar kind of solutions:

$$\Delta y = 0, \, \Delta x = -2\sigma/\delta, \begin{cases} \gamma > 2|\sigma| & \text{if } r = \rho = 0, \\ \gamma\rho/r = -2\sigma & \text{if } \rho, r > 0, \, \sigma < 0. \end{cases} \qquad (8.18b)$$

Consequently, the case of (8.18a) is possible either if there is no migration of predators at all (motile prey—sessile predator) or if the migration parameters meet a special condition, with migration of the prey having the same superior direction as that of the predators (σ and ρ being of the same sign). Previous equilibrium values x^* are preserved for the prey populations, whereas the predator population is increased by $\Delta y > 0$ in the subsystem for which the migration is stronger (the first one in the above case). Relationship (8.18a) between system parameters thus ensures that the increased predator population will graze the surplus prey out caused by stronger migration to the first habitat; naturally following their prey, the predators increase their own population size. The equation $\alpha\sigma/s = 2\rho$ suggests that a point which characterizes the system in the 3-dimensional space of parameters $\{\alpha, \sigma/s, \rho\}$ belongs to a second-order surface (hyperbolic paraboloid) bounded by the plane $\sigma/s = 1$ in the positive orthant, the condition $\alpha > 2\rho$ being automatically true.

Similarly, the case of (8.18b) is possible, either if there is no migration in prey at all (sessile prey—motile predator), or if the migration parameters satisfy a special relationship, with migration in predators having the superior direction opposite to that of the prey migration (σ and ρ are opposite in sign). The preservation of equilibrium values y^* and the redistribution in x^* can be interpreted—perhaps, more in mathematical than ecological terms—as follows: once an excess of predators emerge because of their stronger migration to the second habitat, it is equalized by a decrease in the prey population caused by the stronger migration of prey in the opposite direction; migration efforts by predators are not rewarded with increase in the population size. As in the case of (8.18a), a point which specifies the system belongs to a similar surface now in the space of the parameters $\{\gamma, \rho/r, -\sigma\}$.

C. Migratory Stabilization: Motile Prey—Sessile Predators

Let us examine the equilibrium stability in the case where two prey-predator pairs are linked by migration among prey with the prevailing direction $2 \rightarrow 1$. System (8.14) then takes on the following form (small superscripts indicate the habitat number rather than a power exponent):

$$
\begin{aligned}
\dot{x}^1 &= \alpha x^1 - \beta x^1 y^1 - (r - \rho)x^1 + (r + \rho)x^2, \\
\dot{y}^1 &= -\gamma y^1 + \delta x^1 y^1, \\
\dot{x}^2 &= \alpha x^2 - \beta x^2 y^2 + (r - \rho)x^1 - (r + \rho)x^2, \\
\dot{y}^2 &= -\gamma y^2 + \delta x^2 y^2.
\end{aligned}
\tag{8.19}
$$

When $\alpha > 2\rho$, there is a positive equilibrium of the type (8.18a), that is,

$$
\dot{x}^{1,2} = x^* = \gamma/\delta, \quad \dot{y}^{1,2} = y^* \pm \Delta y = (\alpha \pm 2\rho)/\beta.
\tag{8.20}
$$

Matrix (8.15) takes on the form

$$
\mathcal{A} = \begin{bmatrix}
-r - \rho & -\beta\gamma/\delta & r + \rho & 0 \\
\delta(\alpha + 2\rho)/\beta & 0 & 0 & 0 \\
r - \rho & 0 & -r + \rho & -\beta\gamma/\delta \\
0 & 0 & \delta(\alpha - 2\rho)/\beta & 0
\end{bmatrix}
\tag{8.21}
$$

with the characteristic equation

$$
\lambda^4 + 2r\lambda^3 + 2\alpha\gamma\lambda^2 + 2\gamma(\alpha r - 2\rho^2)\lambda + \gamma^2(\alpha^2 - 4\rho^2) = 0.
\tag{8.22}
$$

By means of the Routh-Hurwitz criterion it is proved (see the Appendix) that

$$\text{Re } \lambda_i(\mathcal{A}) < 0, \quad i = 1, \ldots, 4,$$

excepting the case where $\rho = r$.

Equilibrium (8.20) is thereby asymptotically stable. If we recall now that in isolated prey-predator pairs the equilibrium $[x^*, y^*]$ is neutrally stable (with the frequency of oscillations $\omega = \sqrt{\alpha\gamma}$), it becomes clear that system (8.19) represents an example of the stabilization effect of migration.

The case of $\rho = r > 0$ implies that migration (with the intensity $2r$) goes only in one direction, namely, from the second subsystem to the first one. The spectrum of $\mathcal{A}$ (the roots of (8.22)) then includes a pair of pure imaginary numbers,

$$\lambda_{1,2} = \pm i\sqrt{\gamma(\alpha - 2\rho)},$$

and a pair of negative numbers,

$$\lambda_{3,4} = -r \pm \sqrt{r^2 + 2\gamma r - \alpha\gamma}.$$

The equilibrium (8.20) of the united system thus retains the neutrally stable behavior typical of the isolated subsystems, although the frequency of oscillations around the equilibrium (in the linear approximation) being equal to $\sqrt{\gamma(\alpha - 2\rho)}$, i.e., being reduced as compared to the isolated case. If the the lesser frequency of oscillations is believed to make for greater stability of the ecosystem, then, in this sense, we have another example of the stabilizing effect of migration.

D. Migratory Stabilization: Sessile Prey—Motile Predators

In a similar way, one can investigate the system that accounts for migration only among predators, with the prevailing direction $2 \rightarrow 1$:

$$
\begin{aligned}
\dot{x}^1 &= \alpha x^1 - \beta x^1 y^1, \quad (r = \rho = 0; s, \sigma > 0) \\
\dot{y}^1 &= -\gamma x^2 y^1 - (s - \sigma)y^1 + (s + \sigma)y^2 \\
\dot{x}^2 &= \alpha x^2 - \beta x^2 y^2, \\
\dot{y}^2 &= -\gamma y^2 + \delta x^2 y^2 + (s - \sigma)y^1 - (s + \sigma)y^2.
\end{aligned}
\quad (8.23)
$$

At the equilibrium of type (8.18b), which exists now for $\gamma > 2\delta$, we have

$$\overset{*}{y}{}^{1,2} = y^* = \alpha/\beta, \quad \overset{*}{x}{}^{1,2} = x^* \pm \Delta x = (\gamma \mp 2\sigma)/\delta, \quad (8.24)$$

i.e., $\overset{*}{x}{}^1 < \overset{*}{x}{}^2$. Investigation of the spectrum for the matrix of system (8.23) linearized at this equilibrium shows that, just as in the previous case, all Re $\lambda_i(\mathcal{A}) < 0$ except for the case of $\sigma = s > 0$. Thus, equilibrium (8.24) is asymptotically stable, i.e. system (8.23) again shows the stabilizing effect of migration.

The equality $\sigma = s$ implies that migration with the intensity $2s$ goes only from the second subsystem to the first one; as a result, a pair of pure imaginary eigenvalues appears,

$$\lambda_{1,2} = \pm i \sqrt{\alpha(\gamma + 2s)},$$

as well as a pair of negative numbers,

$$\lambda_{3,4} = -s \pm \sqrt{s^2 + 2\alpha s - \alpha\gamma}$$

i.e., unidirectional migration of predators increases the frequency of oscillations around equilibrium as compared with the isolated case, thus reducing, in this sense, the system stability.

E. When Motility in Common Produces Stability in Equilibria

Consider now the instances where migration takes place on both levels, but with parameters being restricted by (8.18a) or (8.18b). As was noted even by V. Volterra,[10] it is hardly probable that parameters of a real system would encounter a condition of the equality type, but for the sake of mathematical completeness, those improbable cases also deserve investigation.

When the prevailing direction of migration is the same for both prey and their predators (from the second subsystem into the first one), system (8.14) takes on the following form:

$$
\begin{aligned}
\dot{x}^1 &= \alpha x^1 - \beta x^1 y^1 - (r - \rho)x^1 + (r + \rho)x^2, \\
\dot{y}^1 &= -\gamma y^1 - \delta x^1 y^1 - (s - \sigma)y^1 + (s + \sigma)y^2, \\
\dot{x}^2 &= \alpha x^2 - \beta x^2 y^2 + (r - \rho)x^1 - (r + \rho)x^2, \\
\dot{y}^2 &= -\gamma y^2 - \delta x^2 y^2 + (s - \sigma)y^1 - (s + \sigma)y^2.
\end{aligned}
\tag{8.25}
$$

The search for equilibrium solutions in the form $\overset{*}{x}^1 = \overset{*}{x}^2$ or $\overset{*}{y}^1 = \overset{*}{y}^2$ shows that equilibria of the types (8.18a) or (8.18b) are the only steady-state solutions which have equal same-level populations in the subsystems (the latter condition being verifiable in experiments).

Linearization of (8.25) at the point (8.20) under the condition that $\alpha\sigma = 2\rho s$, yields the matrix

$$
A = \left[
\begin{array}{cc|cc}
-r - \rho & -\beta\gamma/\delta & r + \rho & 0 \\
\delta(\alpha + 2\rho)/\beta & -s + \sigma & 0 & s + \sigma \\
\hline
r - \rho & 0 & -r + \rho & -\beta\gamma/\delta \\
0 & s - \sigma & \delta(a - 2\rho)/\beta & -s - \sigma
\end{array}
\right],
\tag{8.26}
$$

whose stability can be proved again by the Routh–Hurwitz criterion (see the Appendix). Thus, equilibrium (8.20) is asymptotically stable in system (8.25).

The equilibrium of type (8.18*b*) is possible when the prevailing directions of migration are opposite for the prey and predators. To keep all the parameters positive, let us rewrite system (8.25) in the following form (substituting $-\sigma$ everywhere for σ):

$$
\begin{aligned}
\dot{x}^1 &= \alpha x^1 - \beta x^1 y^1 - (r-\rho)x^1 + (r+\rho)x^2, \\
\dot{y}^1 &= -\gamma y^1 + \delta x^1 y^1 - (s+\sigma)y^1 + (s-\sigma)y^2, \\
\dot{x}^2 &= \alpha x^2 - \beta x^2 y^2 + (r-\rho)x^1 - (r+\rho)x^2, \\
\dot{y}^2 &= -\gamma y^2 + \delta x^2 y^2 + (s+\sigma)y^1 - (s-\sigma)y^2.
\end{aligned} \tag{8.27}
$$

Arguing along the same lines, one can prove that, when linearized at the point

$$
\overset{*}{y}{}^{1;2} = y^* = \alpha/\beta, \quad \overset{*}{x}{}^{1;2} = x^* \pm \Delta x = (\gamma \pm 2\sigma)/\delta \tag{8.28}
$$

$$
(\gamma - 2\sigma > 0, \quad \gamma\rho = 2r\sigma, \rho, \sigma > 0),
$$

system (8.27) has the real parts in its spectrum all negative if only an additional restriction is imposed on the set of admissible values for the model parameters:

$$
(r+s)^2(\alpha\sigma - \gamma\rho) + (r+s)2rs(\rho+\sigma) - \alpha\sigma(\rho+\sigma)^2 > 0. \tag{8.29}
$$

When the left-hand side vanishes in inequality (8.29), the spectrum contains a pair of pure imaginary roots, $\lambda_{1,2} = \pm i\omega$, where

$$
\omega^2 = \alpha\gamma + \frac{2\alpha\sigma(\rho+\sigma)}{r+s},
$$

and a pair of roots, $\lambda_{3,4}$, with negative real parts. Hence, the frequency ω of the oscillating components in the linear approximation of the solution in the neighborhood of equilibrium exceeds the frequency of oscillations in the isolated subsystem.

Thus, when migrations take place at both prey and predator levels with direction-specific intensities and when identical equilibrium values, $\overset{*}{x}{}^1 = \overset{*}{x}{}^2$, for the two subsystems are preserved at the prey level, the migration stabilizes the equilibrium for all admissible values of the model parameters; but when the same population size is preserved at the predator level, the stabilizing effect of migration shows up under the additional restriction (8.29).

To summarize the results of this section, even with the set of models being quite far from general and quite limited by the linear description of migration flows, we can still infer that stabilizing effects of migration upon the community dynamics are quite far from being unambiguous. Recall that, to identify pure spatial effects, the schemes of the present and preceding sections have also presupposed the identity of equations for biological interactions within different subsystems. In this respect, a more general formulation of the problem is considered in the next section.

III. A SMALL-PARAMETER METHOD TO STUDY MIGRATION EFFECTS

The fundamental assumption of general eigenvalue perturbation theory[11] allows the models we are considering to be quite general, although this is achieved at the expense of the migration parameters being rather small when compared with the parameters of biological interactions.

A. A General Model

Let n habitats, generally different but with the same collection of p species, be linked together by migration flows of a known spatial structure. When isolated, the community dynamics within the kth habitat is governed by the system of equations

$$\frac{dN_i^k}{dt} = f_i^k(N_1^k, \ldots, N_p^k), \quad i = 1, \ldots, p,$$

with the superscript indicating the habitat, or subsystem, number; by means of p-vectors

$$\mathbf{N}^k = \left[N_1^k, \ldots, N_p^k\right] \quad \text{and} \quad \mathbf{F}^k(\mathbf{N}) = \left[f_1^k(\mathbf{N}), \ldots, f_p^k(\mathbf{N})\right], \tag{8.30}$$

it can be rewritten in the vector form as

$$\frac{d\mathbf{N}^k}{dt} = \mathbf{F}^k(\mathbf{N}^k), \quad k = 1, \ldots, n. \tag{8.30'}$$

As for the pattern of biological interactions within a habitat, it will only be assumed representable by sufficiently smooth functions f_i^k (which can be different for different habitats) with a nontrivial equilibrium, say $\overset{*}{\mathbf{N}}{}^k$.

Suppose also that the migration process is described by some functions m_i^{ks}—the instantaneous intensities of the ith species migration on the route from the sth habitat to the kth one ($k, s = 1, 2, \ldots, n$)—which depend (in the very general case) upon population sizes of all species in all habitats:

$$m_i^{ks} = m_i^{ks}(\mathbf{N}^1, \ldots, \mathbf{N}^n), \quad i = 1, \ldots, p; \quad k, s = 1, \ldots, n. \tag{8.31}$$

Then the migration impact on variations in the ith species population of, e.g., the first subsystem, is comprised of the inflow of immigrants into the given habitat, $\sum_{s \neq 1}^n N_i^s m_i^{1s}$, and the outflow of emigrants from this habitat to all the rest, $\sum_{s \neq 1}^n N_i^1 m_i^{s1}$. Equations (8.30) then take the following form:

$$\frac{dN_i^k}{dt} = f_i^k(N_1^k, \ldots, N_p^k) + \sum_{s=1, s \neq k}^n (N_i^s m_i^{ks} - N_i^k m_i^{sk}), \tag{8.32}$$

and if we denote by $\mathbf{z} = [\mathbf{N}^1, \ldots, \mathbf{N}^n]$ the np-vector that specifies the state of all the n linked subsystems, then the dynamic equations for the united system can be aggregated into the following vector form:

$$d\mathbf{z}/dt = \mathbf{F}(\mathbf{z}) + \mathbf{M}(\mathbf{z}). \tag{8.33}$$

Here the vector-function

$$\mathbf{F}(\mathbf{z}) = \left[\mathbf{F}^1(\mathbf{z}^1), \ldots, \mathbf{F}^n(\mathbf{z}^n)\right]$$

characterizes biological interactions within all subsystems, while $\mathbf{M}(\mathbf{z})$ determines the structure of migration links and their impact on the population dynamics. In the notation adopted,

$$\mathbf{M}(\mathbf{z}) = \left[\sum_{s=2}^{n}[N_1^s m_1^{1s}(\mathbf{z}) - N_1^1 m^{s1}(\mathbf{z})], \ldots, \sum_{s=2}^{n}[N_p^s m_p^{1s}(\mathbf{z}) - N_p^1 m_p^{s1}(\mathbf{z})], \ldots,\right.$$
$$\left.\sum_{s=1}^{n-1}[N_1^s m_1^{ns}(\mathbf{z}) - N_1^n m_1^{sn}(\mathbf{z})], \ldots, \sum_{s=1}^{n-1}[N_p^s m_p^{ns}(\mathbf{z}) - N_p^n m_p^{sn}(\mathbf{z})]\right] \tag{8.34}$$

Rejecting now the idealistic postulate of isotropy in the medium with respect to migration and also that of constancy in the functions $m_i^{ks}(\mathbf{z})$(which were adopted in Section 8.II), we assume, on the other hand, that the migration processes influence the state of the system to a far less extent than the basic biological interactions. Mathematically, this assumption is expressed by introducing a small parameter $\varepsilon > 0$ into the right-hand side of system (8.33), so that

$$d\mathbf{z}/dt = \mathbf{F}(\mathbf{z}) + \varepsilon\mathbf{M}(\mathbf{z}), \tag{8.35}$$

and studying stability begins with the existence problem for feasible equilibrium.

B. Migration Links Shift Feasible Equilibria

When $\varepsilon = 0$, the migration links vanish and the system disintegrates into n isolated subsystems; it is logical to denote the set of equilibria $\overset{*}{\mathbf{N}}{}^{k}$ in these subsystems by

$$\mathbf{z}^*(0) = [\overset{*}{\mathbf{N}}{}^{1}, \ldots, \overset{*}{\mathbf{N}}{}^{n}].$$

Can one state that there exists a nontrivial equilibrium $\mathbf{z}^*(\varepsilon) > 0$ for any sufficiently small value of $\varepsilon > 0$ in the united system? The answer is proved to be positive by a general implicit-function theorem of analysis (see the Appendix, Statement A8.3). It is positive also in the sense that the proof reveals the explicit form of the equilibrium,

$$\mathbf{z}^*(\varepsilon) = \mathbf{z}^*(0) - \varepsilon J^{-1}\mathbf{M}(\mathbf{z}^*(0)) + o(\varepsilon^2), \tag{8.36}$$

which may show what particular shifts in equilibria are caused by weak migrations under a given scheme $\mathbf{M}(\mathbf{z})$. It also opens the gateway to stability analysis in the linear approximation.

To find the Jacobian matrix, $\mathcal{A}(\varepsilon)$, for system (8.35) linearized at point $\mathbf{z}^*(\varepsilon)$, i.e., the matrix of the linear approximation system

$$\dot{\mathbf{x}} = \mathcal{A}(\varepsilon)\mathbf{x}, \tag{8.37}$$

where $\mathbf{x}(t) = \mathbf{z}(t) - \mathbf{z}^*(\varepsilon)$, one may apply a general differentiation rule and see that

$$\mathcal{A}(\varepsilon) = \left.\frac{\partial \mathbf{F}(\mathbf{z})}{\partial \mathbf{z}}\right|_{\mathbf{z}^*(\varepsilon)} + \left.\varepsilon \frac{\partial \mathbf{M}(\mathbf{z})}{\partial \mathbf{z}}\right|_{\mathbf{z}^*(\varepsilon)}. \tag{8.38}$$

If the (r, s)th entries of the summand matrices are denoted, respectively, by $j_{rs}(\mathbf{z})$ and $d_{rs}(\mathbf{z})$, then the pertinent entry in $\mathcal{A}(\varepsilon)$ is

$$a_{rs}(\varepsilon) = (j_{rs} + \varepsilon d_{rs})(\mathbf{z}^*(\varepsilon))$$

and, after being expanded in powers of ε at point $\varepsilon = 0$, takes on the form

$$\begin{aligned} a_{rs}(\mathbf{z}^*(\varepsilon)) &= j_{rs}(\mathbf{z}^*(0)) + \left.\varepsilon \frac{d}{d\varepsilon} j_{rs}(\mathbf{z}^*(\varepsilon))\right|_{\varepsilon=0} + \varepsilon d_{rs}(\mathbf{z}^*(0)) + \mathcal{O}(\varepsilon^2) \\ &= (j_{rs} + \varepsilon d_{rs})(\mathbf{z}^*(0)) + \left.\varepsilon \frac{\partial j_{rs}(\mathbf{z})}{\partial \mathbf{z}}\right|_{\mathbf{z}^*(0)} \cdot \left.\frac{d\mathbf{z}^*(\varepsilon)}{d\varepsilon}\right|_{\varepsilon=0} + \mathcal{O}(\varepsilon^2). \end{aligned} \tag{8.39}$$

Here $\partial j_{rs}(\mathbf{z})/\partial \mathbf{z}$, the gradient of function $j_{rs}(\mathbf{z})$, represents a matrix of size $1 \times np$, $d\mathbf{z}^*(\varepsilon)/d\varepsilon$ is a column np-vector, that is equal, by (8.36), to

$$\left.\frac{d\mathbf{z}^*(\varepsilon)}{d\varepsilon}\right|_{\varepsilon=0} = -J^{-1}\mathbf{M}(\mathbf{z}^*(0)).$$

Consequently, the desired matrix $\mathcal{A}$ takes on the form

$$\mathcal{A}(\varepsilon) = J + \varepsilon \left[\left.\frac{\partial \mathbf{M}(\mathbf{z})}{\partial \mathbf{z}}\right|_{\mathbf{z}^*(0)} - H \right] + \mathcal{O}(\varepsilon^2), \tag{8.40}$$

where matrix H consists of entries

$$h_{rs} = \left.\frac{\partial j_{rs}(\mathbf{z}^*)}{\partial \mathbf{z}}\right|_{\mathbf{z}^*(0)} \cdot J^{-1}\mathbf{M}(\mathbf{z}^*(0)), \quad r, s = 1, \ldots, np. \tag{8.41}$$

The well-known principle of the small parameter method states[11] that if all the eigenvalues $\lambda_1, \ldots, \lambda_{np}$ of matrix J are pairwise distinct, the eigenvalues of matrix

$\mathcal{A}(\varepsilon) = J + \varepsilon E + \mathcal{O}(\varepsilon^2)$ can be represented in the form of

$$\lambda_i(\varepsilon) = \lambda_i + \varepsilon \lambda_i^{(1)} + \mathcal{O}(\varepsilon^2), \quad i = 1, \ldots, np, \tag{8.42}$$

with the coefficient at ε expressed in terms of the scalar product,

$$\lambda_i^{(1)} = \langle \mathbf{u}_i, E\mathbf{v}_i \rangle, \tag{8.43}$$

where $\{\mathbf{v}_1, \ldots, \mathbf{v}_{np}\}$ is a normalized system of eigenvectors for matrix J, and $\{\mathbf{u}_1, \ldots, \mathbf{u}_{np}\}$ is the same for the Hermitian-conjugate matrix J^H associated with the complex conjugate eigenvalues $\{\overline{\lambda}_1, \ldots, \overline{\lambda}_{np}\}$.

It should be noted that the schemes examined in the preceding section would not satisfy the restrictions of the small-parameter method, because the spectrum of matrix J, formed by the union of spectra of isolated subsystems, did contain n identical sets of eigenvalues.

C. Stability and Stabilization

Based on the representation (8.42)–(8.43), definitions for the *stabilizing effect* and the *equilibrium-stabilizing structure* of migration[1] can be given in a natural way as follows.

Definition 8.1 *The migration according to a scheme* $\mathbf{M}(\mathbf{z})$ *is said to render a stabilizing effect (or to have a stabilizing tendency) if all the coefficients,* $\lambda_i^{(1)}$, *at the first power of* ε *in the expansion (8.42), have their real parts negative* $(Re\ \lambda_i^{(1)} < 0, i = 1, \ldots, np)$, *i.e., the real parts of eigenvalues of the united system are reduced in comparison with those of the isolated subsystems.*

Definition 8.2 *A migration structure* $\mathbf{M}(\mathbf{z})$ *is said to be* equilibrium-stabilizing *at point* $\mathbf{z}^*(0)$ *if the equilibrium* $\mathbf{z}^*(\varepsilon)$ *is asymptotically stable in system (8.35) for any positive, yet small enough* ε, *while for* $\varepsilon = 0$ *it does not possess asymptotic stability.*

Clearly, the situation is only possible when there are no eigenvalues $\lambda_i(0)$ with nonzero real parts, while Re $\lambda_i(\varepsilon) < 0$, $i = 1, \ldots, np$, i.e., migration improves, in this case, the stability properties of equilibrium, turning the neutral stability of $\mathbf{z}^*(0)$ into the asymptotic stability of $\mathbf{z}^*(\varepsilon)$.

According to (8.43), to see whether or not a migration structure $\mathbf{M}(\mathbf{z})$ has a stabilizing effect on community dynamics and whether $\mathbf{M}(\mathbf{z})$ generates an asymptotically stable equilibrium $\mathbf{z}^*(\varepsilon)$, one has to find eigenvectors $\{\mathbf{u}_1, \ldots, \mathbf{u}_{np}\}$ and $\{\mathbf{v}_1, \ldots, \mathbf{v}_{np}\}$ for the matrix

$$\mathcal{A}(0) = J = \left. \frac{\partial \mathbf{F}(\mathbf{z})}{\partial \mathbf{z}} \right|_{\mathbf{z}^*(0)} = \mathrm{diag}\{A_1(\overset{*}{\mathbf{N}}{}^1), \ldots, A_n(\overset{*}{\mathbf{N}}{}^n)\}, \tag{8.44}$$

whose diagonal blocks $A_k(\overset{*}{\mathbf{N}}{}^k)$ have size $p \times p$ in common and represent the matrices of isolated subsystems linearized at their equilibrium points $\overset{*}{\mathbf{N}}{}^k$. The spectrum of J is therefore composed of n different sets of eigenvalues $\alpha_1^1, \ldots, \alpha_p^k$ belonging to the matrices

$$A_k(\overset{*}{\mathbf{N}}{}^k), \quad k = 1, \ldots, n.$$

Hence, to put the spectrum of $\mathcal{A}(\varepsilon)$ in order and to make it comparable with pertinent spectra of isolated subsystems, one may put

$$\lambda_{(k-1)p+j}(0) = \alpha_j^k, \quad j = 1, \ldots, p,$$

in the representation (8.42).

If, furthermore, $\{e_1^k, \ldots, e_p^k\}$ is a normalized system of (p-dimensional) eigenvectors associated with the eigenvalues $\alpha_1^k, \ldots, \alpha_p^k$, then the following np-vectors:

$$
\begin{aligned}
\mathbf{v}_{(k-1)p+1} &= [0, \ldots, e_1^k, 0, \ldots, 0], \\
\mathbf{v}_{(k-1)p+2} &= [0, \ldots, e_2^k, 0, \ldots, 0], \ldots \\
\mathbf{v}_{kp} &= [0, \ldots, e_p^k, 0, \ldots, 0], \quad k = 1, \ldots, n,
\end{aligned}
\tag{8.45}
$$

make up a normalized system of eigenvectors for matrix J. In an analogous way, the system of vectors $\{\mathbf{u}_1, \ldots, \mathbf{u}_{np}\}$ can be constructed, after which one can find the sign of the expression

$$\operatorname{Re} \lambda_i^{(1)} = \operatorname{Re} \langle \mathbf{u}_i, E \mathbf{v}_i \rangle,$$

where

$$E = \left. \frac{\partial \mathbf{M}(\mathbf{z})}{\partial \mathbf{z}} \right|_{\mathbf{z}^*(0)} - H, \tag{8.46}$$

and matrix H is defined in (8.41).

D. An Example: Asymmetric Exchange between Unequal Habitats

Consider the migration between two predator-prey subsystems, which may take place at both levels. To modify the system (8.25) of the preceding section in such a way that the subsystems become different, assume that

$$
\begin{aligned}
f_x^1(x, y) &= x - xy, \quad f_x^2(x, y) = 2f_x^1(x, y); \\
f_y^1(x, y) &= -y + xy, \quad f_y^2(x, y) = 2f_y^1(x, y).
\end{aligned}
$$

The equations of the united community then takes the following form:

$$\begin{aligned}
\dot{x}^1 &= \alpha x^1 - \beta x^1 y^1 - (r - \rho)x^1 + (r + \rho)x^2, \\
\dot{y}^1 &= -\gamma y^1 - \delta x^1 y^1 - (s - \sigma)y^1 + (s + \sigma)y^2, \\
\dot{x}^2 &= 2(\alpha x^2 - \beta x^2 y^2) + (r - \rho)x^1 - (r + \rho)x^2, \\
\dot{y}^2 &= 2(-\gamma y^2 - \delta x^2 y^2) + (s - \sigma)y^1 - (s + \sigma)y^2,
\end{aligned} \qquad (8.47)$$

now without any restrictions on the signs of ρ and σ but with $r, s > 0$ (and with the superscripts indicating, as before, the habitat number rather than a power exponent).

In terms of the present section, we have

$$\mathbf{z} = \begin{bmatrix} x^1 \\ y^1 \\ x^2 \\ y^2 \end{bmatrix}; \qquad \mathbf{M}(\mathbf{z}) = \begin{bmatrix} -(r - \rho)x^1 + (r + \rho)x^2 \\ -(s - \sigma)y^1 + (s + \sigma)y^2 \\ (r - \rho)x^1 - (r + \rho)x^2 \\ (s - \sigma)y^1 - (s + \sigma)y^2 \end{bmatrix};$$

$$\frac{\partial \mathbf{F}(\mathbf{z})}{\partial \mathbf{z}} = [j_{rs}(\mathbf{z})] =$$

$$\begin{bmatrix}
1 - y^1 - (r - \rho) & -x^1 & r + \rho & 0 \\
y^1 & (x^1 - 1) - (s - \sigma) & 0 & s + \sigma \\
r - \rho & 0 & 2(1 - y^2) - (r + \rho) & -2x^2 \\
0 & s - \sigma & 2y^2 & 2(x^2 - 1) - (s + \sigma)
\end{bmatrix}.$$

Equilibria in isolated subsystems coincide: $\overset{*}{\mathbf{N}}{}^1 = \overset{*}{\mathbf{N}}_1 = [1, 1]^T$, so that $\mathbf{z}^*(0) = [1, 1, 1, 1]^T$ and the corresponding matrices,

$$A_1 = \begin{bmatrix} 0 & -1 \\ 1 & 0 \end{bmatrix}, \qquad A_2 = \begin{bmatrix} 0 & -2 \\ 2 & 0 \end{bmatrix},$$

give the set of eigenvalues for $J = \text{diag}\{A_1, A_2\}$:

$$\{\lambda_1, \lambda_2, \lambda_3, \lambda_4\} = \{i, -i, 2i, -2i\},$$

justifying the neutral stability of equilibrium $\mathbf{z}^*(0)$. The pertinent eigenvectors are

$$\mathbf{v}_1 = \mathbf{u}_1 = \frac{1}{\sqrt{2}} \begin{bmatrix} 1 \\ -i \\ 0 \\ 0 \end{bmatrix}, \qquad \mathbf{v}_2 = \mathbf{u}_2 = \frac{1}{\sqrt{2}} \begin{bmatrix} 1 \\ i \\ 0 \\ 0 \end{bmatrix},$$

$$\mathbf{v}_3 = \mathbf{u}_3 = \frac{1}{\sqrt{2}} \begin{bmatrix} 0 \\ 0 \\ 1 \\ -i \end{bmatrix}, \qquad \mathbf{v}_4 = \mathbf{u}_4 = \frac{1}{\sqrt{2}} \begin{bmatrix} 0 \\ 0 \\ 1 \\ i \end{bmatrix}.$$

Having calculated the vector

$$
J^{-1} \cdot \mathbf{M}(\mathbf{z}^*(0)) =
\left[
\begin{array}{cc|cc}
0 & 1 & & \\
-1 & 0 & \multicolumn{2}{c}{\mathbf{0}} \\
\hline
& & 0 & 1/2 \\
\multicolumn{2}{c|}{\mathbf{0}} & -1/2 & 0
\end{array}
\right]
\cdot
\left[
\begin{array}{c}
2\rho \\
2\sigma \\
-2\rho \\
-2\sigma
\end{array}
\right]
=
\left[
\begin{array}{c}
2\sigma \\
-2\rho \\
-\sigma \\
\rho
\end{array}
\right],
$$

we find the equilibrium (8.36) in the united system,

$$
\mathbf{z}^*(\varepsilon) =
\left[
\begin{array}{c}
1 - \varepsilon 2\sigma \\
1 + \varepsilon 2\rho \\
1 + \varepsilon \sigma \\
1 - \varepsilon \rho
\end{array}
\right]
+ \mathcal{O}(\varepsilon^2),
$$

which shows how the asymmetry in migration intensities shifts the equilibrium populations of isolated habitats.

Next, in order to determine the matrix (8.46) we find

$$
\left. \frac{\partial \mathbf{M}(\mathbf{z})}{\partial \mathbf{z}} \right|_{\mathbf{z}^*(0)}
=
\left[
\begin{array}{cccc}
-(r - \rho) & 0 & r + \rho & 0 \\
0 & -(s - \sigma) & 0 & s + \sigma \\
r - \rho & 0 & -(r + \rho) & 0 \\
0 & s - \sigma & 0 & -(s + \sigma)
\end{array}
\right];
$$

calculated at point $\mathbf{z}^*(0)$, the gradients of the entries to matrix $\partial \mathbf{F}/\partial \mathbf{z}$ are

$$
\begin{aligned}
\partial j_{11}/\partial \mathbf{z} &= [0, -1, 0, 0], & \partial j_{12}/\partial \mathbf{z} &= [-1, 0, 0, 0], \\
\partial j_{21}/\partial \mathbf{z} &= [0, 1, 0, 0], & \partial j_{22}/\partial \mathbf{z} &= [1, 0, 0, 0], \\
\partial j_{33}/\partial \mathbf{z} &= [0, 0, 0, -2], & \partial j_{34}/\partial \mathbf{z} &= [0, 0, -2, 0], \\
\partial j_{43}/\partial \mathbf{z} &= [0, 0, 0, 2], & \partial j_{44}/\partial \mathbf{z} &= [0, 0, 2, 0],
\end{aligned}
$$

while all the rest of $\partial j_{rs}/d\mathbf{z}$ are zero. This results, by (8.41), in

$$
H =
\left[
\begin{array}{cccc}
2\rho & -2\sigma & 0 & 0 \\
-2\rho & 2\sigma & 0 & 0 \\
0 & 0 & -2\rho & 2\sigma \\
0 & 0 & 2\rho & -2\sigma
\end{array}
\right],
$$

and, by (8.46), in

$$
E =
\left[
\begin{array}{cccc}
-(r + \rho) & 2\sigma & r + \rho & 0 \\
2\rho & -(s + \sigma) & 0 & s + \sigma \\
r - \rho & 0 & -(r - \rho) & -2\sigma \\
0 & s - \sigma & -2\rho & -(s - \sigma)
\end{array}
\right].
$$

Now the coefficients at ε in the estimate (8.42), calculated as $\lambda_j^{(1)} = \langle \mathbf{u}_j, E\mathbf{v}_j \rangle$, are the following:

$$\lambda_{1,2}^{(1)} = [-(r+\rho) - (s+\sigma)]/2 \mp \mathrm{i}(\rho + \sigma);$$

$$\lambda_{3,4}^{(1)} = [-(r-\rho) - (s-\sigma)]/2 \mp \mathrm{i}(\rho + \sigma).$$

Hence, if $(r+\rho)+(s+\sigma) > 0$ and $(r-\rho)+(s-\sigma) > 0$, both conditions being fairly unrestrictive, then for small enough $\varepsilon > 0$ the equilibrium $\mathbf{z}^*(\varepsilon)$ is asymptotically stable. Therefore, if at least one of the species migrates from habitat 2 to habitat 1 and at least one of the species from 1 to 2, then the migration structure $\mathbf{M}(\mathbf{z})$ is equilibrium-stabilizing at point $\mathbf{z}^*(0) = [1, 1, 1, 1]$ for the interaction (8.47).

Thus, coupled with the heterogeneity in space, the anisotropy in the migration medium essentially extends the parameter domain where migration brings about stabilizing effects on population dynamics.

Appendix

Lemma 8.1 *Let a matrix $M_n = [m_{ks}]$ have a zero eigenvalue and let $\mathbf{a} = [a_1, \ldots, a_n]^T$ be an eigenvector associated with the zero eigenvalue. If $\mathbf{x} \in \mathbb{R}^p$ is a column eigenvector of matrix A with an eigenvalue λ, then the same value λ is associated with the column eigenvector $\mathcal{X} = [a_1\mathbf{x}^T, \ldots, a_n\mathbf{x}^T]^T \in \mathbb{R}^{np}$ of matrix $\mathcal{A} = I_n \otimes A + M_n \otimes D$, too.*

Proof: Using the technique to tackle block matrices of consistent structures,[12] we have for the matrix $\mathcal{A}$ and vector $\mathcal{X}$:

$$\mathcal{A}\mathcal{X} = \mathcal{A}(\mathbf{a} \otimes \mathbf{x}) = \begin{bmatrix} (A + m_{11}D)a_1\mathbf{x} + m_{12}Da_2\mathbf{x} + \cdots + m_{1n}Da_n\mathbf{x} \\ m_{21}Da_1\mathbf{x} + (A + m_{22}D)a_2\mathbf{x} + \cdots + m_{2n}Da_n\mathbf{x} \\ \cdots\cdots\cdots\cdots\cdots\cdots\cdots\cdots\cdots \\ m_{n1}Da_1\mathbf{x} + m_{n2}Da_2\mathbf{x} + \cdots + (A + m_{nn}D)a_n\mathbf{x} \end{bmatrix}$$

$$= \begin{bmatrix} a_1 A\mathbf{x} + (m_{11}a_1 + m_{12}a_2 + \cdots + m_{1n}a_n)D\mathbf{x} \\ a_2 A\mathbf{x} + (m_{21}a_1 + m_{22}a_2 + \cdots + m_{2n}a_n)D\mathbf{x} \\ \cdots\cdots\cdots\cdots\cdots\cdots\cdots\cdots\cdots \\ a_n A\mathbf{x} + (m_{n1}a_1 + m_{n2}a_2 + \cdots + m_{nn}a_n)D\mathbf{x} \end{bmatrix}.$$

The terms in parentheses vanish due to vector $\mathbf{a}$ belonging to a zero eigenvalue, whereby

$$\mathcal{A}\mathcal{X} = \begin{bmatrix} a_1\lambda\mathbf{x} \\ a_2\lambda\mathbf{x} \\ \vdots \\ a_n\lambda\mathbf{x} \end{bmatrix} = \lambda\mathcal{X},$$

which was stated.   ■

Lemma 8.2 *Let X and Y be $p \times p$ and $n \times n$ matrices, respectively, with eigenvalues λ_i $(i = 1, \ldots, p)$ and μ_j $(j = 1, \ldots, n)$. Then the eigenvalues of matrix $I_n \otimes X + Y \otimes I_p$ are given by all possible sums of the form $\lambda_i + \mu_j$.*

Proof: Let $\mathbf{x}^i$ and $\mathbf{y}^j$ be eigenvectors of matrices X and Y, respectively, associated with the eigenvalues λ_i and μ_j. Consider the matrix $(I_n + Y) \otimes (I_p + X)$. The eigenvalues of the Kronecker product are all possible pairwise products of eigenvalues of its co-factor matrices,[13] which are respectively $1 + \mu_j$ and $1 + \lambda_i$. Associated with these products are eigenvectors of the form

$$\mathcal{Z}^{ji} = [y_1^j \mathbf{x}^i, y_2^j \mathbf{x}^i, \ldots, y_n^j \mathbf{x}^i]^T.$$

Since, furthermore,

$$\begin{aligned}
(I_n + Y) \otimes (I_p + X) &= I_n \otimes I_p + I_n \otimes X + Y \otimes I_p + Y \otimes X \\
&= I_{np} + (I_n \otimes X + Y \otimes I_p) + Y \otimes X,
\end{aligned}$$

we have

$$(1 + \mu_j)(1 + \lambda_i)\mathcal{Z}^{ji} = \mathcal{Z}^{ji} + (I_n \otimes X + Y \otimes I_p)\mathcal{Z}^{ji} + \mu_j\lambda_i\mathcal{Z}^{ji}.$$

Elementary laying out then shows that

$$(\lambda_i + \mu_j)\mathcal{Z}^{ij} = (I_n \otimes X + Y \otimes I_p)\mathcal{Z}^{ij},$$

which was stated.   ∎

Statement A8.1 (localization of eigenvalues for matrix (8.21)) *If $r > \rho$, then matrix (8.21) is stable.*

Proof: makes use of the well-known criterion of Routh–Hurwitz.[14] The Hurwitz matrix for characteristic polynomial (8.22) takes on the form

$$H = \begin{bmatrix} 2r & 2\gamma(\alpha r - 2\rho^2) & 0 & 0 \\ 1 & 2\alpha\gamma & \gamma^2(\alpha^2 - 4\rho^2) & 0 \\ 0 & 2r & 2\gamma(\alpha r - 2\rho^2) & 0 \\ 0 & 1 & 2\alpha\gamma & \gamma^2(\alpha^2 - 4\rho^2) \end{bmatrix},$$

whereby the Hurwitz determinants will be

$$\begin{aligned}
\Delta_1 &= H \begin{pmatrix} 1 \\ 1 \end{pmatrix} = 2r, \\
\Delta_2 &= H \begin{pmatrix} 1, & 2 \\ 1, & 2 \end{pmatrix} = 2\gamma(\alpha r + 2\rho^2), \\
\Delta_3 &= H \begin{pmatrix} 1, & 2, & 3 \\ 1, & 2, & 3 \end{pmatrix} = 16\gamma^2\rho^2(r^2 - \rho^2), \\
\Delta_4 &= \det H = \gamma^2(\alpha^2 - 4\rho^2)\Delta_3.
\end{aligned} \qquad (A8.1)$$

The Routh-Hurwitz theorem states that for real parts of all the roots of a real polynomial to be negative it is necessary and sufficient that the conditions

$$a_0\Delta_1 > 0, \quad \Delta_2 > 0, \ldots, \begin{cases} a_0\Delta_n > 0 & \text{for odd } n, \\ \Delta_n > 0 & \text{for even } n, \end{cases} \tag{A8.2}$$

be satisfied (a_0 being the leading coefficient of the polynomial).

Since all the symbols appearing in the formulae of (A8.1) stand for positive parameters, we have Δ_1 and Δ_2 apparently positive; $\Delta_3 > 0$ if $r > \rho$ (by definition, $r \geq \rho$), and $\Delta_4 > 0$ by the condition $\alpha > 2\rho$, the existence condition for a positive equilibrium (8.20). Thus, except for the case where $\rho = r$, it follows that

$$\text{Re } \lambda_i(A) < 0, \quad i = 1, \ldots, 4. \qquad \blacksquare$$

Statement A8.2 (localization of eigenvalues for matrix (8.26)). *If $\alpha > 2\rho$ and $\alpha\sigma = 2\rho s$, then matrix (8.26) is stable.*

Proof: Routine algebra shows the characteristic polynomial of matrix (8.26) to be

$$\lambda^4 + 2(r+s)\lambda^3 + 2(\alpha\gamma + 2rs)\lambda^2 \quad + \quad 2\gamma[\alpha(r+s) - 2\rho(\rho - \sigma)]\lambda$$
$$+ \quad \gamma^2(\alpha^2 - 4\rho^2) + 4\alpha\gamma(rs - \rho\sigma). \tag{A8.3}$$

The Hurwitz determinants[14] for this polynomial are equal to

$$\begin{aligned}
\Delta_1 &= 2(r+s), \\
\Delta_2 &= 2[(r+s)(\alpha\gamma + 4rs) + 2\gamma\rho(\rho - \sigma)], \\
\Delta_3 &= 16\{(r+s)[\alpha\gamma\sigma(r\sigma + s\rho) + (r+s)\gamma^2\rho^2] - \gamma^2\rho^2(\rho - \sigma)^2\}, \\
\Delta_4 &= [\gamma^2(\alpha^2 - 4\rho^2) + 4\alpha\gamma(rs - \rho\sigma)]\Delta_3.
\end{aligned}$$

Evidently, $\Delta_1 > 0$. Positiveness of Δ_2 is also evident if $\rho - \sigma \geq 0$, while otherwise, since $\sigma \leq s$ and $2\rho s = \alpha\sigma$, we have

$$\begin{aligned}
\Delta_2/2 &= (r+s)(\alpha\gamma + 4rs) + 2\gamma\rho(\rho - \sigma) \geq (r+s)(\alpha\gamma + 4rs) - 2\gamma\rho s \\
&= (r+s)(\alpha\gamma + 4rs) - \alpha\gamma\sigma = r(\alpha\gamma + 4rs) + 4rs^2 + \alpha\gamma(s - \sigma) > 0.
\end{aligned}$$

Positiveness of Δ_3 is equivalent to the condition

$$(\rho + \sigma)^2 \quad < \quad \frac{r+s}{\gamma^2\rho^2}[\alpha\gamma\sigma(r\sigma + s\rho) + (r+s)\gamma^2\rho^2]$$
$$= \quad \frac{\alpha\sigma}{\gamma\rho^2}(r+s)(r\sigma + s\rho) + (r+s)^2,$$

which is true since from restrictions $0 < \rho \le r$ and $0 < \sigma \le s$ it follows that $(\rho + \sigma)^2 \le (r + s)^2$. Positiveness of Δ_4 is then equivalent to the condition

$$\gamma^2(\alpha^2 - 4\rho^2) + 4\alpha\gamma(rs - \rho\sigma) > 0,$$

which holds true by virtue of the above-mentioned restrictions and the condition that $\alpha - 2\rho > 0$.

Thus, all the Hurwitz determinants are positive and, by the Routh–Hurwitz criterion, matrix (8.26) is stable along with equilibrium (8.20) in system (8.25). ∎

Statement A8.3. *In the united system (8.35) composed of n subsystems whose isolated dynamics are governed by systems (8.32), there always exists an equilibrium* $\mathbf{z}^*(\varepsilon) > 0$ *for any sufficiently small* $\varepsilon > 0$.

Proof: Consider the mapping

$$\varphi\begin{bmatrix} \mathbf{z} \\ \varepsilon \end{bmatrix} = \mathbf{F}(\mathbf{z}) + \varepsilon\mathbf{M}(\mathbf{z}),$$

which maps a certain neighborhood of the point

$$\begin{bmatrix} \mathbf{z}^*(0) \\ 0 \end{bmatrix}$$

in the coordinate space $\mathbb{R}^{np+1}$ to the space $\mathbb{R}^{np}$. As functions $\mathbf{F}(\mathbf{z})$ and $\mathbf{M}(\mathbf{z})$ are supposed to be sufficiently smooth, the mapping φ and its partial derivatives are continuous with respect to all arguments. Besides, according to the definition of equilibrium points, we have

$$\varphi\begin{bmatrix} \mathbf{z}^*(0) \\ 0 \end{bmatrix} = \mathbf{0},$$

and require that the Jacobian matrix of the multivariate function φ, calculated at point

$$\begin{bmatrix} \mathbf{z}^*(0) \\ 0 \end{bmatrix}$$

by the formula

$$J = J(\mathbf{z}^*(0)) = \frac{\partial\varphi}{\partial\mathbf{z}}\Bigg|\begin{bmatrix} \mathbf{z}^*(0) \\ 0 \end{bmatrix} = \frac{\partial\mathbf{F}(\mathbf{z})}{\partial\mathbf{z}}\Bigg|_{\mathbf{z}^*(0)},$$

be nonsingular. The latter is equivalent to nonsingularity of the Jacobian matrices for all subsystems linearized at their equilibrium points $\overset{*}{\mathbf{N}}{}^1, \ldots, \overset{*}{\mathbf{N}}{}^n$.

Now the implicit-function theorem (see, e.g., Korn and Korn[15]) ensures that there exists an interval $-\delta < \varepsilon < \delta$, $\delta > 0$, and a function $\mathbf{z}^*(\varepsilon)$ such that

$$\varphi \left[\begin{array}{c} \mathbf{z}^*(\varepsilon) \\ \varepsilon \end{array} \right] = 0,$$

for any value of ε in the interval.

Thus, $\mathbf{z}^*(\varepsilon)$ proves to be an equilibrium for system (8.35) and, moreover,

$$\mathbf{z}^*(\varepsilon) = \mathbf{z}^*(0) - \varepsilon J^{-1}\mathbf{M}(\mathbf{z}^*(0)) + \mathcal{O}(\varepsilon^2). \tag{A8.4}$$

It follows that if all components of the previous equilibrium populations in the isolated subsystems, i.e., the components of $\mathbf{z}^*(0)$, were positive, then for small enough values of ε the positiveness is also retained in the equilibrium $\mathbf{z}^*(\varepsilon)$ of the united system. ∎

Additional Notes

To 8.I. The idea of *segmentation*, i.e., representing a spatially distributed system as a combination of a number of points, or *segments*, is typical not only for migrational models but for other ecological models as well. The bulk of lake modeling, for instance, relies upon such segmentation,[16–17] while the term "box" models originates mainly from conventional system flow diagrams, where segments are often depicted as boxes.

Although "box" approach is not a unique way to account for a migration in population models, it has the distinct advantage of permitting a comparison within the same kind of mathematical apparatus, namely, ordinary differential equations (ODEs). Incorporating a space dimension explicitly leads to partial differential equations (PDEs) and, as a result, to some kinds of model phenomena, like "dissipative structures" and "traveling waves" in the so-called "reaction-diffusion" equations (see, e.g., Svirezhev[18]), which have no analogy in ODEs. Other kinds, like "diffusion instability" or "diffusion stabilization" in pertinent PDEs,[1] still permit a comparison with the effects observable in the proper ODEs.

Similar conclusions about "symmetric" migration being unable to stabilize the dynamics of a biocommunity were also obtained in more specific cases of biological interactions.[19] In Lotka–Volterra competition communities, however, situations are found where migration may provide for the coexistence of otherwise competitively exclusive species,[20] thus exemplifying one more possible attenuation for the competition exclusion principle. In the framework of difference equations of population dynamics equations the migration effect turned out to be sensitive to whether the migration is considered to act before or after the biological interactions.[21]

Another approach to the mathematical study of stabilizing effects of migration[22–23] considers again a system of interconnected habitats, whose isolated dynamics eventually results in species extinction. Uniting habitats by migration guarantees a regional coexistence, as far as computer simulation technique may ever guarantee a general conclusion.

To 8.II. The effect of unidirectional migration of predators appears to be qualitatively the same as that of weak competition at any of two trophic levels, which was discovered in a system of n prey-predator pairs linked by competition either among prey or predators: the competition among prey decreases, whereas that among predators increases, the frequency of oscillations around neutrally stable equilibria.[24]

Particular formulations of problems on the stabilization effects of migration are numerous. Due to analytical difficulties, however, the majority of these problems are studied by computer simulations. In some of the studies the migrating populations are averaged over habitats,[25] others account for migrations occurring only at certain stages of cyclic subsystem dynamics[26] or incorporate a dependence on habitat-specific environmental conditions to determine the level of correlation among habitats that yields the strongest stabilization effect of migration.[27]

To 8.III. The line of reasoning expounded in the section generalizes, for the case of arbitrary interactions within the habitat, a small-parameter method proposed earlier for n prey-predator pairs.[28] Studying a particular two-habitat, two-species model in greater detail reveals that there exist the equilibrium $\mathbf{z}^*(\varepsilon)$ and the nontrivial sufficient conditions for it to be stable—not only in the regular case (where the Jacobian matrix is nonsingular) but in some nonregular cases too.[29]

REFERENCES

1. Svirezhev, Yu. M. and Logofet, D. O. *Stability of Biological Communities* (revised from the 1978 Russian edition), Mir Publishers, Moscow, 1983, 319 pp., Chap. 8.

2. Logofet, D. O. Can migration stabilize an ecosystem? (A mathematical aspect), *Zhurnal Obshchei Biologii* (*Journal of General Biology*), 39, 122–128, 1978 (in Russian).

3. Huffaker, C. B. Experimental studies on predation: Complex dispersion factors and predator-prey oscillations, *Hilgardia*, 27, 343–383, 1958.

4. Huffaker, C. B., Shea, K. P. and Herman, S. G. Experimental studies on predation: Complex dispersion and levels of food in acarine predator-prey interaction, *Hilgardia*, 34, 305–329, 1963.

5. Sakai, K., Narise, T., Hiraizumi, Y. and Iyama, S. Studies on competition in plants and animals. IX. Experimental studies on migration in *Drosophila melanogaster*, *Evolution*, 12, 93–101, 1958.

6. Southwood, T. R. E. Migration of terrestrial arthropods in relation to habitat. *Biol. Rev.*, 37, 171–214, 1962.

7. Luchinbill, L. S. Coexistence in laboratory populations of *Paramecium aurelia* and its predator *Didinium nasutum*, *Ecology*, 54, 1320–1327, 1973.

8. Hutchinson, G. E. Homage to Santa Rosalie, or why are there so many kinds of animals? *Amer. Natur.*, 145–159, 1959.

9. Levin, S. A. Population dynamic models in heterogeneous environments, *Ann. Rev. Ecol. Syst.*, 7, 287–310, 1976.

10. Volterra, V. *Leçons sur la Théorie Mathématique de la Lutte pour la Vie*, Gauthier-Villars, Paris, 1931, 214 pp., Chap. 3.

11. Lancaster, P., *Theory of Matrices*, Academic Press, New York, 1969, Chap. 7.

12. Gantmacher, F. R. *The Theory of Matrices*, Chelsea, New York, 1960, Chap. 2.

13. Marcus, M. and Minc., A. *A Survey of Matrix Theory and Matrix Inequalities*, Allyn and Bacon, Boston, 1964 Chap. 2.15.

14. Gantmacher, F. R. *The Theory of Matrices*, Chelsea, New York, 1960, Chap. 16.

15. Korn, G. A. and Korn, T. M. *Mathematical Handbook*, 2nd edn., McGraw-Hill, New York, 1968, Chap. 4.5–7.

16. Chen, C. W. and Orlob, G. T. Ecological simulation for aquatic ecosystems. In: *Systems Analysis and Simulation in Ecology*, Vol. 3, Patten, B. C., ed., Academic Press, New York, 1975, 475–588.

17. Voinov, A. A. Lake Plesheyevo—A case study. In: *Decision Support Technique for Lakes and Reservoirs* (Water Quality Modelling, Vol. 4), Henderson-Stellers, B., ed., CRC Press, Boca Raton, 1991, 115–151.

18. Svirezhev, Yu. M. *Nonlinear Waves, Dissipative Structures and Catastrophes in Ecology*, Nauka, Moscow, 1987, 368 pp., Chaps. 6,7 (in Russian).

19. Hilborn, R. The effect of spatial heterogeneity on the persistence of predator-prey interaction, *Theor. Popul. Biol.*, 8, 346–355, 1975.

20. Kishimoto, K. Coexistence of any number of species in the Lotka–Volterra competitive system over two patches, *Theor. Popul. Biol.*, 38, 149–158, 1990.

21. Allen, J. C. Mathematical models of species interactions in time and space, *Amer. Natur.*, 109, 319–342, 1975.

22. Maynard Smith, J. *Models in Ecology*, Cambridge University Press, Cambridge, 1974, Chap. 6.

23. Hastings, A. Spatial heterogeneity and the stability of predator-prey systems, *Theor. Popul. Biol.*, 12, 37–48, 1977.

24. Logofet, D. O. Investigation of a system of n "predator-prey" pairs linked by competition, *Soviet Math. Dokl.*, 16, 1246–1249, 1975.

25. Jones, D. D. Stability implications of dispersal linked ecological models, Research Memorandum RM-75-44, IIASA, Laxenburg, Austria, 1975, 46 pp.

26. Ziegler, B. P. Persistence and patchiness of predator-prey systems induced by discrete event population exchange mechanisms, *J. Theor. Biol.*, 67, 687–713, 1977.

27. Dombrovsky, Yu. A. and Tyutyunov, Yu. V. Habitat structure, motility of individuals, and the population persistence, *Zhurnal Obshchet Biologii (Journal of General Biology)*, 48, 493–498, 1987 (in Russian).

28. Chewning, W. C. Migratory effects in predator-prey models, *Math. Biosci.*, 23, 253–262, 1975.

29. Freedman, H. I. and Waltman, P. Mathematical models of population interaction with dispersal. I. Stability of two habitats with and without a predator, *SIAM J. Appl. Math.*, 32, 631–648, 1977.

Epilogue

This "one more book on stability..." would risk remaining one more unfinished book if the purpose were to give comprehensive answers to stability problems of ecology, even within their mathematical framework. Fortunately, the purpose is to show how these problems can be formulated and treated in terms of matrices and graphs, or how these mathematical tools—clear and appealing even for the nonmathematician—may be of help in stability analysis, which is probably the most mathematized aspect of theoretical ecology.

The present-day invasion of computers, armed with their seemingly endless varieties of user-friendly software products ready to solve almost any problem, can hardly inspire anybody to seek a deeper understanding and comprehension of "boring" mathematical theories, but this surfeit can, no doubt, endanger the "population" of modelers who comprehend the mathematical nature of the models themselves, including the computer-aided models. The author, therefore, when drafting and writing this book, was inspired by the hope that it might contribute to "conservation measures" for that population and, in particular, for the subpopulation consisting of those capable of solving on their own a system of linear equations, i.e., of finding equilibrium in a simple model of a few-species community.

As far as "equilibrium" is concerned, an ecological-minded reader probably might be annoyed with how many times this word has been used throughout the book, for even our earliest education at school taught us that nothing is less relevant to living systems than static equilibrium. The author's only excuse is the fact that the "equilibria" dealt with in the book, i.e., nontrivial equilibria in mathematical models of population dynamics, have nothing to do with thermodynamic equilibria, which actually imply death. Instead, these model equilibria are just mathematical abstractions capable of revealing some features of reality—to the extent that models are capable of reflecting reality at all.

Stability analysis is performed with models, but we understood that the very mathematical nature of modeling limits any computer simulation to only partial results, in contrast to theoretical modeling which produces general conclusions. In mathematics, the greater is the generalization of the problem formulated, the more difficult will it be to solve. Fortunately or unfortunately, we also face these problems in stability analysis, where we use matrices and graphs that greatly facilitate our formulating the problems and interpreting the results.

As this book has shown, the most general way to interpret the results involves the roles that the structure and functioning of an ecosystem play in the dynamic stability of the system. Some of the examples have been chosen to be purely hypothetical to illustrate nontriviality, if any, in the mathematics; some are quite practical, dealing with real ecological objects. But what may be the "practical" meaning of particular estimates or recommendations drawn from such highly theoretical models as Lotka–Volterra or other types of population equations? Of course, they can hardly be applied immediately to a field study, but they can specify a general tendency in the relations between structural properties and stability in dynamical behavior that might be exemplified in proper field or laboratory studies; should the matter be about any quantitative inference from the model, it has to be thought of as a minimal restriction ensuing only from the dynamic organization of interactions within the system, and thus to be restrained further in more specific models of the case study.

Evidently, whether and to what an extent the dynamic stability is itself a cause of ecosystem organization can hardly be studied within the context of mathematics alone.

List of Notation

$x \in S$	an element x belongs to a set S
$S \ni x$	a set S contains an element x
$\mathbb{R}^n$	a linear space of n-dimensional vectors (Euclidean space)
$\mathbb{R}^n_+$	the positive orthant of $\mathbb{R}^n$
int $\mathbb{R}^n_+$	the interior of $\mathbb{R}^n_+$
dim L	the dimensionality of (sub)space L
$\|x\|$	the absolute value of (real or complex) number x
$\|\mathbf{x}\|$	a norm of vector $\mathbf{x}$
$\langle \mathbf{x}, \mathbf{y} \rangle$	the scalar, or inner, product of vectors $\mathbf{x}$ and $\mathbf{y}$
diag$\{\mathbf{a}\}$	the diagonal matrix with the elements of vector $\mathbf{a}$ in the main diagonal
I	an identity matrix, $I = \text{diag}\{1, 1, \ldots, 1\}$
$A \oplus B$	the direct sum of matrices, $A \oplus B = \text{diag}\{A, B\}$
$L \otimes S$	the Kronecker product of matrices L and S
det A	the determinant of a square matrix A
$p(\lambda)$	the characteristic polynomial of a matrix
Re λ	the real part of a complex number λ
Im λ	the imaginary part of a complex number λ
$\lambda(A)$	an eigenvalue of a square matrix A
$r(A)$	the dominant (Frobenius) eigenvalue of a nonnegative matrix
$h(A)$	the index of imprimitivity, or Frobenius index, of matrix A
$\Lambda(A)$	the highest Lyapunov exponent for a system defined by matrix A
g.c.d.$\{\mathbf{v}\}$	the greatest common divisor for the components of a vector $\mathbf{v}$
l.c.m.$\{\mathbf{v}\}$	the least common multiple for the components of a vector $\mathbf{v}$
$i = j(\text{mod} h)$	number i differs from j by a multiple of h
$D(A)$	the directed graph (digraph) associated with a matrix A
$F_S[D]$	the factorgraph of a graph D with respect to a subdivision S
$L(\mathcal{A})$	the age factorgraph for a block-structured matrix $\mathcal{A}$
$S(\mathcal{A})$	the status factorgraph for a block-stuctured matrix $\mathcal{A}$
$D_1 \wedge D_2$	the Kronecker product, or conjunction, of digraphs D_1 and D_2
$\mathcal{O}(\varepsilon)$	a quantity or variable of the same order of smallness as ε
$o(\varepsilon)$	a quantity or variable of a higher order of smallness than ε
inf $\{\mathbb{M}\}$	the exact lower boundary for a set $\mathbb{M}$
$\underline{\lim}_{t \to \infty} N(t)$	the lower limit of function $N(t)$ as $t \to \infty$

$\mathbb{D}_n^+$ the set of diagonal $n \times n$ matrices with all positive diagonal entries

$q\mathbb{D}$ the set of quasi-dominant $n \times n$ matrices

$q\mathbb{R}$ the set of quasi-recessive $n \times n$ matrices

$s\mathbb{S}$ the set of sign-stable $n \times n$ matrices

■   end of proof

□   end of example

INDEX